AF798138

# Boston Studies in the Philosophy and History of Science

**Founding Editor**
Robert S. Cohen

Volume 346

**Series Editors**
Jürgen Renn, Max Planck Institute for the History of Science, Berlin, Germany
Lydia Patton, Virginia Tech, Ona, USA

**Associate Editor**
Peter McLaughlin, Department of Philosophy, Universität Heidelberg, Heidelberg, Germany

**Managing Editor**
Lindy Divarci, c/o Divarci, Max Planck Institute for the History of Science, Berlin, Germany

**Editorial Board Members**
Kostas Gavroglu, University of Athens, Athens, Greece
Thomas F. Glick, Department of History, Boston University, Boston, USA
John Heilbron, University of California, Berkeley, CA, USA
Diana Kormos-Buchwald, Dept. of Humanities and Social Sciences, California Institute of Technology, Pasadena, USA
Agustí Nieto-Galan, Dept. de Filosofia, Edifici B, Universitat Autonoma de Barcelona, Bellaterra (Cerdanyola del V.), Spain
Nuccio Ordine, Campus di Arcavacata - Dip. di Filologia, Universitá della Calabria, RENDE, Italy
Ana Simões, Department of History and Philosophy of Science, FCUL, Universidade de Lisboa, Lisboa, Portugal
John J. Stachel, Boston University, Brookline, USA
Baichun Zhang, Inst for the History of Natural Sciences, Chinese Academy of Science, Beijing, China

The series *Boston Studies in the Philosophy and History of Science* was conceived in the broadest framework of interdisciplinary and international concerns. Natural scientists, mathematicians, social scientists and philosophers have contributed to the series, as have historians and sociologists of science, linguists, psychologists, physicians, and literary critics.

The series has been able to include works by authors from many other countries around the world.

The editors believe that the history and philosophy of science should itself be scientific, self-consciously critical, humane as well as rational, sceptical and undogmatic while also receptive to discussion of first principles. One of the aims of Boston Studies, therefore, is to develop collaboration among scientists, historians and philosophers.

*Boston Studies in the Philosophy and History of Science* looks into and reflects on interactions between epistemological and historical dimensions in an effort to understand the scientific enterprise from every viewpoint.

Stefano Gattei • Roberta Corvi
Editors

# Feyerabend in Dialogue

## Critical Essays

Springer

*Editors*
Stefano Gattei
Department of Sociology
and Social Research
University of Trento
Trento, Italy

Roberta Corvi
Department of Philosophy
Catholic University of Milan
Milan, Italy

ISSN 0068-0346 ISSN 2214-7942 (electronic)
Boston Studies in the Philosophy and History of Science
ISBN 978-3-031-71937-0 ISBN 978-3-031-71938-7 (eBook)
https://doi.org/10.1007/978-3-031-71938-7

© The Editor(s) (if applicable) and The Author(s), under exclusive license to Springer Nature Switzerland AG 2024

This work is subject to copyright. All rights are solely and exclusively licensed by the Publisher, whether the whole or part of the material is concerned, specifically the rights of translation, reprinting, reuse of illustrations, recitation, broadcasting, reproduction on microfilms or in any other physical way, and transmission or information storage and retrieval, electronic adaptation, computer software, or by similar or dissimilar methodology now known or hereafter developed.
The use of general descriptive names, registered names, trademarks, service marks, etc. in this publication does not imply, even in the absence of a specific statement, that such names are exempt from the relevant protective laws and regulations and therefore free for general use.
The publisher, the authors and the editors are safe to assume that the advice and information in this book are believed to be true and accurate at the date of publication. Neither the publisher nor the authors or the editors give a warranty, expressed or implied, with respect to the material contained herein or for any errors or omissions that may have been made. The publisher remains neutral with regard to jurisdictional claims in published maps and institutional affiliations.

This Springer imprint is published by the registered company Springer Nature Switzerland AG
The registered company address is: Gewerbestrasse 11, 6330 Cham, Switzerland

If disposing of this product, please recycle the paper.

Paul K. Feyerabend (1924–1994)

# Introduction

This volume is meant to celebrate the centenary Paul K. Feyerabend's birth: he was born in Vienna, on 13 January 1924, and died after a short illness in Genolier, Switzerland, on 11 February 1994.

Feyerabend is a well-known philosopher, and it may seem superfluous to offer a portrait, however cursory, of him as a thinker. In fact, Feyerabend is often remembered for a few of his jokes, which are regarded as slogans, whereas the depth of his thought, which he himself liked to conceal behind provocations, is not always appreciated. His sharp and penetrating outlook was the result not only of his personal qualities, which were undoubtedly exceptional, but also of a varied and diverse cultural background, as well as of life experiences that, as he himself recounts in his autobiography (completed shortly before his death), forged his character and oriented his research. In particular, we have to keep in mind that the years of his youth were marked by the rise of Nazism, both in Germany and in Austria, and World War II, in which he participated as a soldier, sustaining wounds whose consequences afflicted him for the rest of his life. All this notwithstanding, he earned a degree in physics, and then a doctorate in philosophy, in 1951. In 1952, he began his collaboration with Popper at the London School of Economics and Political Sciences, later moving first to Bristol, where he had obtained a teaching position, and then (in 1958) to the United States, at the University of Berkeley, where he taught until 1990, alternating in his later years with a position at the University of Zurich (for more detailed information on Feyerabend's life and academic career, see his autobiography, *Killing Time*, published posthumously in 1995).

The aim of this volume is to remember Feyerabend in the way he would perhaps have liked best, namely, by attempting to reconstruct the dialogue he held not only with his contemporaries, to whom the third and last part of the volume is dedicated, but also with those protagonists of the past—philosophers or scientists—who were a source of inspiration for him, and with whose works he engaged most. Indeed, the first part is devoted to some significant figures in the history of philosophy, while the second part includes two chapters, dedicated to Galileo and Bohr respectively. Both these great scientists appear prominently in Feyerabend's works: Bohr in his early years, when Feyerabend chiefly tackled problems in quantum theory, and Galileo at a later stage, when Feyerabend focused on methodological issues and the notion of rationality. As Stefano Gattei points out, whereas Feyerabend's provocative picture of Galileo as a charlatan who succeeded in undermining the received tradition by appealing to rhetorical and propagandistic weapons, as well as psychological tricks, is usually considered an exaggeration and deemed misleading at best, Feyerabend's unconventional reading of Galileo's *Dialogue* has engaged philosophers and historians alike, breaking grounds for new, more nuanced interpretations as well as questioning contemporary philosophical accounts that take the received clash between Galileo's spotless rationalism and the Church's dogmatic obscurantism for granted. In his contribution to Niels Bohr and quantum physics, John Heilbron moves from Feyerabend's support of David Bohm's challenge to the 'Copenhagen interpretation', based on Popper's criticism of Bohr's complementarity and the theory of propensities, to Feyerabend's later defence of Bohr's views against Popper's criticism. Heilbron—a renowned historian of quantum physics and the author of a challenging biography of Galileo—closes his contribution by critically considering what he deems Feyerabend's unpersuasive account of Galileo, pointing to the way in which such an account was used by Feyerabend both to establish his case against method and to offer insightful objections to Kuhn's model for the growth of scientific knowledge.

The focus on Feyerabend's intellectual relationships with the main interlocutors of the past and present allowed a more multifaceted portrait of the man and philosopher to emerge, avoiding flattening on easy labels such as 'anarchist', 'irrationalist', 'relativist', 'enemy of science' and the like. The idea of dealing with Feyerabend's thought by the relationship he maintained with philosophers and scientists who influenced him arises from the consideration that he did not think in a systematic way, and references to many protagonists of Western culture are scattered throughout Feyerabend's works. It seemed useful to attempt at least a partial arrangement of his 'influences'. We are aware that many names are missing, some even important and significant ones, but given the great variety of readings and interests that characterise Feyerabend, it would have been impossible to be exhaustive within the limits of a project like this. We hope that this first piece can be followed by others, until the complete puzzle is put together, bringing out the richness, depth and varied character of Feyerabend's philosophical proposal, which is not limited to epistemological themes but extends to political, ethical and even aesthetic issues.

A comparison with the authors with whom Feyerabend dialogued, either in person or at a distance, always reveals a chiaroscuro relationship: for, he did accept some of their theses, but never all the way through, always differing in some aspect or disagreeing with their consequences, or else taking the same theses to their extremes. Each of the authors taken into consideration presents some thesis Feyerabend shared, but the convergence is always partial, so that a thinker, however appreciated from a certain point of view, may well be criticised or even contradicted on other issues, or even on the same one.

The thesis that seems to be the pivot around which Feyerabend's thought is built is, as is evident from many contributions, pluralism, which was inspired to him not only by his philosophical readings but also by his association with Brecht, whose emphasis on experimentation and correction of errors is similar to Feyerabend's ideal in philosophy of science (Val Dusek). According to Roberta Corvi, Feyerabend's pluralism comes from Popper's teaching, which leads him to argue that to control empirical theories it is necessary to formulate alternative hypotheses that are not conditioned by observation, and hence metaphysical. However, it must be acknowledged that the need for pluralism also has a political matrix, which may be traced back to Aristotle, who pointed out the advantages of pluralism for democracy (Gonzalo Munévar), and even more so in Mill, whom Feyerabend himself believed was at the origin of his thesis on the proliferation of methods. To those who point out that Mill defended pluralism against religious dogmatism and not against science, Elisabeth Lloyd reminds us that the English philosopher advocated tolerance in general and invoked nonconformity and freedom of opinion, principles that can also be coherently applied to science, just as Feyerabend did.

Obviously, the strenuous defence of pluralism is also present in Feyerabend's political proposal, which fights the hegemony of science, like any other hegemony; it is therefore surprising to find frequent references to Marx and Marxism in his works, which, according to Rory Kent and Ian J. Kidd, must be considered more as provocations against Popper, on the one hand, and the cultural climate that characterised the West during the Cold War, on the other. Indeed, Feyerabend could not accept the basic tenet of Marxism, as pluralism, in his eyes, requires the permanent coexistence of very many systems of thought; for Marx, by contrast, in order to realise a communist society, a single-minded ideology is needed.

Another central theme in Feyerabend's thought concerns the theoretical character of observation, as has been noted in a number of contributions (Corvi's, Munévar's, Preston's). In this instance, too, Corvi identifies a Popperian matrix, which generates a critique of empiricism, the roots of which go back as far as Aristotle's views. In this regard, as Munévar notes, Feyerabend, an admirer of Galileo, seems closer to Plato than to Aristotle, since the latter advocated a naïve form of empiricism, which considered observation neutral. The theme of theory-ladenness of observation is what Feyerabend also shares with Kuhn, whom he criticised for his advocation of 'normal science', one of the cornerstones of Kuhn's theory.

That this is a fundamental idea for the theorist of methodological anarchism is also proven by his assiduous search for precursors, which leads him, for example, to attribute it to Mach, misunderstanding him, according to John Preston. Preston points out that while before the 1970s Feyerabend considered Mach a positivist, he later saw him as a historical oriented thinker, whose fallibilism and evolutionary epistemology Feyerabend appreciated. Moreover, the relationship between Feyerabend and Logical Positivism is anything but linear, as also observed by Friedrich Stadler, who notices that Feyerabend was strongly influenced by the members of Vienna Circle and by groups of Viennese derivation that moved to the United States after World War II. Although Feyerabend disguised his closeness to logical positivism, his Viennese roots would become more visible after his appointment in Zurich. Stadler thinks that throughout his whole life Feyerabend was actually a Viennese philosopher, even if he himself may not have subscribed to this statement. Vasso Kindi also agrees on this point, as she believes that Feyerabend's upbringing was actually logically positivistic, as testified by his doctoral dissertation supervised by Victor Kraft. Kindi underlines that Wittgenstein's idea of meaning had a decisive influence on Feyerabend's concept of incommensurability, not to mention the prize of wonder and the sense of mystery, which unites both philosophers and leads them to reject scientism, but not science. In fact, Feyerabend's polemic is not about science, but rather about the philosophers of science, each of whom claims to codify the method that scientists must follow.

Therefore, although Feyerabend is not an enemy of science, he might be seen as an enemy of an algorithmic account of rationality: he rejected the very notion of rationality, and more specifically scientific rationality, in the name of creativity, intuition and tacit inference. But he did not inquire whether there is an alternative notion of it, according to which these abilities are not irrational at all, but may be part of a broadly conceived rationality, extending well beyond logical-mathematical reason.

We cannot help mentioning another constituent element of Feyerabend's thought: his relativism, which derives from pluralism in conjunction with the theoretical character of observation. According to Helmut Heit, it is inevitable if we admit pluralism. In order to turn to relativism, Feyerabend's reconstruction of the origin of the Western rational tradition was decisive, in particular the path from Protagoras to Nietzsche. Heit appreciates Feyerabend's relativism, because, in his opinion, it does not imply that all perspectives are symmetric and equally valid. He considered relativism as the natural and appropriate epistemic attitude of a confident and conscious scientific time, keeping in mind that we are inevitably caught in a net of beliefs, and that there are more and different such nets, which derive from our contingent human nature.

Feyerabend was (and still is, to some extent) a controversial personality and author, as emerges from Joseph Agassi's retrospect, at the end of the volume. Agassi and Feyerabend knew each other well: both were pupils of Popper's in the 1950s, and seated 'at the feet of the great thinker', as Agassi recalled in his autobiography (*A Philosopher's Apprentice: In Karl Popper's Workshop*, 1993 and 2008[2]); both appreciated each other's sharp intelligence, and did not spare harsh words towards one another. Both appealed to reason and rationality, although Feyerabend's provocative downplay of rationalism led to their intellectual estrangement. According to some, Feyerabend enjoyed fame and recognition he not always deserved, while according to others, the author *of Against Method* managed to intelligently provoke Western philosophy and the mainstream, pointing out the conformism that pervades not only society, but science itself. According to this interpretation, there are more continuities than ruptures in his biography, and more consistencies than contradictions in the theoretical aspects of Feyerabend's philosophical work.

Feyerabend, moreover, is also a complex personality, as his relationships with reference authors reveal. Suffice it for all, the example of Kuhn: 'Feyerabend and Kuhn are often seen as sitting in the same boat, as if their opinions in the philosophy of science were identical' (Paul Hoyningen-Huene). Actually, Feyerabend criticised Kuhn not only as far as historical and methodological approach is concerned, but also with respect to his mode of presentation; particularly he doubted the very existence of normal science, and eventually turned his back on incommensurability, famously introduced by both Kuhn and Feyerabend in 1962. In the last years of his life, however, Feyerabend himself partly changed his mind, acknowledging that his and Kuhn's positions were not so far apart.

Even towards Lakatos, who was a very good friend of his, Feyerabend had stinging words: he criticised research programmes taken as normative criteria for progress and accused him of being an anarchist 'in disguise'. Actually, as Kadvany points out, 'Feyerabend and Lakatos fundamentally agreed that no knowledge has certain foundations, including mathematics, nor are foundations needed', but provided different answers to Hume's problem of how the sceptic lives scepticism.

For all these reasons, the dispute whether Feyerabend belongs to one school or another seems entirely sterile: Feyerabend is no more Popperian than he is Wittgensteinian or Marxist, or positivist. He dialogued with his contemporaries as with the authors of the past, taking cues, reworking their suggestions, taking certain assumptions to their extreme consequences, constructing an original, complex thought, not devoid of inconsistencies and contrasts—because reality, or rather life, is rich, complex, dynamic, unpredictable and does not allow itself to be imprisoned in a static model that claims to explain everything and be valid in any possible situations. Feyerabend's great lesson, the reason why we remember him on the centenary of his birth and thirty years after his untimely death, is that he taught us that in philosophy, as in life, figures never add up: there is always at least some decimal point out of place, which leaves room for mystery and creativity.

Catholic University of Milan
Milan, Italy
Roberta Corvi

University of Trento
Trento, Italy
Stefano Gattei

# Editorial Note

We wish to thank all authors for accepting our invitation and for their kind availability. All contributions to the volume are original, except for Elisabeth A. Lloyd's, which was first published as 'Feyerabend, Mill, and Pluralism', *Philosophy of Science*, 64 (Supplement), 1997, pp. S396-S407 (and reprinted in *The Worst Enemy of Science? Essays in Memory of Paul Feyerabend*, edited by John Preston, Gonzalo Munévar and David Lamb, New York-Oxford: Oxford University Press, 2000, pp. 115-124), and for Friedrich Stadler's, which appeared as 'Paul Feyerabend and the Forgotten "Third Vienna Circle"', in *Vertreibung, Transformation und Rückkehr der Wissenschaftstheorie: Am Beispiel von Rudolf Carnap und Wolfgang Stegmüller*, edited by Friedrich Stadler, Vienna-Münster: LIT, 2010, pp. 169-187. We thank Elisabeth Lloyd and Friedrich Stadler for their permission to reprint their works.

Sadly, two contributors passed away before the volume reached the press: Joseph Agassi (7 May 1927–22 January 2023) and John L. Heilbron (17 March 1934–5 November 2023). Their works, extending over several decades, significantly influenced and shaped their respective fields, philosophy and history of science. We are very glad to include here their last contributions.

# Contents

# Editors and Contributors

## About the Editors

**Stefano Gattei** teaches history and philosophy of science at the University of Trento, Italy. He has worked extensively on key issues and authors of contemporary philosophy of science, as well as on the history of early modern astronomy and cosmology. He is the author of *Thomas S. Kuhn's "Linguistic Turn" and the Legacy of Logical Positivism: Incommensurability, Rationality and the Search for Truth* (Routledge, 2008), *Karl Popper's Philosophy of Science: Rationality without Foundations* (Routledge, 2009), and *On the Life of Galileo: Vincenzo Viviani's 'Historical Account' and Other Early Biographies* (Princeton University Press, 2019). He co-edited, with Joseph Agassi, the fourth volume of Paul Feyerabend's collected philosophical papers: *Physics and Philosophy* (Cambridge University Press, 2016).

**Roberta Corvi** teaches Theories of knowledge and Philosophy of mind at the Catholic University of Milan. Her research is mainly concerned with English-speaking contemporary epistemology, and several of her papers deal with issues related to rationality and the complexity of knowledge, chiefly focusing on post-positivist epistemology and American pragmatism, so as to identify a transdisciplinary path that emerges in the recently published, *Frontiere aperte: Verso un'epistemologia transdisciplinare* (Scholé, 2023). Her other monographs include *I fraintendimenti della ragione: Saggio su P.K. Feyerabend* (Vita e Pensiero, 1992), *An Introduction to the Thought of Karl Popper* (Routledge, 1997), and *Ritorno al pragmatismo: L'alternativa Rorty-Putnam* (Mimesis, 2017).

## Contributors

**Joseph Agassi** Tel-Aviv University, Tel Aviv, Israel
York University, Toronto, ON, Canada

**Roberta Corvi** Department of Philosophy, Catholic University of Milan, Milan, Italy

**Val Dusek** Department of Philosophy, University of New Hampshire, Durham, NH, USA

**Stefano Gattei** Department of Sociology and Social Research, University of Trento, Trento, Italy

**John L. Heilbron** University of California at Berkeley, Berkeley, CA, USA
California Institute of Technology, Pasadena, CA, USA

**Helmut Heit** Kolleg Friedrich Nietzsche, Klassik Stiftung Weimar, Weimar, Germany

**Paul Hoyningen-Huene** Institute of Philosophy, Leibniz University, Hannover, Germany

**John Kadvany** Policy & Decision Science, Menlo Park, CA, USA

**Rory Kent** Department of History and Philosophy of Science, Cambridge University, Cambridge, UK

**Ian James Kidd** Department of Philosophy, University of Nottingham, Nottingham, UK

**Vasso Kindi** Department of History and Philosophy of Science, National and Kapodistrian University of Athens, Athens, Greece

**Elisabeth A. Lloyd** Department of History and Philosophy of Science and Medicine, Indiana University Bloomington, Bloomington, IN, USA

**Gonzalo Munévar** Lawrence Technological University, Southfield, MI, USA

**John Preston** Department of Philosophy, University of Reading, Reading, UK

**Friedrich Stadler** Institute Vienna Circle, University of Vienna, Vienna, Austria

# Part I
# Feyerabend and the History of Philosophy

# Chapter 1
# Feyerabend, Protagoras and Nietzsche

**Helmut Heit**

**Abstract** This paper reconstructs Paul Feyerabend's dialogue with two other representatives of allegedly self-contradictory relativism: Protagoras and Nietzsche. I argue that such a comparison mutually clarifies Feyerabend's arguments in favor of a certain kind of relativist anti-dogmatism. Unlike the widely accepted image, a closer reading reveals that neither Feyerabend nor Nietzsche or Protagoras adopt the so-called equality-thesis. All three authors argue against the absolutism and dogmatism of their time. They subject human reason and scientific activity to a critical analysis and strive for a fundamental conceptual and cultural transformation. They share the suspicion that the concrete scientific practice and its theoretical self-understanding are not identical, and that it could be possible and useful to enlighten science about certain prejudices about itself.

**Keywords** Relativism · Absolutism · Dogmatism · Man is measure · Homo Mensura · Self-contradiction · Post-truth

> 432. *Investigators and experimenters*. – There are no scientific methods which alone lead to knowledge! We have to tackle things experimentally, now angry with them and now kind, and be successively just, passionate, and cold with them (Nietzsche 1997[1881], §432).

Hardly any history of relativism fails to mention the ancient Greek sophist Protagoras, the nineteenth century German thinker Friedrich Nietzsche and the Austrian philosopher Paul Feyerabend. "The first known statement of a relativist position in Western philosophy is the famous dictum 'man is the measure of all things' by the Sophist Protagoras" (Baghramian 2004, 16), while Nietzsche is

H. Heit (✉)
Kolleg Friedrich Nietzsche, Klassik Stiftung Weimar, Weimar, Germany
e-mail: helmut.heit@klassik-stiftung.de

© The Author(s), under exclusive license to Springer Nature Switzerland AG 2024

S. Gattei, R. Corvi (eds.), *Feyerabend in Dialogue*, Boston Studies in the Philosophy and History of Science 346,
https://doi.org/10.1007/978-3-031-71938-7_1

"possibly the most influential single philosopher of relativism in recent history" (Baghramian 2010, 45). Feyerabend's position in the history of relativism is similarly undisputed, not the least since he—unlike his two famous predecessors—explicitly adopts and defends certain versions of relativism. Despite the fact, however, that Protagoras counts as the inventor of relativism and Nietzsche as its most explicit modern figure-head, their relation to the infamous "worst enemy of science" (Theocharis and Psimopoulos 1987, 596) was never addressed in detail. This paper aims to start filling that desideratum. My hope is that such a comparison can contribute to a mutually improved understanding of all three authors. Assuming that the readers of this book are more familiar with Feyerabend as with Protagoras or Nietzsche, I put more emphasis on the reconstruction of their thoughts.

In addition to this reconstructive and comparative ambition, I argue that such a more detailed historical study clarifies the systematic arguments in favor of a certain kind of relativism. Unlike the widely accepted negative image, a closer reading reveals that Feyerabend is as little an enemy of the sciences as Nietzsche or—for that matter—Protagoras were; rather, here as there, criticism is confused with enmity. All three authors subject human reason and our scientific endeavor to a critical analysis and strive for a fundamental conceptual and cultural transformation. They argue in favor of modesty against certain versions of absolutism and dogmatism of their time. They share the suspicion that the theoretical self-understanding does not match the actual scientific practice, and that it is possible and useful to enlighten science about certain prejudices about itself. Among these prejudices they include a number of metaphysical assumptions and unifying dogmatic methodologies that could be removed from the sciences of the future. Moreover, they question the cultural task of the sciences, whose inviolable hegemony Feyerabend compares to the social function of a church. They ask what role sciences can and should play in our world. Relating Feyerabend to Protagoras and Nietzsche helps to clarify and strengthen his reasons for relativism.[1] It also clarifies the scope of these relativist position. A careful reading evidences that all three argue for dependency and pluralism: there is a variety of human beliefs and all are relative to human epistemic systems and practices. None of these three, however, proposes epistemic symmetry or equal value of any given belief. This is particularly informative because if these three godfathers of relativism do not argue for the indistinguishably equal validity of propositions, who does? In contrast to stereotyping clichés of irresponsible and self-contradictory defeatism towards reason, this paper argues that Feyerabend's dialogue with Protagoras and Nietzsche provides good arguments to adopt relativism as a productive and virtuous epistemic attitude.

[1] This paper continues a line of reasoning I proposed in "Reasons for Relativism: Feyerabend on the 'Rise of Rationalism' in Ancient Greece" (Heit 2016), where I argue that Feyerabend's reconstruction of the origin of the Western rational tradition is decisive for his turn to relativism.

## 1.1 Protagoras and the Human Measure

Feyerabend refers to Protagoras of Abdera (490–411 BC) and certain tropes associated with this early Greek sophist at various occasions in his published works and regularly discussed Plato's *Theatetus* in lectures and seminars, but he never devoted a detailed analysis to him. Towards the end of the preface of his two volumes of philosophical papers published in 1981, he points out that the "reader will notice that some articles defend ideas which are attacked in others. This reflects my belief (which seems to be held by Protagoras) that good arguments can be found for the opposite sites of any issue" (Feyerabend 1981, xiv). In *Against Method* he states: "Protagorean relativism is reasonable because it pays attention to the pluralism of traditions and values. And it is civilized for it does not assume that one's own village and the strange customs it contains are the navel of the world" (Feyerabend 1993, 226, cf. 1978a, 28). On this position, he elaborates more extensively in his *Notes on Relativism* in *Farewell to Reason* (Feyerabend 1987, 42-53). This text distinguishes various dimensions of his understanding of relativism and argues that Protagoras captures better than Plato or Popper what scientists are actually doing. "What we have, therefore (apart from grandiloquent promises and superficial popularizations), is a variety of approaches based on a variety of models and successful in restricted domains, i.e. what we have is a Protagorean practice" (Feyerabend 1987, 53). Similar consideration can be found in *Science in a Free Society* or in the opening sections of *Three Dialogues on Knowledge*. His last work, *Conquest of Abundance*, again contains a lengthy discussion of the notion of sophistic *antilogiké* in a chapter on Parmenides and the logic of being (Feyerabend 1999, 72 f.).

While his references to Protagoras are noted occasionally, I am not aware of a detailed analysis. John Preston devotes a few pages on the topic and comes to the conclusion that Feyerabend's "actual relativist pronouncements are not well thought through. [...] His defense of Protagorean relativism either leaves Protagoras behind, or falls victim of the same kind of objections Plato leveled against Protagoras' views" (Preston 1997, 209). Martin Kusch suggests that "Feyerabend's intriguing reinterpretation of the Protagoras of Plato's *Theaetetus* seems worthy of a more detailed analysis in a different context; it makes contact with some of the best recent literature on the topic by presenting Protagoras as an infallibilist rather than a relativist" (Kusch 2016, 109). However, he does not devote any further analysis to this relation. Instead, Kusch introduces nine dimension of what he considers the 'standard model' of relativism and moves on to check their applicability on Feyerabend's later texts. Instead of asking what Feyerabend means by relativism, Kusch concludes from the difference between his model and Feyerabend's theses that he is either unclear or describes considerations as 'practical relativism' or 'epistemic relativism' that are not relativism at all. I think this method neither fits well with Feyerabend's historically informed arguments nor does it do justice to his thoughts. In contrast to formal definitions, in the following I would like to take Feyerabend's actual arguments for his special relativism as a starting point, by means of a closer look at Protagoras.

Protagoras worked primarily in Athens as a successful teacher, political advisor, and author. His writings were influential and important components in all central fields of contemporary philosophical debate, with the exception of natural philosophy, in which he showed little interest. Although almost everything Protagoras ever taught and wrote is lost, he counts among the most distinctive and provocative figures in early Greek philosophy. We hear that his books were collected and burned in Athens during his lifetime after a trial for impiety (*asébeia*) (80A1).[2] Superficially, he was sued for the sober agnostic statement that for human beings the question of the divine could not be answered (80B4). In the background, as in the similar case of his contemporary Anaxagoras (59A1, 59A17), there might have been the political motive to harm his protégé Pericles with the trial against Protagoras. The poor source situation in the matter of Protagoras, however, is not because all copies of his writings were destroyed. Plato, Aristotle, Sextus Empiricus and others were able to access his texts for a long time. Diogenes Laërtius mentions books on dispute arts (*Téchnē eristikōn*), constitutions (*Perì politeías*), general knowledge (*Perì tōn mathēmátōn*), virtue (*Perì aretōn*), and other topics (80A1). That we do not have ten fragmentary sentences of all this today mainly result from the fact that the doxographic perspective of Aristotle and his successors strongly determined early philosophical historiography. Our present source situation is the sediment of a value judgment, carried forward over centuries, according to which we need as much as possible from Plato and almost nothing from Protagoras.

The lasting fame of Protagoras derives primarily from statements that seem to identify him as a radical critic of absolute concepts of knowledge. Nothing can be known about the nature of the gods (80B4), contradictory statements can be made about everything (80B6a), the weaker case should be made stronger (80B6b) and—most famously—that man is the measure of all things (80B1). Especially through the *Homo Mensura Statement* (hereafter: HMS) Protagoras has had considerable impact; that is, he has caused offense. It is hardly an exaggeration to count HMS among the most successful philosophical catchphrases of all time (Berkel 2013: 38), even in comparison to Feyerabend's famous 'anything goes'. The phrase is said to have been the prelude of a writing *On Truth* or with the title *Crushing Speeches* and is handed down by Plato as well as by Sextus Empiricus and by Diogenes Laërtius. While the three versions are not quite identical, the differences relate only to the grammatical form of individual words, which in Plato's case is probably due to the sentence order of the quotation in his text. The version of Sextus Empiricus seems to correspond most closely to the Protagorean wording: "The measure of all things is man, of the being that/as they are, of the not being that/as they are not (*pantōn chremátōn métron estìn ánthrōpos, tōn mèn óntōn hōs éstin, tōn dè ouk óntōn hōs*

[2] I cite texts of early Greek philosophy according to the standard edition by Hermann Diels and Walther Kranz (eds.): *Die Fragmente der Vorsokratiker*. 3 vols. 6th ed., Berlin 1952. I use the established system of abbreviations: The first number stands for the respective philosopher (80 for Protagoras), the character 'A' refers to a testimony or paraphrase while 'B' denotes (supposedly) original quotations, and the second number counts the single fragments. References to Plato and Aristotle are also via standard abbreviations.

*ouk éstin*)" (80B1). What exactly this sentence means and what consequences result from it has always been disputed and eludes a final clarification already because the original context is missing. Various physiological, linguistic, sociological or epistemological considerations are possible to give the sentence concreteness and plausibility, but we do not know the justifications and defenses of Protagoras. Accordingly, the readings are full of variations. In addition, each of the words used is in need of clarification. Thus, it makes a considerable difference whether *anthrōpos* refers to a single individual, to socialized man in community, to all humankind, or to 'the human' in the collective singular. Interpretations can accordingly vacillate between simplistic subjectivism, intersubjectivity models, and an ultimately downright Kantian understanding of *anthrōpos* as a transcendental subject. Similar problems affect the other side of the relation. Does *panta chremata* refer to all respective individual things, or to the totality of things, or to their interrelation as a world? Is it only about material objects or also about cognitive objects? Do we speak of the existence of the things themselves or of their properties and qualities? On the answer to these questions also depends how the unclear speech of the *metron* is to be understood. Does it make man a factual truth-maker, or an epistemic criterion, or is it to emphasize that knowledge is always situated in a way to which man also belongs? The terminology of being (*estin*) and not-being (*ouk estin*) of Being (*ontōn*) and Not-Being (*ouk ontōn*) was introduced by Parmenides and opens up another field of problems, especially since Protagoras seems to position himself at a critical distance from the "extremism" of Eleatic ontology (Schiappa 2003: 121). Interestingly, many of these problems already come up in the little word *'hōs'*, since it possibly translates both as 'that' and as 'how'. When translated as 'that', the being or non-being of things seems to depend on man as the measure. A translation with 'how', on the other hand, might relate the properties or qualities of things to a human measure rather than the things themselves. Against the background of these ambiguities, to reduce the name of Protagoras simply to a label for inconsistent relativism evidences a regrettable stereotyping of thought. To achieve a better understanding of Protagoras and the way Feyerabend adopts his thought, I suggest having a closer look at the sources.

The earliest detailed discussion of Protagoras position is laid out by his intellectual and political opponent Plato in his epistemological dialogue called *Theaítētos*—a text Feyerabend regularly and extensively discussed in seminars, lectures and publications. Plato equates the HMS with a sensualist theory of knowledge. After Socrates challenged his interlocutor, the young mathematician Theaetetus, to give a definition of knowledge, the latter hesitantly proposes: "knowledge (*epistēmē*) is nothing other than perception (*aisthēsis*)" (Plato Tht., 151e). Socrates praises this proposal, noting that Protagoras had presented quite the same idea in a different form with his notion of *metron anthrōpon*. The connection between human measure and perception is explained by an example of wind. While to one person the same wind appears warm, to another it appears cold. According to this explanation, Protagoras' theorem seems to refer to the relation between qualitative properties and individual sensations. The quality of the wind is measured by the particular person and her/his sensations. A particular wind is not intrinsically warm or cold, but these attributes make sense only in relation to a specifically sensing subject. Although this idea seems harmless at first, Plato argues that Protagoras' proposition ultimately cancels itself out because it is self-contradictory.

Assuming that knowledge is identical with perception, and assuming that two different propositions cannot be true at the same time and in the same respect, the HMS seems to contradict itself not only in its application but also formally. In particular, the HMS seems to undermine its own claim to validity. For if Protagoras is right in that things are in truth as they appear to one in each case, it follows that the opposite judgment is also true, namely, if to another person Protagoras' proposition appears to be false. "Thus he would admit that his own opinion is false if he admitted that the opinion of those who hold that he is mistaken is true" (Plato Tht., 171b). If it follows from Protagoras' statement that the opposite of this statement is also true, then it is self-contradictory. According to Sextus Empiricus, already Democritus uses this kind of argument against Protagoras (68A114), which was made more precisely by Aristotle (Met., 1062bff) and has been used ever since against various versions of relativism. A self-contradiction occurs when the implications of one's claim refute that claim. When Protagoras says that all truth is relative to man, this seems to imply, 'It is an absolutely true statement that there are only relatively true statements.' This would make the self-contradiction clear, for these two statements are mutually exclusive. The second part of the sentence refutes the first. However, I argue in accordance with the majority of recent scholarship that this formal objection is hardly effective against Protagoras (Schiappa 2003, 126–130; Zilioli 2007, 134). This does not render Protagorean relativism a philosophically valid position but it points to the need for a more detailed refutation.

The most perceptive reconstruction of this kind of refutation can be found, in my impression, in two essays by Myles F. Burnyeat from 1976, which Feyerabend read very carefully.[3] The first paper argues that Sextus Empiricus, following Aristotle, wrongly conceived of Protagoras as a crude subjectivist and thus created an easy opponent for refutation (Burnyeat 1976a). The second paper argues that the Platonic reconstruction is more accurate and subtle, which at the same time makes his refutation more valid. According to Burnyeat, the HMS contains in particular the thesis: "P1. If it seems to x that p, then it is true for x that p" (Burnyeat 1976b: 178). One may wonder if the reduction of a content-free ancient fragment to such a formalism really captures the essential parts of the text. It should at least be noted that Protagoras did not use the word "true" in the sentence in question. With this reading, however, Burnyeat translates Protagoras' phrase into a conditional determination of truth and then states:

> P1 is indeed Protagoras' major contribution to the extreme empiricist epistemology which the dialogue elaborates out of Theaetetus' definition of knowledge as perception. It guarantees that, no matter how 'wild' and variable (from the common-sense point of view) a person's experience may be, the judgments he is led to will be true for him, giving him certain knowledge of how things are for him. If, as Protagoras holds, all things are for each person

[3] Feyerabend's heavily annotated copy of these articles is kept in the philosophical archive in Konstanz with the number PF 6-6-18.

> exactly as they appear to him, no one can be mistaken about the reality that confronts him, of which he is the sole authoritative judge (measure). (Burnyeat 1976b: 178)

Burnyeat proposes an individualist reading of *anthrōpos*, associates *panta chremáta* with the reality of all things and translates *métron* into the sole authority of certain knowledge. This is the most extreme interpretation of the statement in question. Moreover, he silently introduces sociological, ontological and epistemological criteria, whose validity are well worth discussing. The rhetorical assumption of a presumably unproblematic social 'common-sense perspective' in contrast to the 'wild' and 'changing' experiences of individuals actually deserves to be bracketed. Distrust of the notoriously unreliable common-sense of the common people and their everyday opinions is virtually constitutive of early Greek philosophy, and even more so of Plato. Ultimately, the pejorative juxtaposition of 'wild' vs. 'common sense' rather confirms Protagoras' theorem by pitting the 'normal' human as the authoritative perspective against the idiosyncratic (philosophical) individual. Even more problematic, however, are the assumptions that a Protagorean relativist 'cannot be wrong' because she/he is the 'sole authoritative judge' with respect to 'the reality that confronts him'. Burnyeat thus not only postulates a stable, objective, external reality about which it is possible to be mistaken, he also implies that this reality confronts us in a clear and direct way. The ontological status and accessibility of this reality, however, is precisely the critical object of debate. The HMS emphasizes that the reality of things does not simply confront us, but that it has to be examined, interpreted and developed by human beings according to human, all-too-human capacities and needs. Thereby Protagoras does not even have to deny the ontological existence of a subject-independent reality. It suffices to argue that this reality is never given to us without human measures. More importantly, such a cautious and self-reflexive view on the relation between man and world most naturally combines with a modest epistemology. It is hardly plausible to assume 'certain knowledge' and 'sole authority' for any judgment, if HMS is taken serious. The sentence of Protagoras does not result in a self-righteous carte blanche, according to which one can never be mistaken, but on the contrary in the prudent warning that we probably do not have the simple truth at our disposal.

Burnyeat reproduces an argumentative trick already used by Plato. Protagoras fails not because of inconsistent implications of his own statement, but because his opponents apply standards whose validity and usefulness he puts into question. Plato, too, in preparing his critique of Protagoras, had noted in passing: "Perception (*aísthēsis*) is probably always one of being (*toũ óntos*) and infallible (*apseudès*), if it is then knowledge (*epistēmē*)" (Plato Tht., 152c). This is how an absolutist criterion of truth is taken as a basis in advance and determined as an adequate relation between a belief and reality. By guilelessly accepting this requirement in the course of the dialogue, the verdict on Theaetetus' first proposition (and also on the following ones) is spoken. By making the infallible recognition of a purely objective being the standard of knowledge, the Platonic Socrates postulates a criterion of knowledge, which is not only diametrically opposed to HMS, but must swamp any empirical or sensualistic epistemology. Moreover, if truth is defined as infallible correspondence with the thing, it cannot, of course, be understood without contradiction as relative to man. In a not particularly dialogical spirit, this criterion is set quite implicitly without open debate about its justification and fruitfulness. In *Conquest of Abundance*, Feyerabend points to this questionable Platonic strategy. While Socrates only apparently aims at a definition of knowledge in an open spirit, he already applies criteria such as infallibility and stability. "Yet Socrates argues as if the term had already been defined and comes to a clear and unambiguous conclusion. Thus it was not the argument that produced the conclusion (i.e. the rejection of Theaetetus's thesis that knowledge is perception) but the conclusion (the rejection) produced the argument" (Feyerabend 1999: 75).

Protagoras obviously suspends this Platonic requirement when he conceives of man as the measure of the qualities and how-ness of things. His central insight seems to be that we only ever have the putative properties of the 'reality that confronts us' in the medium of human appropriations and translations, the relative validity and credibility of which can be debated. He does not contradict himself but rather avoids an over-ambitious and not applicable philosophical ideal of absolute authoritative expert-knowledge. Ironically, Plato's absolutist criterion is too demanding not only for the first proposal of Theaetetus, but also for today's textbook definition, according to which knowledge consists in true and justified belief (Plato Tht., 202c). Socrates also explicitly rejects the definition of knowledge as 'justified true belief', for the criterion of truth remains dependent on the criterion of good justification and vice versa. It is unclear how to prove that the truth-criterion is matched seperately from the standards of good justification, and we do not know what we mean by a good justification if it is not true (Plato Tht., 210ab). The dialogue therefore closes without reaching its goal, leaving us with the advice "prudently not believe to know what you do not know" (Plato Tht., 210c).

With this insight into the limitedness of human knowledge, Plato reaches a conclusion ultimately more devastating than Protagoras, who constantly defends the idea that virtue is teachable and that improved knowledge is possible. In contrast to the Abderite, however, the Platonic Socrates seems to oscillate between two extremes, according to which knowledge is either absolutely certain or none at all. If one were to impute an eristic intention to Plato, the *Theaítētos* appears to be virtually a pointed debate that, starting from a false alternative, systematically undermines confidence in the human capacity for knowledge. Protagoras avoids this exclusive disjunction. With regard to the Platonic ideal of infallible knowledge of what exists, he may have advocated a similarly moderate agnosticism as with regard to the gods: "Of the gods I am unable to know anything, either that they are or that they are not; for much hinders us in this knowledge, both the obscurity of the matter and the brevity of human life" (80B4). Protagoras could not have assumed that a perception is valuable and real knowledge only if it grasps infallibly that which exists in itself. It is the essence of perception to refer to the changeable world of experience, and not to a being that might underlie these experiences. Consequently, it can never be infallible, but always only more or less clear and credible.

This approach also fits well with two other famous Protagorean statements, which also resonate with Feyerabend: "About every thing there are two opposing arguments (*logoi*)" (80B6a); and, "It is necessary to make the weaker argument stronger (*ton hēttō logon kreittō poiein*)" (80B6b). Much again depends on how exactly these phrases are to be understood and how the words *hēttō* and *kreittō* in particular are to be translated. In a pejorative reading, as suggested by Aristophanes, Plato, and Aristotle, *hēttō* appears as the untrue and unjust thing and *kreittō* as the good and true. Thus, the maxim of Protagoras would amount to an invitation to fraud that purposefully and deceptively confuses right and wrong. What is irritating about such an interpretation is that Protagoras, according to everything else we know about him, was a honorable man with moral integrity (Schiappa 2003, 107). If, on the other hand, we take into account that Protagoras taught his students to reprove and praise the same person (80A21), we can see the sentence more as a maxim for balanced judgment. It fits with the observation that *hēttō* and *kreittō* in the fifth century are probably not to be translated as 'good' and 'bad', but rather as 'weaker' and 'stronger' (Schiappa 2003, 113). According to this, it would not be a matter of exchanging right and wrong, but of taking sides argumentatively for what at first sight seems to be the less promising cause. This art of weighing can be understood either rather didactically, or as an expression of the conviction that alternative interpretations can never be completely ruled out. Understood in this way, Protagoras's maxim would not be an expression of destructive belligerence, but a useful exercise in theory and an important skill in practice. It may prove useful as a method of cognition to pursue ostensibly futile hypotheses and thus test the potential dogmatism of the established majority opinion. It could be that the weaker argument proves to be the stronger one under changed problem constellations. Epistemic relativism thus combines with the idea of theory proliferation. This is exactly, how Feyerabend understood that dictum, when he argued in favor of counterinductive procedures. Actually, he states that such a "*pluralistic methodology*" is not only advisable, but rather defines scientific practice today: "The task of a scientist, however, is no longer to 'search for the truth' or 'to praise god', or to 'systematize observations', or 'to improve predictions'. These are but side effects of an activity to which his attention is now mainly directed and which is '*to make the weaker case stronger*', as the sophists said, *and thereby to sustain the motion of the whole*" (Feyerabend 1993, 21).

One may think of Feyerabend's assault against established theories and positions in philosophy of science as applications of such a sophistic counter-rule. It is also worth noting how he applies it to his own philosophical enterprise as well. In the preface of the first two editions of *Against Method*, he states: "There may, of course come a time, when it will be necessary to give reason a temporary advantage and when it will be wise to defend its rules of exclusion of everything else. I do not think that we are living in such a time today." (Feyerabend 1975, 22). While revising this book again for the third English edition in 1993, Feyerabend remarkably adds: "This was my opinion in 1970 [...] Times have changed. Considering some tendencies [...] I think that reason now should be given greater weight not because it is and always was fundamental but because it seems to be needed, [...] to create a more human approach" (Feyerabend 1993, 13). This passage not only evidences, how Feyerabend thought twice about central issues, it also shows that even within a relativist framework, the balance between weaker and stronger cases can change. Actually, to speak about weaker and stronger cases or arguments already proves, that Feyerabend and Protagaros did not assume that any given proposition is equally valid in any given context or situation, or that no qualitative evaluations can be made.

The assumption that man is the measure does not render attempts futile to compare and evaluate different views in the hope to reach justified agreement. This is well illustrated by an anecdote that has been handed down by Plutarch, among others. After a boy was killed by a spear on a sports field, Protagoras and Pericles discussed whether the thrower, the spear, the boy, or the supervisor should be held responsible for the incident. For a whole day they are said to have wrestled over "the right judgment", the *orthótaton lógon* (80A10). The problem is not artificial and the answer is of high political, social and legal importance. If Protagoras and Pericles debated it for a whole day, one may assume three things: first, they obviously take the problem seriously; second, even clever people find it difficult to give a simple and unambiguous answer. Third and most importantly, they think it possible and worthwhile to engage in such discourse even if man is measure. The practice of weighing correctness takes the place of purely objective truth: "The notion of ‚correctness' (*orthon*) is the criterion which enables one to unravel the problem: given that all perspectives make a legitimate claim to truth and validity, the problem is to find the solution which best fits the situation without pretending to be offering a univocal and abstract answer "(Bonazzi 2013, 28). The right *logos* is not the (unavailable) ideal and true *logos*, but it is the relatively best *logos* according to human judgment, appropriate to the situation. The difference to Plato consist in the fact that Protagoras denies the recognizability of a purely objective and at the same time absolutely true *logos*. If one considers that today hardly anyone takes Platonic optimism of truth to be justifiable, the position of Protagoras does not only appear theoretically in a different light; it also has other consequences. Protagoras takes on a complex task as educator and advisor, which Plato – according to the seventh letter – withdrew from out of annoyance with the political circumstances. This is also how Feyerabend perceived Protagoras as political relativist and democratic philosopher: "'You and I,' Protagoras seems to say, 'our physicians, artists, artisans know many things and we live as we do because of this knowledge. Now these philosophers call our knowledge opinions based on shiftless experience and contrast 'the many', i.e. people like us, with the enlightened few, i.e. themselves and their strange theories" (Feyerabend 1987, 50).

## 1.2 Nietzsche and the Perspectival Condition

The debunking assault against prevalent absolutism and dogmatism connects Protagoras and Feyerabend with the third most infamous anti-rationalist. A copy of Nietzsche's collected works is kept among the books of the Feyerabend-collection at the philosophical archive in Konstanz. In *Killing Time* he mentions that he had read *Thus Spoke Zarathustra* in the 1940s and was fallen for the same "bombast" for a while (Feyerabend 1995, 50). This fits with letters he exchanged with Babette Babich in which Feyerabend admitted, that he had read Nietzsche but that he did not like him (Babich 2010, 142). And indeed, Nietzsche appears in Feyerabend's works even less than Protagoras. Gordon Bearn, however, was right to point to central similarities in their philosophical modes and argumentative strategies regarding relativism. Rather than stating relativism as a (probably self-contradictory) doctrine, they both produce debunking polemics against absolutism, describe deep disagreements between competing perspectives, and exemplify the advantages of modesty. "These are the voices of relativism" (Bearn 1986, 151). Moreover, at one point Feyerabend explicitly associates his views with those of Nietzsche. In a German essay with the title: *Die 'Rationalität' der Forschung* he distinguishes four positions: A naïve rationalism, represented by Descartes, Kant and Popper, for example, and a context-dependent rationalism, which he identifies with Marxists and anthropologists. He confronts these two views with exceptions and counterexamples, thus rejecting their implicit or explicit claim to universal validity. He sees a third position in naïve anarchism, with which his philosophy of science is often confused. To counter such confusion and to clarify "my own 'position'", whereby he does not neglect to put 'position' in quotation marks, Feyerabend mentions "precursors" of this approach, "for instance Kierkegaard's *Concluding Unscientific Epilogue*, Niels Bohr, and also Nietzsche, if only he would not take himself so seriously" (Feyerabend 1978a, 343).

In contrast to naïve anarchists, Feyerabend's places his position actually more down to earth. From the fact that all rules and standards have their limits, he does not derive the conclusion that one must disband rules and standards whatsoever, or that one can even practically do without them (Feyerabend 1978a, b, 344). Rather, he demands that "every standard that guides a course of action is itself made a part of the course of action" (Feyerabend 1978a, b, 345). According to this view, the rules and methodological standards by which we conduct scientific research or other rational investigation are not simply applied, but are shaped and modified, invented and altered in use. We apply measures to evaluate the quality of propositions, but these measures are also relative to the epistemic practices and goals. They are inseparable from the course of action. Feyerabend calls such a practice a science for free people (Feyerabend 1978a, b, 349) and it resembles the Protagorean quest for a right judgment. In the following I will show that this view indeed connects Feyerabend with Nietzsche, and that he thereby is correct to count him among his predecessors.

Nietzsche was an intent and thoughtful observer of the discussions about human knowledge, its limits and prospective in the later nineteenth century. During his readings he encountered the concept and idea of relativism at various occasions (Heit 2019). Authors such as Schopenhauer, Comte, Mill, Lange, Guyau, Spir and others opposed absolutist aspirations and insisted on the conditional and contextual character of human knowledge. Some of them, namely Comte, Lange und Guyau, introduced the term 'relativism' in a positive way to denote the liberating anti-absolutist and post-Kantian philosophy of their time. According to Comte, the beneficial validity of "relativisme" not only applies to the entire intellectual domain, where its general acceptance is no longer seriously disputed, but also to the practical and moral order (Comte 1855, 23).[4] According to Comte, relativism is not only the informed, reflexive, and modest mind-set of modern science, it is also the only intellectual attitude that ensures human sympathy and cultural progress. Nietzsche owned a German translation of introductory parts of Comte's *Cours de philosophie positive* (1839). The word 'relativism' does not appear in it, but the very idea is present. Comte associates the impression of "unconditioned knowledge" with the theological and metaphysical age of humankind, while the mind of our scientific stadium finally renounces from such ambitions and "realizes the impossibility to achieve absolute knowledge" (Comte 1880, 3).[5] According to John Stuart Mill's sympathetic explication, this is the core doctrine of Comte's positivism: "We have no knowledge of anything other than phenomena, and our knowledge of phenomena is relative, not absolute" (Mill 1874, 4).[6] We only know the relation between facts, their successions and similarities, the essence of things is unknown and unknowable to us. Nietzsche carefully read and annotated this passage and he might have come to the conclusion, that Kantians, positivists and even inductivists of Mill's persuasion acknowledge some inevitable relativism.

The idea and the term 'relativism' appears at several occasions in Lange's *History of Materialism*, first published in 1866, and even more prominent in the following revised editions. This book belongs to the most important philosophical influences on Nietzsche. Lange draws a line of relativist thought from the ancient Greek materialists and sophists via Thomas Hobbes and Baron d'Holbach to contemporary philosophy and science and mainly grants it to the philosophical rigor of Kant and the intellectual talent of French thinkers "that today the exact science in all domains of experience no longer set up absolute truths, but only *relative* ones; that the conditions of the acquired knowledge are always recalled, and that the accuracy of all doctrine is justified on the *reservation of the progress of knowledge*" (Lange 1866, 244). The combination of empiricism and fallibism with a hypothetico-deductive understanding of science leads Lange to a certain concept of relativism. This concept not only means to provide an appropriate understanding of human knowledge and science; it is also taken as a better, more promising, and self-conscious epistemic attitude. When Nietzsche entered the world of philosophy, the collapse of

[4] Maria Baghramian fails to note the use of relativism in Comte. Referring to the *Oxford English Dictionary*, she writes that John Grote was the first to introduce 'relativism' as a technical term in 1865 (Baghramian 2004, 11). Grote indeed claims that the "notion of the mask over the face of nature is […] what I have called 'relativism'" (Grote 1865, 229). Comte, however, spoke of 'relativisme' 10 years earlier in 1855.

[5] Nietzsche owned a copy of a German translation of Comte's *Cours*, which is kept in the Herzogin Anna Amalia Bibliothek in Weimar. The reference number is C 246.

[6] The reference number of Nietzsche's heavily annotated copy of this book in his private library is C 713-b. Friedrich Albert Lange suggested (in a treatise most likely unknown to Nietzsche) that Mill "appears to me, namely in a purely formal sense, as a master in the handling of the relativism which forms the soul of all the exact sciences" (Lange 1866, IV). This is one of the first appearances of the term "Relativismus" in a German text.

absolutism was obvious and hardly anybody defended it. Anti-absolutism was basically a fundamental assumption of later nineteenth century philosophers.

Accordingly, Nietzsche took relativism of one kind or another to be the most advanced and widely accepted contemporary epistemological view. The opening remarks of *Beyond Good and Evil* state that "dogmatism of all types standing sad and discouraged," it "is in its last gasps" and basically failed to conquer truth (Nietzsche 2001[1885], 3). This negative diagnosis regarding dogmatism is presented as a matter of fact, as the widely accepted view. Nietzsche does not mean to be original here, he basically takes it for granted. The defeat of dogmatism is not Nietzsche's main task, he is more interested in the causes of its downfall. The failure of dogmatism, Nietzsche assumes in a moderate claim on originality, results from its overly solemn approach, from prejudices, false assumptions, and over-ambitious expectations. Traditional philosophers tried to erect eternal systems, but the allegedly "sublime and unconditional philosophical edifices that the dogmatists used to build" actually rested on "folk superstition" on a "seduction of grammar, or an over-eager generalization from facts that are really very local, very personal, very human-all-too-human" (Nietzsche 2001[1885], 3). Far from being unconditioned or absolute, these philosophical systems were relative to their culture, their common-sense, their language, and their perspectives. According to Nietzsche, ideas such as 'pure spirit' or 'the Good in itself' are "the worst, most prolonged, and most dangerous of all errors to this day", they meant "standing truth on its head and disowning even the *perspectival*, which is the fundamental condition of all life" (Nietzsche 2001[1885], 4). The institutionalized quest for objective and unconditioned truth itself lead to the self-undermining insight that we human beings might be peculiar animals, we are even able to comprehend the perspectival condition of life, but we do not escape it.

One may wonder if such views qualify for relativism. If one, however, assumes an exclusive disjunction between absolutism and relativism, Nietzsche is relativist. Occasionally, he is even singled out as the godfather of modern relativism: "The saying, 'There are no facts, only interpretations' could serve as a motto for the relativist movement. It comes from late notes of Friedrich Nietzsche, probably the greatest figurehead of that tendency since Protagoras" (Blackburn 2005, 90). The—wrong cited—notorious quote is taken from Nietzsche's late notes written down in spring 1887. It is a private jotting, a memo not meant for publication and reads as follows:

> Against positivism which halts at the phenomenon 'There are only facts' I would say: no, precisely facts do not exist, only interpretations. We cannot determine any fact 'in itself': perhaps it is a nonsense to want such a thing. 'Everything is subjective', you say: but that itself is an *interpretation*, the 'subject' is not a given, but an added-on-fiction, tucked-behind. – Is it at last necessary to posit the interpreter behind the interpretation? Even that is poetry, hypothesis. Inasmuch as the word 'knowledge' ['Erkenntnis'] has any meaning at all, the world is knowable: but it is variously *interpretable*; it has no meaning behind it, but countless meanings. 'Perspectivism'. [...] (Nietzsche 2003[1887], 139).

Despite its provisional character, the text is clear enough. The apparent denial of facts is not simply stated as a doctrine but introduced as an anti-thesis and anti-dote to a positivist assumption, that there are clear, plain, and obvious facts alone. Moreover, the text actually admits the phenomenon of 'facts', but requires and invites further critical examination. In opposition to positivism Nietzsche states that there are no facts without interpretation, i.e. without selection, valuation, adoption, contextualization, simplification, reduction of complexity, etc.. The note continues with a warning that the significance of interpretation should not be confused with subjectivism. The world is knowable to us, in a way according to us, to our capacities and needs. The alternative to objectivism is not subjectivism, but a certain kind of anthropocentrism. Human beings form interpretations of the world we live in, which must—for the sake of survival—be somewhat successful. In consequence, Nietzsche argues that interpretations could be good or bad, despite their inevitable under-determination by data or text. They are ordering simplifications of empirical abundance relative to specific human standards. These standards accord to the needs and values of living human beings and different standards and different successful interpretations, i.e. different worlds of experience are possible, they were and still are actually real.

The idea of manifest alternative worldviews is actually the most relevant concept Feyerabend adopts from Nietzsche. To make that point, however, Feyerabend does not refer to any of the texts mentioned above, but rather to a passage from Nietzsche's early essay *On Truth and Lies in a Extra-Moral Sense*. Nietzsche wrote this text in 1872 but never published it himself and Feyerabend quotes a certain passage at least at three occasions during a time-span of 20 years. The quotation first appears in 1963 in a letter to the Australian philosopher Jack Smart. Feyerabend is concerned with genuine alternative worldviews: "Yet I think we shall have to admit, following a careful analysis of the structure of the worldviews which have been connected with different cultures, that they are genuine alternatives, genuinely different manifest images" (Feyerabend 2016, 212). He points to the Hopi and to Ernst Cassirer and his theory of symbolic forms to exemplify his idea. "I wish I could present some examples at least to make this assertion more plausible. I shall certainly do this in one of my future papers. I always wanted to write a paper about the nature of myth in order to show that they are fully blown worldviews" (Feyerabend 2016, 212 f.). The first chapters of his *Philosophy of Nature*, written in first years of the 1970 but only published posthumously, are dedicated to this task. The earlier letter to Smart instead continues with an astounding remark:

> All I can do here is quote some Nietzsche. In his earlier life Nietzsche was very much interested in the philosophy of nature and he intended at some time to write a systematic treatise on this topic. He adopted for some time Mach's point of view and he interpreted this point of view as showing that the world in which we live (our perceptual world) is a mental construction, and that different kinds of such mental constructions are possible. This brings him at some places rather close to Wittgenstein (Feyerabend 2016, 213).

It is fascinating that Feyerabend already associates Nietzsche with Ernst Mach and Ludwig Wittgenstein, since his relation to the intellectual contexts of the Vienna

Circle was only acknowledged much later (Fischer 1999; Brusotti 2009).[7] The main incentive of Nietzsche's unpublished early essay is to argue that language is a conventional tool for coping, not for copying, that the nature of language is therefore intrinsically metaphorical system of signs, and that our social rules to use these signs are subject of change. Towards the end of the text Nietzsche suggests that "the waking day of a mystically excited people as are the older Greeks is indeed more similar to a dream than is the day of a sober mind; and this is due to the always present miracle as assumed by the myth" and that "like in a dream everything is possible at any moment and the whole of nature surrounds man as if it were only a masquerade of the gods" (Nietzsche 1979[1873], 89). Feyerabend translated the full passage himself and he later used the same quote and the same thought in his *Philosophy of Nature* (written in 1976) and again 20 years after his letter to Smart in his essay *Science as Art* (1984). In his *Philosophy of Nature* he provides an explication of this Nietzschean idea that illuminates not only his relation to the nineteenth century philosopher, but also clarifies his understanding of cultural, political, ontological, and epistemological relativism:

> The concept of reality, which impacts people's actions, feelings, and perceptions within a certain period of time, thus depends in part on psychological, social, and cultural factors. The factors change partly under the pressure of external circumstances, and partly as a result of conscious human efforts (rationalists tend to grossly overestimate the second part). Their impact is more or less "rational," and "reason" itself is determined either by these factors themselves or by the factors of a competing ideology. "Absolute" standards, that is, standards independent of all factors, may exist in Heaven, but here on Earth with its diverse forms of reality there are none (Feyerabend 2016, 78).

## 1.3 Conclusion: On the Epistemic Virtue of Relativism

Relativism is often defined as a composition of (at least) three theses: A relativity- or dependency-thesis, a plurality-thesis, and a equality- or symmetry-thesis (Baghramian 2004, 138; Boghossian 2006, 73). While Protagoras, Nietzsche and Feyerabend clearly subscribe to the first and second, they do not accept the third. They state that man is the measure of epistemic systems and practices and they assume a diachronic and synchronic plurality of manifest world-views. They deny, however, explicitly or in practice, that all perspectives are symmetric and equally valid. If absolutism and relativism are exclusive alternatives, by virtue of being anti-absolutist, they must be relativist. If any given knowledge-claim must either be relatively well justified or absolutely true, they certainly deny the possibility of the latter. Experiencing and cognizing are active processes, the characteristics of the biological, social, cultural and individual subject contribute to their emergence and to their justification. But if one assumes that an epistemic relativist must entertain the view, that any statement is equally valid or appropriate in any given context or

[7] See also the chapters on Feyerabend's relation to Mach and to Wittgenstein in this volume.

situation, neither Protagoras and Nietzsche nor Feyerabend are relativists. They all express respect for the ontological assumptions, methods, practices, and results of the sciences. They rather seem to consider relativism as the natural and appropriate epistemic attitude of a confident and conscious scientific time. But it is really relativism?

In his reconstruction of relativism in Feyerabend's later works, Martin Kusch states that relativism not only implies dependence, plurality, exclusiveness and notional confrontation—all of which Feyerabend assumes in accordance with Protagoras and Nietzsche—but also "Symmetry: Epistemic systems and practices must not be ranked [...] Most characterizations of relativism – by friends and foes alike – take Equality to be the natural consequence of Non-Neutrality and thus the best way to spell out Symmetry" (Kusch 2016, 107). In a footnote Kusch briefly considers the idea, that "the relativist deny that there is a uniquely best system or practice while at the same time allowing that there are better and worse social practices" but moves on with the clarification that according to "the model of relativism suggested here, this position would not amount to a full-blown relativism. While the denial of the unique best system or practice is a relativistic element, the (presumably neutral) ranking of systems or practices into better and worse is not" (Kusch 2016, 107). By means of a careful analysis of Feyerabend's *Notes on Relativism*, Kusch concludes with some disappointment that "the *Symmetry as Equality* involved here is not that all cultures are equally correct but that all views have – to use modern jargon – important and, ultimately not-to-be-ignored, undefeated defeaters" (Kusch 2016, 110). Who is to blame for the mismatch: Feyerabend or the model? If neither the ancient inventor, nor the nineteenth century godfather, nor our anarchist philosopher defend such a notion of symmetry, one may wonder, whom among the 'friends and foes' of relativism actually do. Given that all three try to make the weaker case stronger and employ our changing and context-dependent human standards to pursue reasonable conversions, the request that a relativist must assume that all perspectives are equally valid, is not fulfilled. As it seems, relativism is either a strawman or the view proposed by its allegedly most prominent representatives.

Drawing on Protagoras and Nietzsche, Feyerabend approaches the issue of relativism from a broader perspective and points to a number of arguments in favor of it. He assumes that accepting relativism can help to bring about a more human approach. His diagnosis of relativism assumes that we are inevitably caught in a net of beliefs, and that there are more and different such nets. Not only our conceptual interpretation, but already sensual perception is an active process, produced by our contingent human nature. Sensory organs simplify and interpret the world, but they respond to the same unqualified world of becoming and they express similar human needs. Our intellect is an instrument of survival, not of truth-acquisition. Regarding observation, however, members of the human species are equipped with similar sensory organs and therefore captured within overlapping perspectives. The same applies to our language, which serves a social function embedded in evolutionary needs. Language is no representation of plain facts, but a translation into human metaphors. Every propositional worldview is a translation into constructed signs and human grammar, only by which we can speak and think. The social character of language and grammar, however, requires us to use the established metaphors, we cannot do without. This explains why the philological art of careful reading remains important to Nietzsche as did the new developments in physics to Feyerabend. Feyerabend allows a hierarchy of interpretations by means of non-absolute standards of evaluation. Not only our cultural certainties and idiosyncrasies, even epistemic categories are subject to change and the development of knowledge is not only progress but change of (epistemic) interests, values, orientations. Relativism results from a self-reflection and self-overcoming or our will to truth; it is not meant to be an objectively true doctrine, but denotes a reasonable epistemic attitude. The absolutist alternative, according to which we could form beliefs which carry no signs of such nets, is considered untenable. The denial of such relativism therefore either leads to ethnocentric parochialism or to an universalist self-misunderstanding. Both attitudes are the product of weak, bounded, and delicate conditions. Regarding the prospects of cultural transformation, the latter is ultimately even worse, since it denies the possibility of justified or promising different views from the outset. Relativism is a liberating and advantageous epistemic attitude of educated, strong, and cultured people. Unlike the cliché of a blatant dogmatist, who takes whatever pleases him for a sacred and unquestionable truth, these criticisms of absolutism require cautious reserve. The dialogue between Protragoras, Nietzsche and Feyerabend suggests relativism as a liberating and advantagous epistemic attitude of educated und cultured people.

## Literature


Babich, Babette. 2010. Das 'Problem der Wissenschaft' oder Nietzsches philosophische Kritik der wissenschaflichen Vernunft. In *Der Tod Gottes und die Wissenschaften. Zur Wissenschaftskritik Nietzsches*, ed. Carlo Gentili and Catrin Nielsen, 123–169. Berlin; New York.

Baghramian, Maria. 2004. *Relativism. London/New York* 2004.

———. 2010. A Brief History of Relativism. In *Relativism. A Contemporary Anthology*, ed. Michael Krausz, 31–50. New York: Columbia University Press.

Bearn, Gordon. 1986. Nietzsche, Feyerabend, and the Voice of Relativism. *Metaphilosophy* 17 (2/3): 135–152.

Berkel, Tazuko A. van. 2013. Made to Measure: Protagoras' *mètron*. In *Protagoras of Abdera: The Man, His Measure*, ed. Johannes M. van Ophuijsen, Marlein van Raalte, and Peter Stork, 37–67. Leiden: Brill.

Blackburn, Simon. 2005. *Truth. A Guide for the Perplexed*. London: Penguin.

Boghossian, Paul. 2006. *The Fear of Knowledge. Against Relativism and Constructivism*. Oxford: Oxford Scholarship Online.

Bonazzi, Mauro. 2013. Universals before Universals: Some Remarks on Plato in His Context. In *Universals in Ancient Philosophy*, ed. Riccardo Chiaradonna and Gabriele Galluzzo, 23–40. Pisa: Edizioni della Normale.

Brusotti, Marco. 2009. Wittgensteins Nietzsche. Mit vergleichenden Betrachtungen zur Nietzsche-Rezeption im Wiener Kreis. *Nietzsche-Studien* 38: 335–362.

Burnyeat, Myles F. 1976a. Protagoras and Self-Refutation in later Greek Philosophy. *The Philosophical Review* LXXXV (1): 44–69.

———. 1976b. Protagoras and Self-Refutation in Plato's Theaetetus. *The Philosophical Review* LXXXV (2): 172–195.

Comte, Auguste. 1855. *Appel aux Conservateurs*. Paris.
———. 1880. *Einleitung in die positive Philosophie*. Leipzig: Deutsch von H.G. Schneider.
Feyerabend, Paul K. 1975. *Against Method. Outline of an Anarchistic Theory of Knowledge*. London (first edition).
———. 1978a. Die 'Rationalität' der Forschung. In *Der wissenschaftstheoretische Realismus und die Autorität der Wissenschaften*, ed. Paul K. Feyerabend, 339–350. Braunschweig.
———. 1978b. *Science in a Free Society*. London.
———. 1981. *Realism, rationalism, and scientific method*. New York.
———. 1987. *Farewell to Reason*. London.
———. 1993. *Against Method*. 3rd ed. London.
———. 1995. Killing Time. *The autobiography of Paul Feyerabend*. Chicago.
———. 1999. In *Conquest of Abundance. A Tale of Abstraction versus the Richness of Being*, ed. Bert Terpstra. Chicago.
———. 2016. In *Paul Feyerabend: Philosophy of Nature*, ed. Helmut Heit and Eric Oberheim. Cambridge.
Fischer, Kurt Rudolf. 1999. Nietzsche and the Vienna circle. In *Nietzsche. Theories of Knowledge and Critical Theory*, ed. Babette E. Babich, 119–128. Dordrecht.
Grote, John. 1865. *Exploratio Philosophica. Rough Notes on Modern Intellectual Science*. Cambridge.
Heit, Helmut. 2016. Reasons for Relativism: Feyerabend on 'the Rise of Rationalism' in Ancient Greece. *Studies in History and Philosophy of Science* 57: 70–78.
———. 2019. Nietzsche et le 'Relativisme Épistémique. In *Nietzsche et la Relativisme*, ed. Olivier Tinland and Paolo Stellino, 37–67. Montpellier.
Kusch, Martin. 2016. Relativism in Feyerabend's Later Writings. *Studies in History and Philosophy of Science* 57: 106–113.
Lange, Friedrich Albert. 1866. *Geschichte des Materialismus und Kritik seiner Bedeutung in der Gegenwart*. Iserlohn.
Mill, John Stuart. 1874. Auguste Comte und der Positivismus. Aus dem Englischen übersetzt von Elise Gomperz. In *John Stuart Mills Gesammelte Werke. Band IX*, ed. Theodor Gomperz. Leipzig.
Nietzsche, Friedrich. 1979[1873]. On Truth and Lies in a Nonmoral Sense. In Ed. and transl. by Daniel Breazeale, *Philosophy and Truth. Selections from Nietzsche's Notebooks from the early 1870's*, 79–91. London.
———. 1997[1881]. Ed. by Maudemarie Clark and Brian Leiter, transl. by R. J. Hollingdale, *Daybreak. Thoughts on the prejudices of Morality*. Cambridge.
———. 2001[1885]. Ed. by Hans Peter Horstmann, transl. by Judith Norman, *Beyond Good and Evil. Prelude to a Philosophy of the Future*. Cambridge.
———. 2003[1887]. Ed. by Rüdiger Bittner, transl. by Kate Sturge, *Writings from the Late Notebooks*. Cambridge.
Preston, John. 1997. *Feyerabend. Philosophy, Science and Society*. Cambridge.
Schiappa, Edward. 2003. *Protagoras and Logos. A Study in Greek Philosophy and Rhetoric*. Columbia: University of South Carolina Press.
Theocharis, Theo, Psimopoulos, Michaelis (1987). *Where science has gone wrong*. Nature 329, 595–598.
Zilioli, Ugo. 2007. *Protagoras and the Challenge of Relativism. Plato's Subtlest Enemy*. London.

# Chapter 2
# Feyerabend, Plato and Aristotle

**Gonzalo Munévar**

**Abstract** It may seem surprising that some of the most radical ideas proposed by Paul K. Feyerabend have important aspects in common with those proposed by two of the most respected philosophers in the history of philosophy, Plato and Aristotle. Nonetheless, they do. Feyerabend's principle of proliferation as developed in *Against Method*, for example, may remind us of Aristotle's arguments for proliferation in making his case for democracy as the best form of government, principally in his *Politics*. And Plato's "Allegory of the Cave," in his *Republic*, gives support to Feyerabend's position on how empirical evidence makes theoretical assumptions. Plato argues that the observer is misguided if he concludes that the real positions and motions of heavenly bodies can be determined by his observations. Plato's approach to education would also be extremely congenial to Feyerabend, as would his use of the "Myth of the Metals" in Book III of the *Republic* that places the greatest obligation on the rulers of a society to ensure that all children get a good chance to develop their talents. Extraordinarily, this obligation is extended to women, as he shows in Book V.

**Keywords** Paul Feyerabend · Plato · Aristotle · Democracy · Proliferation · Education · Evidence · Theoretical assumptions

## 2.1 Introduction

Throughout most of his career, Paul Feyerabend was viewed with suspicion by "respectable" philosophy of science. Indeed, some of his ideas were considered not just anarchistic but shocking. In this chapter I will concentrate on some important

G. Munévar (✉)
Lawrence Technological University, Southfield, MI, USA
e-mail: gmunevar@ltu.edu

© The Author(s), under exclusive license to Springer Nature Switzerland AG 2024

S. Gattei, R. Corvi (eds.), *Feyerabend in Dialogue*, Boston Studies in the Philosophy and History of Science 346,
https://doi.org/10.1007/978-3-031-71938-7_2

points of convergence between several of Feyerabend's most cherished proposals and the insights of the two most respected and "respectable" Classical Greek philosophers: Plato and Aristotle.

## 2.2 Aristotle and Feyerabend's Principle of Proliferation

Aristotle developed excellent arguments in favor of the notion that ordinary citizens are qualified to challenge the experts, a notion cherished by Feyerabend. Indeed, Aristotle's arguments led to his demonstration (in his *Politics*) that democracy is generally the most reasonable form of government. Since, Feyerabend insisted on the democratization of science, as well as on the advantages of proliferation of scientific views (pluralism) over the rigid following of methodological rules, Aristotle's arguments are of particular relevance to his approach.

In Book III, Chapter 11 of the *Politics* (2005b [c. 330 BC]), Aristotle offers an interesting argument that supports Feyerabend's suggestion that non-experts may criticize well-recognized experts. When considering Plato's notion that ruling a society should be assigned to the most virtuous people, Aristotle begins to make a case for pluralism and democracy. "For the many," he says, "who are not as individuals excellent men, nevertheless can, when they have come together, be better than the best few people, not individually but collectively, just as feasts to which many contribute are better than feasts provided at one person's expense" (1281b, 387). Aristotle points out that the multitude have "many parts of virtue and practical wisdom" as well as many character traits, so they are like a human being with multiple senses and experiences that can be used also to judge better such things as music and poetry, for it is as if the multitude brought together what is scattered and separate into one (1281b, p. 387). This may give the multitude an advantage even over an exceptional man. Of course, as Aristotle concedes, there are cases in which the multitude proves inferior to the excellent few, or else "beasts" would be superior to their owner. Nevertheless, giving the multitude "no share and not to allow them to participate at all would be cause for alarm." For as Plato points out in Book VIII of the *Republic*, "a state in which a large number of people are excluded from office and are poor must of necessity be full of enemies" (1281b, p. 388).

For Aristotle the key question here is whether experts should be evaluated only by other experts: doctors by doctors, geometers by geometers, and ship captains by ship captains. Aristotle expands his previous argument, "since each may be a worse judge than those who know, but a better or no worse one when they all come together" (1282a, p. 388). And thus, in some crafts, "those who do not possess knowledge of the craft nevertheless have knowledge of its products." Therefore, "the maker of a house is not the only one who has some knowledge about it; the one who uses it is an even better judge … A captain, too, judges a rudder better than a carpenter, and a guest, rather than the cook, a feast" (1282a, p. 388).

In this way, Aristotle's arguments seemingly support Feyerabend's claim to the effect that "in a democracy local populations not only will, but also should, use the sciences in ways most suitable to them" (1999, p. 251). Feyerabend justifies his proposed relationship between scientific experts and citizens by offering a case for democracy and pluralism. Aristotle works in the other direction: arguments about why ordinary citizens may criticize experts help him develop a case for democracy and pluralism.

In Chapters 13 and 15 of Book III of the *Politics*, Aristotle emphasizes that even in the question of possessing virtue the ability of the many, when coming together, may be greater than that of the excellent few. For example, the many are more incorruptible than the few, for it is easier to corrupt a society in which only a few are in authority, say, by bribing those few. In Chapters 15 and 16, he introduces two important considerations. Should a society be ruled by universal rules or by men? Aristotle acknowledges that written laws are beneficial, and that it is important that the rulers possess universal reason. Nevertheless, as he points out, a law is the mean: it provides guidance in the most likely circumstances, but not in all. The law, "having educated the rulers for this special purpose [the mean] hands over the rest to be decided and managed in accordance with the most just opinion of the rulers" (1287a, p. 394). That is, experience and good sense allow the rulers to make corrections where those lead to improvements.

It is crucial, then, to decide who should be those rulers, since exceptions to the law should be expected when dealing with particulars: "a human being will deliberate better about particulars" (better than the law could have foreseen, that is, since, in creating the law, we aim for the mean) (1286a, p. 393). Aristotle's conclusion is that the multitude should be chosen, as we will see below.

Whatever the merits of the procedures of abstraction which lead to universal laws or rules, as Feyerabend points out, "the details did not therefore cease to exist, just as people don't cease to have a nose when being weighed" (p. 14). Aristotle's emphasis on particulars, which again enhances the need for pluralism, is very congenial to Feyerabend's approach.

It is precisely on this key point that the diversity of experience and expertise of the many will serve the society best. In a very small society, perhaps, an individual of exceptional virtue should be king, but as the society becomes larger, "it is perhaps no longer easy for any other constitution to arise besides democracy" (1286b, p. 394). Even in those cases where a monarchy makes sense, there is a need for pluralism: "monarchs provide themselves with many eyes, ears, hands and feet," for it would be very strange, e.g., that someone judging with just one pair of eyes could see better than many people with many pairs (1287b, p. 395).

In Chapter 17, Aristotle makes it clear how strongly he feels about the advantages that pluralism gives to democracy:

> But it is surely evident from what has been said in a case where people are similar and equal, it is neither beneficial nor just for one person to control everything. This holds whether there are no laws except the king himself, or whether there are laws; whether he is a good person ruling good people, or not a good one ruling not good ones; and even whether he is their superior in virtue (1288a, p. 395).

An exception would exist in the case of an individual (or family, or small group) "whose virtue is so superior as to exceed that of all the others [as a whole]" (1288a, p. 396), but of course in a large society that is practically impossible.

Aristotle's defense of pluralism offers interesting parallels to Feyerabend's scientific pluralism. As Feyerabend argues in *Against Method* (1975) and other books, just as universal laws of a society strive for the mean and are thus bound to have exceptions, so should universal methodological rules in science. Universal laws, as well as methodological rules, are abstractions, either because they aim for the mean, as Aristotle pointed out, or because they ignore particulars by not taking the many relevant natural contexts into account, as we have seen in Feyerabend. That is why we may have theories or models of the formation of a solar system, but we expect every solar system we encounter to be different from all others, for their formation would have included a unique combination of factors (e.g., the presence of other stars in the cosmic neighborhood). Feyerabend understands that for several practical purposes (e.g., to achieve a ballpark approximation) it may be useful to ignore such variety in favor of abstraction. As he says in *Conquest of Abundance*, "Anyone who tries to make sense of a puzzling sequence of events…is forced to introduce ideas that are not in the events themselves, but put them in perspective" (p. 12). As he explains further, "*understanding a subject means transforming it*, lifting it out of a natural habitat and inserting it into a model or a theory or a poetic account of it" (p. 12). But he and Aristotle, however, object to the notion that the views that result from abstraction constitute "reality." Not only could there be different abstractions, but the universal laws, or universal methodological rules, may ignore features that may be very important to us and be considered "real" for different (practical) reasons. Aristotle criticizes Plato's inclination for universal ideas in the *Nicomachean Ethics*, Bk. 1, Ch. 6 (2005a [c. 332 BC]):

> …it is a puzzle to know what the weaver or carpenter will gain for his own craft from knowing this Good Itself, or how anyone will better at medicine or generalship from having gazed on the Idea Itself. For what the doctor appears to consider is not even health [universally] but human health, and presumably the health of this human being, even more since he treats one particular patient at a time (1097a, 258–9).

According to Aristotle in the *Politics*, the correct forms of government (constitutions) turn out to be those that look to the common good, whereas those that look to the benefit of the rulers are mistaken (1279a, p. 384), and eventually, in Book 4, Chapters 11 and 12, he comes to favor a democracy in which the middle class is in the majority as the best constitution in practice, for that society is the one most likely to emphasize benefitting as many citizens as possible.

## 2.3 Plato and Aristotle on Happiness and Knowledge

A scientific view of the world should not be abandoned just because some of the predictions it makes turn out to be false. This is an important lesson we learn from Thomas Kuhn, against the dictates of the falsificationist method, or so Feyerabend tells us in his article "Consolations for the Specialist" (1970). A paradigm, as Kuhn calls it in *The Structure of Scientific Revolutions* (1970), should be given time to develop, to prove its value. Commitment to a view, the "Principle of Tenacity" in Feyerabend's terms, is of extreme importance to science, as long as it is applied by several competing groups working on those ideas that they find particularly promising (which Feyerabend calls the "Principle of Proliferation," i.e., pluralism). Kuhn errs, according to Feyerabend, when he expects the whole discipline to be committed to just one point of view or paradigm. Moreover, Feyerabend tells us, the principle of proliferation, and within its action, the principle of tenacity, leads to greater human happiness. Those two principles also create the conditions for fruitful change and improvement. Thus, both humanity and science are the better for their presence.

This point is important to Feyerabend because he believes that science, as a human activity, must be justified ultimately by the difference it makes to human lives, including the lives of those who practice science. According to him, the most important question about science is "to what extent the happiness of individual human beings, and to what extent their freedom, has been increased?" (p. 209) It seems to him that "the happiness and the full development of a human being is now, as ever, the highest possible value" by which we should probe the sciences of today (p. 210). This value, he adds, "does not exclude the values which flow from institutionalized forms of life (truth, valour, self-negation, etc.)" (p. 210). On the contrary, "It rather encourages them *but only* to the extent to which they can contribute to the advance of some individual" (p. 210). What is excluded, he says, "is the use of those institutionalized values for the condemnation, or perhaps even the elimination, of those who prefer to arrange their lives in a different way" (p. 210). And, Feyerabend emphasizes, "What is excluded is the attempt to 'educate' children in a manner that makes them lose their manifold talents so that they become restricted to a narrow domain of thought, action, emotion" (p. 210).

It may seem surprising, but it is here that Plato makes important contributions to philosophy in what we may call the "spirit of Feyerabend."

In Book III of Plato's *Republic* (2005 [c. 385 BC] Socrates, the main character in Plato's dialogues, mentions what is now called the "Myth of the Metals" (415 a-c, 128–129). According to the Myth, the god mixed different metals into different people as he was forming them: gold into those who should be rulers, silver into future soldiers, and iron and bronze into farmers and other craftsmen. Children tend to resemble their parents, but nevertheless some gold children will be born of iron parents and vice versa. The most important command from the god to the rulers of a city is to watch carefully the mixture of metals in the souls of their offspring and to honor each mixture in a manner appropriate to its nature. By metals Plato means talents. Different people are born with different talents, and the society's most important obligation is to make sure that every child has an opportunity to explore and develop his talents. A just society will also ensure that every job is assigned to the person most qualified. A society that falls short of these two demands will be not only inefficient but unjust.

According to the myth's insight emphasized by Plato, the reality in practically all past and present societies is that the chances of children doing well as they grow up depend on the status of their parents. This bias leads to a great waste of human potential and much unhappiness and conflict. Children of poor parents do not even have a chance to a decent education in many countries, to say nothing of the opportunity to become rulers of their society. The situation of women tends to be particularly distressing. In many countries they risk their lives if they wish to have any education at all, and in some other countries they are not even allowed to drive a car. Many jobs are forbidden to them. Even in presumably enlightened countries like the U.S. it was not that long ago, in the 1960s, that graduate programs in science had rules against accepting women as students. In Book V, Plato is adamant that women should be given equality of opportunity. Girls who are born with the talent to be rulers or doctors or scientists should be given the opportunity and the education to reach their potential, if they wish to. Indeed, there is no reason to suppose that women would not be capable to rule the society. Philosophy Queens, we might say, shall join Philosopher Kings in Plato's ideal society, a point that he emphasizes again in Book VII.

To the objection that men and women obviously have different natures and, therefore, should not be assigned the same jobs, Plato replies that the differences and similarities that matter are those pertinent to the pursuit in question; otherwise, we would have "to ask ourselves whether the natures of bald and long-haired men are the same or opposite. And once we agree that they are opposite, it is open to us to forbid the long-haired ones to be shoemakers, if that is what the bald ones are to be, or vice versa" (454c, p. 151). That would be absurd, of course. But is it not the case, Plato then asks, that whereas "a male and a female whose souls are suited for medicine have the same nature" … "a male doctor and a carpenter have different ones?" (454d, 151). Plato then demands to know from the opponents of equality for women what is "the precise craft or pursuit, relevant to the organization of the city, for which a woman's nature and a man's are not the same but different" (455a, p. 151). For it seems that, for most if not all pursuits, "they differ in this respect alone, that the female bears the offspring while the male mounts the female" (454e). Denying women equality of opportunity because of such a difference would be a travesty.

Of course, if you are going to employ males and females for the same tasks, you must give them the same opportunity for education. And education, as it turns out, is the proper method by which to determine what talents (metals) children have. Like Feyerabend, Plato was against compulsory education, as he makes clear in Book VII: "Because a free person should learn nothing slavishly. For while compulsory physical labors do not harm the body, no compulsory instruction remains in the soul" (536e). Plato proposes, instead, an approach that would also be agreeable to Feyerabend: "…use play instead. That way you will also be able to see better what each of them is naturally suited for" (537a, p. 200). This will not be a quick decision either. The young citizens will have plenty of opportunity to self-select into their future professions, for the play-based education will last most of their youth. Education may take on a more specialized and formal aspect around the age of 20. The most important consequence of Plato's suggested system of education is that the society (we may also read science) will benefit by allowing individuals to follow the inclinations that interest them the most. This was pretty much Feyerabend's point as well.

We might also conclude that this is the sort of arrangement that would increase the happiness of the individuals involved, since they would lead more fulfilling lives. Individual happiness is also an important concern for Aristotle, and he manages to complement Plato's account, given the relationship he finds between an individual's development of his talents and happiness.

Aristotle points out in Chapter 7 of Book I of the *Nicomachean Ethics* that the good of man is an end in itself, complete and self-sufficient, and only happiness can be such an end (1097b, p. 259). We do many things so as to achieve our goal to become happy, but it makes little sense even to ask why we want to be happy. It is obvious. Happiness, however, as Aristotle argues in Book X, cannot consist merely in pleasure. Hedonism is mistaken. For it is clear that "no one would choose to live with a child's [level of] thought for his whole life, taking as much pleasure as possible in what pleases children" (1174 a, p. 351). Pleasure still plays a key role in happiness, but the pleasure that truly matters is that which completes the activities to which we ought to devote our lives. It is the pleasure that comes from fulfillment: "… the best activity is [that] of the subject in the best condition in relation to the best object of [his] capacity. This activity will also be the most pleasant" (1174 b, 1175a-b, pp. 351–54).

Plato's ideal scheme from the Myth of the Metals would then produce the conditions under which individual citizens will have the best chances to flourish, and thus of being happy (as long as what Aristotle calls the "externals" are there to a reasonable degree: good health, love, family, friends, etc., 1179a, p. 357).

Furthermore, and in accordance with Feyerabend's insight, the proper pleasure of the activity is the one that helps us improve (and then leads to even greater pleasure, and thus greater happiness). Therefore, when we devote ourselves freely (Plato and Feyerabend) to those activities for which we have the most talent, we get more out of life, that is, we are more likely to achieve happiness.

To sum up, if a society allows its citizens to freely determine and develop their true talents, as required by Plato's Myth of the Metals, then Aristotle explains why the majority of its citizens will be happy: they will achieve their potential, and the proper pleasure of their activities will satisfy them more than anything else (as long as the external circumstances are normal). It seems to me that this provides a plausible and strong foundation for Feyerabend's argument to the effect that scientists should have the freedom to develop those ideas that fascinate them the most. Scientists will be happier by doing so, will lead more fulfilling lives, and will make greater contributions to science.

Plato's position is not entirely aligned with Feyerabend's view of science, though. As he says in Book IV (419 a—d, 130), the happiness of the entire community deserves the greatest priority. His worry is that in his ideal society the philosopher-kings (and philosopher-queens) are the only people truly qualified to rule, but only love and gratitude to their society will make them sacrifice the exciting life of the mind to devote themselves to the well-being of their society. In Feyerabend's view of science, the pursuit of individual happiness will result in the well-being of science.

As for Aristotle, he comes to believe that the greatest happiness comes from study (1177a, 355), but this conclusion is inconsistent with his own examples. Human beings may truly (and exemplary) fulfill themselves in life pursuits that differ greatly from those that would fulfill other human beings. When the extraordinary footballer Messi finally won the World Cup recently, should we have informed him that he would have been better off to devote himself to the life of study? And should Chopin have abandoned the piano for philosophy in order to strive for greater happiness?[1]

## 2.4 The Tower Argument and the Allegory of the Cave

From the point of view of empiricism, the great victory of the Copernican Revolution should be credited to the scientific method championed by Galileo in his defense of Copernicus. The standard empiricist conception of the scientific method lies in specifying the procedures by which sensory evidence determines the worth of scientific theories. For example, falsificationism tells us that theories in disagreement with the facts should be rejected. Inductivism places emphasis on how sensory evidence provides evidence for the theories that agree with it, and indeed may even require that such theories be derived from that evidence. Of course, Feyerabend's main claim to fame, or infamy, is his rejection of all such empiricist methodologies.

It thus comes as a shock that adhering to this empiricist approach would result in the strong and clear refutation of the Copernican claim that the Earth moves, and more specifically of the claim that the Earth's rotation accounts for its day/night cycle. Galileo, then, had to drive a dagger through the heart of empiricism in order to save the Copernican Revolution.

The powerful refutation of the motion of the Earth was made explicit in Aristotle's Tower Argument, and in "The Second Day" of his *Dialogues Concerning the Two Chief World Systems*, Galileo acknowledges that the Aristotelians had observation and experiment on their side when they argued that the Earth did not move: "As the strongest reason of all is adduced that of heavy bodies, which, falling down from on high, go by a straight and vertical line to the surface of the earth" (1989 [1623],72). If the Earth rotated and a stone were allowed to fall from the top of a tall tower, the tower, being carried by the motion of the Earth, "would travel many hundreds of yards in the east in the time the stone would consume in its fall, and the stone ought to strike the earth that distance from the base of the tower." But the stone strikes the earth next to the tower, which refutes the notion that the Earth rotates.

[1] Feyerabend's anti-intellectualism, or rather his taking exception to the contempt intellectuals have often expressed towards ordinary people (the "vulgar"), would lead him to agree with the author on this point.

Salviati, Galileo's spokesman in the *Dialogues*, then adds extremely powerful thought experiments to these devastatingly clear perceptions of the stone's path. For example, two equal cannons shooting one west and the other east (if the Earth rotated, the respective cannon balls should travel significantly different distances, but we are certain that they do not); or shooting them north or south (the paths should deviate from a straight north/south direction, but we know they do not); or shooting the cannon ball straight up (it should fall at a large distance from the cannon, but we would not expect it to and would move out of the way, Salviati might add). The empirical or experimental support offered by Aristotle's Tower experiment for the immovable Earth, and made even more obvious by Salviati, prompts Simplicio, the Aristotelian spokesman, to proclaim in admiration that it would appear to be "an impossible feat to contradict such palpable experiences." If these experiments were false, Simplicio asks, "What true demonstrations were ever more elegant?" (p. 73).

Nevertheless, as we know, Galileo defeated the Tower Argument's conclusion that the Earth could not move. Did he produce new observations or experimental results? No. He offered a theoretical argument instead. He began by asking what may seem to be a simple-minded question: How do we know that the rock falls vertically? We see it, obviously, as Simplicio points out ("by means of the senses"). This answer seems eminently sensible, as Aristotle had reasonably argued that a normal person observing the world under normal conditions was likely to form a realistic impression of what the world was like.[2] In the case of the Tower Argument the relevant visual perceptions seemed to be about as reliable as we could possibly wish.

But not every one in Classical Greece was of the same mind. In his famous Allegory of the Cave (Book VII of his *Republic*) Plato told us a story of chained men who spent their lives looking at the shadows of objects projected on the wall of a cave and guessing what those objects were. Plato equated sensory perceptions with those shadows. One day a man is able to free himself from his chains and walks out of the cave into the real world illuminated by the sun. He then sees what those objects are really like and how extraordinary the real world is. Reality goes well beyond the sensory appearances. Plato applies his allegory to the practice of astronomy. In 530a,b,c, he tells us, a "real astronomer" will not be compelled to conclude that the motions of the stars as well as other "phenomena" such as the ratio of night and day are always the same as he has observed. Only very advanced mathematics (e.g., solid geometry) will begin to give us some means "to grasp the truth about them." He then suggests that "We will leave the things in the heavens alone," i.e., our mere perceptions of them.

In a rather Platonic spirit, then, Galileo tries to cast doubt on the sensory perceptions from which Aristotle concludes that the Earth cannot move. He asks a simple question. What would be the motion of the stone if the Earth did rotate? He has the

[2] For a sophisticated account of Aristotle's views on this subject see Anna Marmodoro, *Aristotle on Perceiving Objects*, Oxford University Press, 2014.

Aristotelian Simplicio supply the answer: The rock would move with a compound of two motions, "one with which it measures the tower, and the other with which it follows it." The real motion would thus be a compound of a vertical and a circular motion. Of course, it is implied, we only observe the vertical motion, since we share with the rock and the tower the circular motion of the Earth. A few pages earlier Galileo had pointed out that any motions that may be attributed to the Earth "must necessarily remain imperceptible to us… for as inhabitants of the earth, we consequently participate in the same motions" (p. 69).

It follows, then, that from *seeing* the motion of the stone "you could not say for sure that it described a straight and perpendicular line, *unless you first assumed the earth to stand still*" (p. 77, my italics). But whether the Earth stands still is precisely what is in question. The evidence adduced to show that the earth stands still *assumes* that the earth stands still! Aristotle, the great logician, has committed the fallacy of begging the question.

In a few pages, then, and without providing one single piece of new empirical evidence, Galileo disposes of the main objection against the very possibility that the Earth rotates, by offering an argument much in the spirit of Feyerabend's advice to imagine "*a dream-world in order to discover the features of the world we think we inhabit*" (1993, p. 32). By relinquishing the evidence of experiments for the sake of a dream of his own devising, he was able not only to discover important features of the world we thought we inhabited, but eventually to show that such a world was itself a dream. Galileo thus creates the stage for the eventual triumph of the Copernican Revolution.

Galileo's overthrow of Aristotle's Tower Argument is probably the most important historical example of a feature of science: that in determining the presumed empirical basis we make theoretical assumptions. The critique of such assumptions may introduce new ones and lead to a different empirical basis. Galileo illustrates this process beautifully. Instead of the simple Aristotelian motions, he introduces compound motions that result in "true" parabolic motions. Those are the "facts" of the new, Copernican world. He opens the door to this new set of facts by demonstrating that claims about the "real" motions of objects make assumptions about theories of the Earth (whether it moves or not). Thus Galileo strongly supports a third infamous thesis defended by Feyerabend: the theoretical character of observation and experiment (or in the dreadful parlance of analytic philosophy, the thesis that "observational languages are theory-laden"). In this case, by removing the crucial objection to the motion of the Earth, it led to the most important revolution in the history of science.

This revolution did not take place in a philosophical vacuum. Aristotle's empiricism was defeated by the Platonic turn favored by Galileo. This does not mean, of course, that Plato would have become a Copernican, for he would likely have expected a still deeper mathematical exploration in our search for reality. Many different views and approaches might then have much to recommend them to those researchers willing to develop them, just as Feyerabend told us with his principle of proliferation.

In the history of science, this example is the most significant, but it is not the only one by any means. Time and again views have been "conclusively refuted" only to be rehabilitated at some point in the future thanks to the introduction of new theories. A discussion of this matter goes beyond the scope of this chapter, but I hope the reader may wish to consider a look at my recent book *A Theory of Wonder: Evolution, Brain and the Radical Nature of Science* (2021), which is in great part devoted to the elucidation and consequences of Feyerabend's challenging work. In this chapter, my purpose has been to acknowledge some of the convergence between that work and some of the great philosophical insights offered by Plato and Aristotle.

## References

Aristotle. 2005a. Nicomachean Ethics (T. Irwin, Trans.). In *Classics of Moral and Political Theory*, ed. M.L. Morgan, 4th ed. Indianapolis: Hackett Publishing Co.. (First published c. 332 BC).

———. 2005b. Politics (C.D.C. Reeve, Trans.). In *Classics of Moral and Political Theory*, ed. M.L. Morgan, 4th ed. Indianapolis: Hackett Publishing Co.. (First published c. 330 BC).

Feyerabend, P.K. 1970. Consolations for the Specialist. In *Criticism and the Growth of Knowledge*, ed. I. Lakatos and A.E. Musgrave, 197–230. Cambridge: Cambridge University Press.

———. 1975. *Against Method: Outline of an Anarchistic Theory of Knowledge*. London: New Left Books.

———. 1993. *Against Method*. 3rd ed. London: Verso.

———. 1999. In *Conquest of Abundance: A Tale of Abstraction Versus the Richness of Being*, ed. B. Terpstra. Chicago: University of Chicago Press.

Galileo. 1989. Dialogue Concerning the Two Chief World Systems. In *The Scientific Background to Modern Philosophy*, ed. M.R. Mathews, 61–81. Indianapolis: Hackett Publishing Company. (First published in 1632).

Kuhn, T. 1970. *The Structure of Scientific Revolutions*. 2nd ed. Chicago: Chicago University Press.

Marmodoro, A. 2014. *Aristotle on Perceiving Objects*. Oxford University Press.

Munevar, G. 2021. *A Theory of Wonder: Evolution, Brain and the Radical Nature of Science*. Wilmington: Vernon Press.

Plato. 2005. Republic (C.D.C. Reeve, Trans.). In *Classics of moral and political theory*, ed. M.L. Morgan, 4th ed., 75–251. Indianapolis: Hackett Publishing Co. (First published c. 385 BC).

# Chapter 3
# Feyerabend and Marx

**Rory Kent and Ian James Kidd**

**Abstract** In this chapter we explore Feyerabend's relationship to Karl Marx, with a focus on their respective views on politics. Throughout Feyerabend's writings, there are references to Marx and his written work, to the historical Marxist tradition and to later scholars of Marxist philosophy. After a brief discussion of the intellectual-historical, disciplinary and personal contexts behind these engagements with Marxism, we develop a comparison of their views on politics. We identify two points of major divergence in their views, namely those regarding the dynamics of political change and those regarding the ultimate end of politics. We conclude by offering several themes for 'dialogue' between Marx and Feyerabend, suggesting the possibility of future work.

**Keywords** Paul Feyerabend · Karl Marx · Science · Ideology · Politics · Tradition

## 3.1 Introduction

As this volume documents, Paul Feyerabend was famously eclectic in his intellectual influences and sympathies. Unfortunately, Feyerabend was also unreliable and vague about his influences, sometimes even downplaying them in later works such as in the case of Karl Popper (Collodel 2016). A further difficulty is that Feyerabend would often praise certain figures with the apparent intention of provoking his audiences. In some cases, there may have been genuine inspiration—such as with the Dadaists—while in others, a desire to provoke alone seems to explain why certain

R. Kent (✉)
Department of History and Philosophy of Science, Cambridge University, Cambridge, UK
e-mail: rdk32@cam.ac.uk

I. J. Kidd
Department of Philosophy, University of Nottingham, Nottingham, UK
e-mail: ian.kidd@nottingham.ac.uk

© The Author(s), under exclusive license to Springer Nature Switzerland AG 2024

S. Gattei, R. Corvi (eds.), *Feyerabend in Dialogue*, Boston Studies in the Philosophy and History of Science 346,
https://doi.org/10.1007/978-3-031-71938-7_3

figures appear, as may be the case with his favourable references to Stalin (Feyerabend 1978, 113). Assessments of Feyerabend's relationship to certain figures should therefore be sensitive to the context in which he cites them.

In this chapter we explore Feyerabend's relationship to Karl Marx with a focus on their respective views on politics. Throughout Feyerabend's writings, there are references to (a) Marx and his written work; (b) the historical Marxist tradition and some of its key figures, including Vladimir Lenin and Leon Trotsky; and (c) later scholars of Marxist philosophy, such as Louis Althusser. As usual, few of these references amount to systematic treatments or discussions, and there are difficulties in trying to work out the points being made.

Consider, for instance, the discussion of the 'unevenness' of the history and practice of science in Chapter 11 of *Against Method*. Feyerabend's supporting footnote contains a long quotation from the 1859 *Introduction to the Critique of Political Economy* (Feyerabend 1993, 107). However, this point about the 'unevenness' could have been supported with an appeal to then-contemporary history and sociology of science, rather than any specific Marxist treatment of science. Moreover, in the main body of the text, Feyerabend implies that he is simply borrowing a piece of Marxist terminology, rather than invoking any more specific theoretical ideas.

Such points could encourage scepticism of our project of exploring how Feyerabend might be in dialogue with Marx. We want to push back on that scepticism. For a start, there are quite a few discussions of Marx in Feyerabend's writings that must be carefully explored. *The Poverty of Philosophy* and *Introduction to the Critique of Political Economy* are both discussed in *Against Method*; the *Remarks on Hegel's Philosophy of Right* is discussed in *Science in a Free Society*; and there are the references to Lenin, Bakunin, Althusser, and others. While Feyerabend's erratic bibliographic practices make it hard to tell what he read and when, it is clear he had a sustained interest in Marxist thought, especially in the 1960s and 70s (after which his references to Marx and Marxists appear to drop off).

## 3.2 Context

We can think about the relationship between Feyerabend and Marx with respect to different contexts. First, the general *intellectual-historical context*. Feyerabend was working during the Cold War. For mainstream audiences in the West, the villains of the time were communists and their 'sympathisers', including and perhaps especially Marxists; in other words, those on the 'wrong side' of the Iron Curtain, be that literally or by association. Feyerabend was well aware of the provocative rhetorical power of citing Lenin, Stalin and Marx in his work. Citing them would have annoyed many Western philosophers—and not just dogmatic champions of Western ways of life. Feyerabend was never particularly engaged in political philosophy in a serious way—many of his own contributions were undertheorised or naïve (Kidd 2016a). But he was still working in a context of ideological opposition, and Marxist thinking was—from the perspective of mainstream Anglo-American culture—on the wrong side.

A second, related context is the narrower *disciplinary* one of mid-twentieth century Anglo-American philosophy of science. As George Reisch and others have argued, the Second World War and the Cold War shaped the agenda of disciplinary philosophy of science (Reisch 2005; Kidd 2016b). Many of its leading figures were wartime émigrés—Popper from Austria, Carnap from Germany, and Lakatos from Hungary—and thus who worked alongside whom was in no small part determined by contemporary geopolitics. Feyerabend, himself an émigré from Austria, worked at different times in London, California, Bristol, New Zealand and Zurich, bringing him into direct contact with Popper, Lakatos (both ex-Marxist critics of Marxism) and others in the field. Moreover, his time at UC Berkeley in the 1960s motivated a formative shift in his thinking due to his encounters with increasing numbers of Black and otherwise racially-oppressed students, and with militant student movements (Feyerabend 1978, 118; Kidd 2013). The shifting culture of post-war philosophy of science makes Feyerabend's engagements with Marx and Marxists especially provocative to a disciplinary audience so intimately affected by geopolitics.

A third, even narrower context is presented by Feyerabend's *personal* life, including his close friendships with Walter Hollitscher and Lakatos. As Feyerabend tells it, it was Hollitscher who encouraged him to take Marxism more seriously as a critique of positivist philosophy of science, and to adopt more concrete approaches to philosophical argumentation (Feyerabend 1978, 112–13; Kent 2022, 53–54). His friendship with Lakatos seems to have influenced him to an even greater degree. There are jocular references to their constructive antagonism in Feyerabend's letters: the book we know as *Against Method* was originally intended as part of a co-authored book, with Lakatos arguing 'for' method, order and reason, while Feyerabend supplied the argument 'against' (Motterlini 1999). In one letter likely intended to poke fun at Lakatos (who had a complex philosophical and political relationship to Hegel and Marx), Feyerabend groups the two of them into the same category as Lenin (Kadvany 2001; Motterlini 1999, 216). Likewise, as Eric Oberheim has suggested, some of Feyerabend's references to Marx in *Against Method* may have been intended to 'insult' his and Lakatos's former teacher, Popper, who perceived Marxism to be a 'pseudoscience' (Oberheim 2008, 33). Thus, the context of Feyerabend's personal relationships is also important when considering his engagements with Marx.

With all this in mind, we will discuss Feyerabend's relationship to Marx with a particular focus on politics. Of course, a more complete discussion will include dedicated attention to other areas, including philosophical anthropology, theory of history and the philosophy of science. As Feyerabend himself points out, 'Marxism … is a *philosophy*' and there is a Marxist philosophy of science that can be contrasted with Feyerabend's views on science (Feyerabend 1978, 155; Dorn 2000; Murray 1988). We warmly welcome the recent and thorough investigation of Feyerabend's discussions of Marx and Marxism by Yuanlin Guo and Chubi Yan (Guo and Yan 2024). Unfortunately, despite Feyerabend's sustained engagements, much of the extant scholarship says nothing at all about Marx—he is hardly mentioned across (Farrell 2003), (Niaz 2020), (Preston 1996) and the papers collected in (Brown and Kidd 2016) and (Bschir and Shaw 2021). Gonzalo Munévar considers various antecedents to Feyerabend's ideas—including Machiavelli and J. S. Mill—but does not include Marx (Munévar 2016). Roy Bhaskar examines Feyerabend's relationship to Gaston Bachelard, who influenced Marxist philosophers such as Althusser, but does not discuss the link to Marx directly (Bhaskar 1975). Jamie Shaw explores the anarchistic elements in Feyerabend's philosophy but does not touch on the contrast with Marxism (Shaw 2017, 2022). Moreover, contemporary Anglophone philosophers of science have generally tended to show little interest in Marx (there are honourable exceptions, such as Railton 1984). Many scholars of Marx seem to return the indifference, even those who write on Marx's views on science (Sheehan 1985 is among the exceptions here). A few Marx scholars do discuss Feyerabend but only as a prescient critic of positivism (Little 1986; Walker 2001). Thus, there is a lot of material that could be covered in a chapter on Marx and Feyerabend. We focus here on political philosophy, however, for the sake of brevity and because we believe this will throw the contrast between their ideas into sharpest relief.

## 3.3 Politics

Feyerabend and Marx diverge most widely on political-theoretic matters. Their basic concepts and motivations are markedly different, as are some of their main influences and aspirations. It is well known that Feyerabend consistently and emphatically praised Mill's *On Liberty*. This 'immortal essay' was for Feyerabend 'the best modern exposition and defence of a critical philosophy' (Feyerabend 1981, ix–x). Crucially, Feyerabend did not laud Mill's epistemology or his philosophy of science, nor indeed any other political works such as *Considerations on Representative Government*—when speaking of Mill, as Feyerabend told Lakatos, he was 'always talking of *On Liberty*' (Motterlini 1999, 239). Marx was far less admiring of Mill, describing his research as 'neither extensive nor profound' (Marx 1976, 221). That said, Feyerabend never probes issues of political economy and social transformation that are central to the disagreements between Marx and Mill (Duncan 1976). Thus, though Feyerabend and Marx diverge in their respective attitudes towards Mill, this doesn't tell us much about their intellectual relationship to each other.

We focus on two issues that to us seem representative of the points of greatest divergence between Marx and Feyerabend. First is their analyses of dynamics of political change. For Marx this is examined through accounts of *class society* and *class conflict*, in which the state has a somewhat secondary role. For Feyerabend, political change is examined through notions of *intercultural exchange* between *traditions*, where the state can take on a diversity of roles but ideally ought to remain neutral between traditions. A second, related point of divergence concerns the end of politics—the abolition of class society for Marx and the establishment of the 'free society' for Feyerabend.

## 3.3.1 *Marx's Politics*

Marx's political views are complicated and there are a great many interpretations available to us. Decades of Marxological debate has led to the appearance of 'many Marxes' (Berman 1999, 19). It is not our aim to adjudicate these debates nor to take sides. Moreover, that might not help us in exploring Feyerabend's relationship to Marx since his views were expressed within a much earlier generations of scholarship. Still, we must say something to get things moving and we focus on two points that we hope are relatively uncontroversial. First, that Marx's political thought rests on a social theory of class society and of class conflict as the dominant motor in the development of human societies. Second, that Marx's political thought contains an account of the state and its activity as secondary to and partisan in class conflict. The state typically works to support the interests of one or another class. Each point is central to Marx's account of the dynamics of political change that, we argue, puts him at odds with Feyerabend.

### 3.3.1.1 Class and Class Conflict

In Marx's social theory, people can be characterised as members of social-economic classes determined ultimately by their relationship to the 'mode of production' characteristic of the societies in which they live. 'The handmill gives you society with the feudal lord', proposed Marx, while 'the steam mill society with the industrial capitalist (Marx 1973, 95; on differing interpretations of this widely quoted remark, see Rigby 1998, 145). A society's mode of production is determined by the dominant processes and technologies by which goods used to meet human needs and desires are produced. In the capitalist mode of production, wage-labour is the dominant production process. In exchange for a wage, *proletarians* (those who do not own any of the 'means of production'—natural resources, tools, machinery, factories, etc.) must sell their labour to *capitalists* (those who *do* own the means of production). Proletarians are directed by their capitalist employers to engage their labour towards the production of goods, which are then sold by the capitalists on exchange markets for profit. In the Marxist vocabulary, these goods thus take the form of 'commodities' (Marx 1976, 125–31). Commodities appear at the same time as things used to meet human needs or desires, and as things to be exchange on the market for other goods of equivalent 'exchange value' (including sums of money). Given that proletarians do not own any of the means of production themselves (and cannot thereby produce things to meet their own needs directly) they are effectively 'forced' to sell their labour to the capitalists (Marx and Engels 1970, 30–47).

Proletarians are thus *dominated* by the system of wage-labour (for a sophisticated analysis of the dominating character of capitalist class society, see Cicerchia 2021).

The capitalist labour process is also conceptualised by Marx as *exploitative* (Marx and Engels 1970, 34). Roughly speaking, wage-labour extracts 'surplus value' from proletarian workers, whose own value comes to be defined in terms of commodity-production (39). To turn a profit, capitalists must pay the workers wages *lower* than the market value of the commodities. Capitalists engage none of their own labour but extract a profit from the labour of proletarians, simply by virtue of owning the means of production: this is the essence of exploitation (Marx 1976, 769; Wood 1981, 232).

Marx argues that modes of production and the class societies they engender emerge out of historical processes (Marx and Engels 1970, 31–32). In practice, Marx only paid substantial attention to the emergence of capitalism out of feudalism in Western Europe the feudal mode of production was replaced when the capitalist class achieved political and economic supremacy during what he calls a 'bourgeois revolution' against the aristocracy (the ruling class in feudal societies). Marx's historical approach to theorising resembles that which Feyerabend also advocated in the context of modern science—scientific culture became established thanks to 'idiosyncratic historical developments' and the assertion of 'a certain form of life' with which it 'resonated' (Feyerabend 1999b, 138, 79). These deep contingencies belie reassuring stories of rational progress and undercut triumphalist claims about science. Deeper, grittier forces are at work that can be exposed by careful historical-sociological inquiry (Feyerabend 1993).

Marx's theory of class society includes a notion of class *conflict*. By virtue of their different relations to the mode of production, different classes have different and sometimes conflicting interests *vis-à-vis* the future development of that mode. Proletarians desire higher wages and better working conditions in order to maintain their subsistence, whereas their capitalist employers aim to sell commodities at the highest possible rate while paying the lowest possible wages in order to extract as much value from the labour process as possible (Marx and Engels 1970, 41–43). The interests of the proletarians and capitalists are thus oppositional and this will manifest in political activity such as wage disputes, disruptive industrial action and boycotts—all of which may be organised by trade unions and other organisations. The exploitative and dominating character of the wage-labour process thus compels capitalist societies towards class conflict. For Marx, class conflict can develop into a higher stage if critical masses of proletarians become sufficiently conscious of their indelibly oppositional relationship to capitalists—they may, for instance, embrace the radical goal of abolition of private ownership through state seizure of the means of production (Marx and Engels 1955, 43–45; Marx 1976, 344, 533, 700). The exact character of the goals and the means to achieving them is, however, disputed among different kinds of Marxists, as is the shape of the society after the abolition of capitalism.

In sum, Marx's politics rests on an account of classes locked in class conflict. Before asking how this compares to Feyerabend's political thought, we will say more about Marx's views on the state.

### 3.3.1.2 Bourgeois and Proletarian States

The state plays an unusually secondary role in Marx's political thought, compared to, say, Plato or Hegel. *The Manifesto of the Communist Party* explains that the 'executive of the modern state is nothing but a committee for managing the common affairs or the whole bourgeoisie' (Marx and Engels 1970, 33). Bourgeois states constitute a system of political, legal and administrative apparatuses that maintain the capitalist class's domination over the proletariat. The set of laws and punitive systems surrounding private ownership is one example (Marx and Engels 1939, 58–62). Marxist social theorists have devoted sustained attention to identifying in greater detail and scope the ways in which institutions of the modern state work to maintain capitalist domination over the mode of production and over the proletarian class; alas we cannot do justice to that literature here (Miliband 1969; Althusser 1971; Hall et al. 1978; Gilmore 2007).

The crucial point for Marx is that the formation and operation of modern nation-states is itself an historical phenomenon tied ultimately to the development and expansion of the capitalist mode of production. The political-economic ascendence of the capitalist class was central to this process: hence the partisan status of the state in Marx's account. Indeed, nothing short of a social revolution in the sphere of production could alter this position, which will itself involve the seizure of the means of production by another class, i.e., the proletariat. There is general agreement among Marxists that any such seizure will require confrontation with and transformation of the bourgeois state; what form these take, whether through 'reform' or 'revolution', is a matter of contention. For Marx, what at least is needed is a transition away from the bourgeois state to one that supports the proletariat, which Marx dubbed 'the class dictatorship of the proletariat' (Marx and Engels 1955, 223). If that sounds ominous, 'dictatorship' did not, for Marx and Engels, have the associations it does today to tyranny or despotism. It meant a socialist state in which the proletarian class controlled the means of production ('dictating' its use, as it were, and steering it away from one predicated on exploitative wage-labour; see Draper 1987 for a detailed analysis of the historical development of the term 'dictatorship' from Marx's day to its contemporary usage). Such a state would also have the features described by Marx and Engels—abolition of property in land and rights to inherited wealth, say, and establishment of free universal education (Marx and Engels 1955, 53–54).

For Marx, therefore, the state is unavoidably partisan. He rejects the liberal ideal of the state as a politically-neutral arbitrator between a plurality of interest groups. The state is not and cannot be independent of the class conflict that drives history onwards. With this account in place, we now move to ask how Marx's politics compares to Feyerabend's political thought.

### 3.3.2 Feyerabend's Politics

The majority of Feyerabend's writings tended to focus on epistemology and the philosophy of science. From the late 1960s, however, he began to broach a range of political, moral and cultural issues. They ranged well-trodden issues like freedom of thought—a reflection of his admiration for Mill—to 'postcolonial' concerns with the complicity of the sciences in the subjugation of the Global South. These were not independent of his philosophical treatments of science. Critique of philosophical conceptions of science is continuous with a critique of the sorts of cultures those conceptions helped legitimate (Kidd 2021). Unfortunately, scholars agree that Feyerabend's ventures into political philosophy were underwhelming. *Science in a Free Society*—arguably containing his fullest attempt at a political theory—is unsystematic, polemical and lacking in evidentiary detail (Feyerabend 1978). Feyerabend later described some of his early 'political' papers, like "Experts in a Free Society" as being written 'in a fit of anger and self-righteousness caused by what I thought were certain disastrous developments in the sciences' (Feyerabend 1970, 1999a, 112). There is also unclarity over the character of his political-philosophical views. Consider the admiration for Mill and the constant celebration of a plurality of forms of life, the stern denunciation of epistemically and culturally oppressive institutions (including the sciences) and the consistent ethos of variety, liberal tolerance and pluralism. At the same time, there is an abiding respect for organically-developed traditions and hostility to attempts to impose abstract visions on human life and practice—the prerogative of philosophical conservatives. As Shaw explains, Feyerabend's pluralism has an unclear axiology (Shaw 2021).

At a more general level, Feyerabend was averse to taking sides and often preferred to play the role of defending unpopular positions for its own sake, an inheritance from Mill (Lloyd 1997, reprinted in this volume). An important consequence is that Feyerabend acted as a sort of dialectical prodder, unlike Marx whose political activity was wholly commensurate with a thoroughgoing commitment to his theoretical positions. Feyerabend seemed to be concerned that the entrenchment of one view would become dogmatic—excluding or derogating alternatives. *Many* views have to be in place. *Many* doctrines need to be competing. The history of human thought involved shifting patterns of tendencies to monism and pluralism. At certain points, a state of dogmatic monism is clearly starting to emerge, and, at that point, one must intervene and stir things up. If so, then it can be hard to imagine Feyerabend taking a side and working hard to establish it as a permanent part of human life.

We think this ambivalence runs through most of Feyerabend's political thinking. It also points to a major difference between him and Marx, for whom there is no such ambivalence. We focus, however, on the themes we established earlier—the dynamics of political change and the ultimate end of politics. In Feyerabend's writings these are explored in his accounts of *inter-tradition exchange* and *the free society*. We are also now in a position to describe contrasts with Marx's political thought throughout.

#### 3.3.2.1 Traditions and Cultural Exchange

Feyerabend does not identify individuals in terms of their membership to socio-economic classes in relation to modes of production. Instead, he sees individuals in terms of their participation with very broad *traditions*. Such traditions include whole conceptions of reality and associated sets of ontological and epistemological commitments. Feyerabend does not provide a clear, systematic account of these traditions. Generally, he offers an abstract account of how traditions can relate, illustrated with more-or-less elaborate case studies (Feyerabend 1978, 16–24). His favourite example was the shift from archaic Greek to presocratic philosophical cosmologies—rehearsed in *Farewell to Reason* and *Conquest of Abundance*, and first conceptualised in the early 1970s in the project posthumously published as the edited book *Naturphilosophie* (Feyerabend 1985, 1999b, 2019).

A few Feyerabend scholars have explored in detail his account of traditions and his historical case studies therein (Clark 2002; Farrell 2003; Kusch 2016). Some of the main themes: traditions are historical systems, broadly but not rigidly integrated, consisting of stable sets of practices and associated values or ideals and a more-or-less articulated worldview or vision of reality. Such traditions can develop over time due to internal dynamics and external influences. Each tradition supplies ontological commitments (about what kinds of objects there are) and epistemological commitments (about the sources and scope of knowledge). Members of a tradition might be reflective and self-aware of the particularities of their own tradition. Not necessarily, though. *Conquest of Abundance* offers a sort of inverted Hegelian narrative—of history since the Greeks as the increasing alienation of human beings, as we become ever-more self-conscious and so entrenched by ever-more elaborate abstractions (Feyerabend 1999b: Part 1). One implicit ideal seems to be a natural immersion in a tradition—uninterrupted by the alienating artifice of a self-consciousness that spoils what was a sort of unforced inhabitation of the world. Feyerabend's admiration for other 'non-scientific' traditions seemed in part to rest on their success, so far, in sustaining these sorts of comportment (Feyerabend 1978, 1999b). That said, elsewhere Feyerabend elevated the notion of 'maturity' with regards to one's own tradition, according to which it is recognised as engendering a partial perspective on the world, one that might be complimented by dialogical engagement with others (Feyerabend 1978,). This is a paradigmatic instance of the ambivalence in Feyerabend's political thought. In any case, Feyerabend has a more *culturalist* social ontology than does Marx, who instead focused on systematic economic processes of production, exchange and consumption and the ways these structured our social existence. Rather, for Feyerabend, our social-political lives are mediated by our relationship to cultural traditions, especially so far as these traditions provide us with the values and standards against which make practical, ethical and epistemic choices.

Beyond this, Feyerabend was mainly interested in *interactions* between traditions. Abstractly speaking, his focus on interactions, i.e., what Helen Longino understands as a properly *social* analysis, suggests a point of resonance with Marx, who is also focused on dynamics of interaction (albeit between social-economic classes) as a motive force in history (Longino 2022). Several articles by Feyerabend from the late 1970s onwards explore the theme: from the experiments with relativism in *Farewell to Reason* to the late essay whose title, "Potentially Every Culture is All Cultures," was one of Feyerabend's favourite aphorisms (Feyerabend 1985; Feyerabend 2019). This interest in inter-tradition interactions raises all sorts of questions, many of them unanswered in his writings. What does it mean to 'belong' to a tradition? Are traditions really as isolated as he sometimes seems to suggest? Can people only really belong to one tradition? What historical or anthropological evidence is needed to justify or confirm such claims? The answers to these questions will depend on how one defines a tradition, a concept which remains elusive in Feyerabend's writings. What we can say is that Feyerabend's ideal was to ensure the ongoing existence of a diverse range of traditions, so that a thousand interactions might bloom, and to condemn whatever attitudes and convictions might threaten this vibrant diversity. This ideal—rough as it is—brings us to Feyerabend's controversial idea of a 'free society' and brings us back to Marx.

#### 3.3.2.2 The Free Society as the End of Politics

The ideal of a free society is widely considered to be poorly-worked-out even if was also well-motivated. Feyerabend never supplied a proper answer to the obvious political questions. Most scholars agree that the details disinterested him, and on a few occasional Feyerabend announced much the same sentiment. Detailed political and economic and infrastructural issues lay outside the scope of his interests and abilities. He might also have wanted to distance his own ideas from Popper's, whose interest in political philosophy was well-known (1945's *The Open Society and its Enemies* was the first, but not the last, of his contributions to political philosophy).

Our question is how Feyerabend's vision of the free society relates to Marx's account of the communist society. Granted, Marx does not say a lot about the precise form of that society, since it lay in the future and would be quite different from hitherto existing societies. Still, he offers a few remarks:

> [I]n communist society, where nobody has one exclusive sphere of activity but each can become accomplished in any branch he wishes, society regulates the general production and thus makes it possible for me to do one thing to-day and another to-morrow, to hunt in the morning, fish in the afternoon, rear cattle in the evening, criticize after dinner, just as I have a mind, without ever becoming hunter, fisherman, herdsman or critic (Marx and Engels 1939: 22)

Such freedom to explore and acquire new skills would only be possible in a society where there is universal education, where general production is 'taken care of' (so to speak), and where no one is subject to exploitation or excessive toil. For Marx, an individual in this communist society can live in ways characterised by *spontaneity*. Without the constraints of domination and exploitation engendered by class society, people can cultivate their epistemic and practical skills and experiment with different kinds of life. Unlike Mill's liberal society, however, this is achieved through social and economic revolution—Marx distinguished his social theory from 'utopian' accounts of socialism, which specified normative ideals without analysis of how they might be achieved through real political and economic transformations.

Feyerabend's accounts of the free society can be described as a pluralistic and rather radical form of egalitarianism. In a free society 'all traditions have equal rights and equal access to the centres of power', and the whole society 'recognizes the value of any particular mode of life' (Feyerabend 1978, 106; 1999a, 112). The overall value of pluralism can be put in consequentialist terms: a plurality of traditions allows for vibrant inter-tradition exchanges to take place on a continual basis, which will in turn help produce an ever-expanding ocean of ideas, practices and yet further traditions from which individual humans might draw. Alongside this genial pluralism there is a further conviction: a free society must not be based on 'any particular creed' and even if one 'creed' prevails there must be equality with 'all other beliefs' (Feyerabend 1999a, 135; 1993, 228). Such equality and access would not be happy accidents, either, but should all be established and actively protected by the institutions of the free state. Feyerabend, alas, fails to supply any specifications for such institutions (although he gives us some remarks on how children's general education ought to proceed; see Kidd 2013).

Two principles seem to underlie the idea of the free society. The *hegemony principle* is that the establishment of any one system of thought, tradition or worldview necessarily constitutes a politically-epistemically restrictive hegemony. Genuine freedom, says Feyerabend, in 'science, philosophy, and civilization' is equated with the absence of any entrenched 'single … system of thought' (1985: vii). Supporting this is the *liberation principle*: that political and epistemic freedom requires a plurality of freely coexisting traditions and systems of thought. To the denizens of a free society, liberation entails freedom and hegemony means oppression. 'The freedom of a society', claims Feyerabend, 'increases as the restrictions imposed on its traditions are removed' (Feyerabend 1999a: 220). In a further claim, Feyerabend also ruled out the entrenchment of those traditions and systems of thought that promise liberation. "How to Defend Society against Science" opens by emphasising there is nothing intrinsically liberatory about the sciences (Feyerabend 1975). They did, can and do serve oppressive ends—a claim developed in feminist and postcolonial science studies. In the same essay, Feyerabend also extends that same accusation to Marxism which has 'deteriorated' into a 'stupid religion'. Even the traditions and systems of thought with genuine liberatory aims can degenerate if become dogmatic. This would happen if (a) they are not subjected to constant critical scrutiny, if (b) they are not regularly put into critical contact with robust alternatives and (c) if they become entrenched in ways that grant to them disproportionate epistemic and political hegemony.

All of this is consistent with the sort of epistemological pluralism Eric Oberheim has identified as a constant theme of Feyerabend's work (Oberheim 2008). We think it also explains why Feyerabend could not in practice endorse Marx's guiding political aims: the establishment of a community society. What Marx appears to aspire to politically and epistemically is a state of what Feyerabend would see as a baleful hegemony—a single system of thought and life, whose main rivals (roughly, forms of bourgeois life and thought) have been removed. Pluralism for Feyerabend requires the *permanent* coexistence of very many ways of thinking and living—including a plurality of *types* as well as *tokens*. For Marx, by contrast, in order to transition away from the capitalist class system and towards a classless, communist society, and organised proletariat must be more single-minded and steadfastly committed to a 'scientific' understanding of revolutionary politics though 'internal' theoretical criticism, i.e., between committed revolutionaries, is viewed by most Marxist as essential to their task). That said, it is left open by Marx whether such single-mindedness will remain after a full socialist transition—perhaps a pluralism of ways of thinking and living will re-emerge and flourish, once general production is 'taken care of'. In any case, Marx would have likely viewed Feyerabend's commitment to radical and persistent pluralism as an instance of 'eclecticism', just as he did Mill's philosophy.

We think Feyerabend's pluralism can be separated from his proposals surrounding the 'free society'. Feyerabend never gave us a blueprint of the free society or an account of how it might be brought about (nor these were part of his intentions). But he broadly retained his commitment to those principles of liberation and hegemony. They are still there, for instance, in the earnest celebrations of cultural and epistemic 'abundance' in the later writings. One can imagine replies that a Marxist might make to Feyerabend. Couldn't one try to protect a system of thought against decline into dogmatism by ensuring criticism? Could we not argue that criticism requires some alternatives, but not *radical* alternatives? Could we not appeal to the intellectual honest and integrity of members of an established and hegemonic scientific-socialist tradition?

We can imagine how Feyerabend might reply to these questions. We suspect he was too committed to the epistemological pluralism Oberheim described to give up so easily on the commitment to criticism through a more radical plurality of traditions—and too hostile to what Ian Hacking called 'singlemindedness':

> What Feyerabend disliked was any form of intellectual or ideological hegemony ... Single-mindedness in the pursuit of any goal, including truth and understanding, yields great rewards; but single vision is folly if it makes you think that you see (or even glimpse) the truth, the one and only truth (Hacking 1994: 28)

More pertinently, Marxists might press Feyerabend on the issue of identifying the political and economic conditions required to establish a free society. How can citizens be expected to experiment with different forms of life and systems of thought if they must spend most of their waking hours selling their labour to capitalists, and thereby without free direction over their own activity? Even further, Marxist-feminist might press Feyerabend on the issue of working-class women's exclusion from public life altogether due to the constraints of bourgeois family apparatuses: for a society to be truly free, those women and any other systematically marginalised people must be liberated and allowed to participate in cultural exchanges on their own terms too. As we have noted, Feyerabend spent almost no time examining the kinds of political institutions or economic systems that would support his free society. This is a serious shortfall, if, as Herbert Marcuse put it, 'society must first create the material prerequisites of freedom for all its members before it can be a free society' (Marcuse 1964, 40).

Does this mean that Marx and Feyerabend are ultimately incompatible? Given the availability of different interpretations of Marx's thought and the ambivalences present in Feyerabend's political writings, we would be too quick to conclude that. Perhaps there are ways in which their ideas can be seen as complementary. One possible route to reconciliation is suggested by our remarks above: that Feyerabend's vision of the free society is a worthwhile goal but can only be achieved by transitioning class societies away from the capitalist mode of production, so that all citizens can be freed from exploitation and to engage in inter-tradition exchanges with spontaneity. Some Marxists might thereby advocate more single-minded revolutionary commitments *in the meantime*, that is, while the communist society is being prepared, so that true freedom of thought can be achieved down the line (how and when 'down the line' will ever be reached is, of course, a matter of serious concern). In the reverse direction, a principled Feyerabendian ought to allow Marxists to develop their 'scientific-socialist' tradition on their own terms, such that it might contribute to the vibrant plurality of ideas in an aspirational free society. It seems easy to say that a free society ought to at least *tolerate* the existence of Marxists and their political-philosophical tradition and provide it equal rights alongside other traditions. Moreover, as Guo and Yan conclude, Dadaists like Feyerabend can (as Feyerabend himself did) make 'opportunistic' uses of Marxist philosophy as part of their own theoretical activity (Guo and Yan 2024, 40). The question of whether Marxists ought to tolerate sharing centres of power with bourgeois, chauvinist or otherwise reactionary traditions is a much harder one.

In this chapter, we have offered some possible themes for a 'dialogue' between Feyerabend and Marx, indicating some points of agreement and disagreement. Developing that dialogue would be an interesting task, especially if it engaged with themes we have not emphasised, such as their conceptions of the nature and value of science, their discussions of creativity in science and human life, or the cultural and political functions of the discipline of philosophy.

**Acknowledgements** We are grateful to the editors, Stefano and Roberta, for their invitation to contribute to this collection. We are also grateful for constructive comments from the collection's anonymous reviewer.

## References

### *Works by Feyerabend*

Feyerabend, Paul. 1970. Experts in a Free Society. *The Critic* 29: 58–69.

———. 1975. How to Defend Society against Science. *Radical Philosophy* 11: 3–8.

———. 1978. *Science in a Free Society*. London: Verso.

———. 1981. *Realism, Rationalism, and Scientific Method, Philosophical Papers*. Vol. 1. Cambridge: Cambridge University Press.
———. 1985. *Problems of Empiricism, Philosophical Papers*. Vol. 2. Cambridge: Cambridge University Press.
———. 1993. *Against Method*. 3rd ed. London: Verso.
———. 1999a. In *Knowledge, Science, and Relativism, Philosophical Papers*, ed. John Preston, vol. 3. Cambridge: Cambridge University Press.
———. 1999b. In *Conquest of Abundance: A Tale of Abstraction versus the Richness of Being*, ed. Bert Terpstra. Chicago: University of Chicago Press.
———. 2019. Potentially Every Culture is All Cultures. *Common Knowledge* 25 (1–3): 13–20.

## *Works by Marx*

Marx, Karl. 1973. *The Poverty of Philosophy*. Moscow: Progress Publishers.
———. 1976. *Capital*. Vol. 1. London: Penguin Books.
Marx, Karl, and Friedrich Engels. 1939. *The German Ideology*. New York, NY: International Publishers.
———. 1955. *Selected Works*. Vol. 1. Moscow: Foreign Languages Publishing House.
———. 1970. *Manifesto of the Communist Party*. Peking: Foreign Languages Press.

## *Secondary Literature*

Althusser, Louis. 1971. Ideology and Ideological State Apparatuses." Translated by Ben Brewster. In *Lenin and Philosophy and Other Essays*, 127–186. London: New Left.
Berman, Marshall. 1999. *Adventures in Marxism*. London: Verso.
Brown, Matthew, and Ian James Kidd. 2016. Reappraising Feyerabend. *Studies in History and Philosophy of Science Part A* 57: 1–8.
Bhaskar, Roy. 1975. Feyerabend and Bachelard: Two Philosophies of Science. *New Left Review* 1 (94).
Bschir, Karim, and Jamie Shaw. 2021. *Interpreting Feyerabend*. Cambridge: Cambridge University Press.
Cicerchia, Lillian. 2021. Why Does Class Matter? *Social Theory and Practice* 47 (4): 603–627.
Clark, Stephen R.L. 2002. Feyerabend's Conquest of Abundance. *Inquiry: An Interdisciplinary Journal of Philosophy* 45 (2): 249–267.
Collodel, Matteo. 2016. Was Feyerabend a Popperian? Methodological Issues in the History of Philosophy of Science. *Studies in History and Philosophy of Science Part A* 57: 27–56.
Dorn, Harold. 2000. Science, Marx, and History: Are There Still Research Frontiers? *Perspectives on Science* 8 (3): 223–254.
Draper, Hal. 1987. *The "Dictatorship of the Proletariat" from Marx to Lenin*. New York, NY: Monthly Review Press.
Duncan, Graeme Campbell. 1976. *Marx and Mill: Two Views of Social Conflict and Social Harmony*. Cambridge: Cambridge University Press.
Farrell, Robert P. 2003. *Feyerabend and Scientific Values*. Dordrecht: Springer.
Gilmore, Ruth Wilson. 2007. *Golden Gulag*. Oakland, CA: University of California Press.
Guo, Yuanlin, and Chubi Yan. 2024. Paul Feyerabend and Marxism. *Filozoficzne Aspekty Genezy* 20 (2): 77–121.
Hacking, Ian. 1994. Paul Feyerabend, Humanist. *Common Knowledge* 3: 23–38.
Hall, Stuart, Chas Critcher, Tony Jefferson, John Clark, and Brian Roberts. 1978. *Policing the Crisis*. London: Macmillan Press.
Kadvany, John. 2001. *Imre Lakatos and the Guises of Reason*. Durham, NC: Duke University Press.

Kent, R. 2022. Paul Feyerabend and the Dialectical Character of Quantum Mechanics: A Lesson in Philosophical Dadaism. *International Studies in the Philosophy of Science* 35 (1): 51–67.
Kidd, Ian James. 2016a. Feyerabend on Politics, Education, and Scientific Culture. *Studies in History and Philosophy of Science Part A* 57: 121–128.
———. 2013. Feyerabend on Science and Education. *Journal of Philosophy of Education* 47 (3): 407–422.
———. 2016b. 'What's so Great about Science?' Feyerabend on Science, Ideology, and the Cold War. In *Science Studies during the Cold War and Beyond*, ed. Elena Aronova and Simone Turchetti, 55–76. Basingstoke: Palgrave Macmillan.
———. 2021. Feyerabend, Science, and Scientism. In *Interpreting Feyerabend: Critical Essays*, ed. Karim Bschir and Jamie Shaw, 172–190. Cambridge: Cambridge University Press.
Kusch, Martin. 2016. Relativism in Feyerabend's Later Writings. *Studies in History and Philosophy of Science Part A* 57: 106–113.
Little, Daniel. 1986. *The Scientific Marx*. Minneapolis, MN: University of Minnesota Press.
Lloyd, Elisabeth. 1997. Feyerabend, Mill, and Pluralism. *Philosophy of Science* 64: S396–S407.
Longino, Helen E. 2022. What's Social about Social Epistemology? *Journal of Philosophy* 119 (4):169–195.
Marcuse, Herbert. 1964. *One-Dimensional Man: Studies in the Ideology of Advanced Industrial Society*. London: Routledge & Kegan Paul.
Miliband, Ralph. 1969. *The State in Capitalist Society*. London: Quartet Books.
Motterlini, Matteo. 1999. *For and Against Method: Including Lakatos' Lectures on Scientific Method and the Lakatos-Feyerabend Correspondence*. Chicago: University of Chicago Press.
Munévar, Gonzalo. 2016. Historical Antecedents to the Philosophy of Paul Feyerabend. *Studies in History and Philosophy of Science Part A* 57: 9–16.
Murray, Patrick. 1988. *Marx's Theory of Scientific Knowledge*. Atlantic Highlands, NJ: Humanities Press.
Niaz, Mansoor. 2020. *Feyerabend's Epistemological Anarchism: How Science Works and its Importance for Scientific Education*. Dordrecht: Springer.
Oberheim, Eric. 2008. *Feyerabend's Philosophy*. Berlin: Walter de Gruyter.
Preston, John. 1996. *Feyerabend: Philosophy, Science, and Society*. Cambridge: Polity.
Railton, Peter. 1984. Marx and the Objectivity of Science. *PSA: Proceedings of the Biennial Meeting of the Philosophy of Science Association* 1984: 813–826.
Reisch, George. 2005. *How the Cold War Transformed Philosophy of Science: To the Icy Slopes of Logic*. New York: Cambridge University Press.
Rigby, S.H. 1998. *Marxism and History: A Critical Introduction*. Manchester: Manchester University Press.
Shaw, Jamie. 2017. Was Feyerabend an Anarchist? The Structure(s) of 'Anything Goes'. *Studies in History and Philosophy of Science Part A* 64: 11–21.
———. 2021. Feyerabend, Funding, and the Freedom of Science: The Case of Traditional Chinese Medicine. *European Journal for the Philosophy of Science* 11: 37.
———. 2022. Feyerabend's Anarchism: A Political and Epistemological Analysis. *Zilsel* 2.
Sheehan, Helena. 1985. *Marxism and the philosophy of science*. London: Verso.
Walker, David M. 2001. *Marx, Methodology, and Science: Marx's Science of Politics*. Aldershot: Ashgate.
Wood, Allen. 1981. *Karl Marx*. London: Routledge and Kegan Paul.

# Chapter 4
# Feyerabend and Mill

**Elisabeth A. Lloyd**

**Abstract** I suggest following Paul Feyerabend's own advice, and interpreting Feyerabend's work in light of the principles laid out by John Stuart Mill. A review of Mill's essay, *On Liberty*, emphasizes the importance Mill placed on open and critical discussion for the vitality and progress of various aspects of human life, including the pursuit of scientific knowledge. Many of Feyerabend's more unusual stances, I suggest, are best interpreted as attempts to play certain roles—especially the role of "defender of unpopular minority opinion"—that are necessary to fulfilling Mill's conditions for rational exchange and optimal human development.

**Keywords** Paul Feyerabend · John Stuart Mill · Liberty · Anarchism · Scientific change · Truth

## 4.1 Feyerabend Was Not an Anti-scientific Irrationalist

Paul Feyerabend had a reputation, among many, for being anti scientific, irrationalist, anti-methodological, anti-reason, a relativist about evidence, and an epistemological anarchist. More generally, he frequently anchors the "extreme relativist" end of many a comparison in philosophy of science. The slogan, *anything goes,* though, has certainly captured the philosophical imagination; as I have emphasized before (Lloyd 1996), this slogan is frequently misinterpreted as being Paul Feyerabend's *methodological recommendation* for conducting scientific research. Not so.

E. A. Lloyd (✉)
Department of History and Philosophy of Science and Medicine,
Indiana University Bloomington, Bloomington, IN, USA
e-mail: ealloyd@iu.edu

© The Author(s), under exclusive license to Springer Nature Switzerland AG 2024

S. Gattei, R. Corvi (eds.), *Feyerabend in Dialogue*, Boston Studies in the Philosophy and History of Science 346,
https://doi.org/10.1007/978-3-031-71938-7_4

As Feyerabend himself said:

> *"anything goes" does not express any conviction of mine, it is jocular summary of the predicament of the rationalist:* if you want universal standards, I say, if you cannot live without principles that hold independently of situation, shape of world, exigencies of research, temperamental peculiarities, then I can give you such a principle. It will be empty, useless, and pretty ridiculous – but it will be a "principle." It will be the "principle" "anything goes" (1978, 188; his emphasis).

Thus, Feyerabend's slogan was essentially a *reductio* against a certain form of rationalism, rather than a statement of his own positive view (Feyerabend 1975). I now suspect that one reason underlying the standard misreading of the slogan involves one of Feyerabend's genuine, positive, beliefs—specifically, his defense of the value of a *proliferation* of views and methods, and his insistence on the *tolerance* that must accompany such proliferation.

Feyerabend consistently traces his arguments for the central importance of proliferation of views and methods—and the appropriate attitudes of openness and tolerance—to John Stuart Mill's essay, *On Liberty.* When introducing Mill's ideas in his most extensive discussion of them, Feyerabend emphasizes and endorses Mill's view that, as Feyerabend puts it, "pluralism is supposed to lead to *truth*" (Feyerabend 1981b, 67, his emphasis). Or, as Feyerabend put it in *Science in a Free Society,* "the only way of arriving at a useful judgement of what is supposed to be the truth, or the correct procedure, is to become acquainted with the widest possible range of alternatives... The reasons were explained by Mill in his immortal essay *On Liberty.* It is not possible to improve upon his arguments" (1978, 86).

## 4.2 Some Constraints of Interpretation: Feyerabend as a Millian

There is, with Feyerabend, a pervasive difficulty of interpretation: his writings frequently exhibit a dialectical structure, some of them explicitly appearing in the form of dialogues. It is thus sometimes difficult to identify many of the views that Feyerabend articulates as being those that he would, in some other context, defend. Those who read his slogan, "anything goes," as "Paul Feyerabend's Positive Methodological Program" ran into just this difficulty. There is thus a serious question whether Feyerabend's regular appeals to Mill are best read as straightforward endorsements, or as moves adopted in particular battles, in which Feyerabend takes on the assumptions and standards of his opponents in order to beat them on their own turf (one of his favorite strategies).

I believe that Feyerabend's appeals to Mill's views were, in fact, genuine endorsements: Feyerabend *never* distanced himself from Mill's position, and he used it repeatedly in arguments aimed against other views. For instance, among his rather obsessive criticisms of Karl Popper, Feyerabend emphasized that the *good* parts of Popper's philosophy of science, which include his selectionist process and the emphasis on proliferation and criticism, were all actually *Mill's* views

(Feyerabend 1981a, 141–142). I do not wish to pursue Feyerabend's attempts to undermine Popper's originality—preferring, as I do, to the position that the good ideas in Popper were articulated and better defended by Charles Peirce and John Dewey—but I do think that the general questions about the pluralism of methods and views are highly relevant to today's philosophy of science.

Specifically, philosophers of science are working today within a context that includes the widespread philosophical rejection of any foundational doctrine of pure empirical content or sense-data, the acceptance of some constrained but very real embeddedness of evidence in theory, and the repeated failure of all attempted formulations of a set of methodological *rules.*

In addition, many thinkers have independently come to endorse some variant of a generally evolutionary picture of scientific inquiry, in which cultivation of variation—of interests, theories, and methods—is complemented by a variety of selection mechanisms to produce a *process of scientific change.* Feyerabend's endorsement of Mill's claim amounts to a claim that the *process of scientific change,* when it occurs within the preferred context of a plurality of views and methods, is one that leads toward truth; it does not simply wander around the space of possibilities, as some have described the biological evolutionary process as doing, nor does it pass from the views of one powerful ruling class to another.

Moving away from this claim in its most general form, when we consider Feyerabend's own arguments regarding specific candidate views and methods, we seem to run into trouble. Feyerabend did, after all, defend witchcraft, astrology, faith healing, Chinese medicine, and other non-scientific ways of understanding the world. It might seem that Feyerabend defeated his own purposes by doing so. Arguing that the widest variety of views and methods should be tolerated, pursued, and even nurtured, by those seeking knowledge of the world, Feyerabend could be seen—and *has* been seen—as displaying disregard, or even contempt, for empirical evidence, for sound reasoning, and for established scientific results, by his enthusiastic defenses of marginal, mythical, or magical systems, explanations, and practices.

To read him this way is to misunderstand what he is doing. To interpret Feyerabend's arguments in the light of John Stuart Mill's essay, *On Liberty,* in contrast, is to illuminate the stage that Feyerabend took himself to be on, and to understand properly the role he so frequently elected to play.

## 4.3 What J.S. Mill Said that Feyerabend Adopted and Embodied

Let us turn to John Stuart Mill. We concentrate on Mill's long essay, *On Liberty,* co-written with his wife, Harriet Taylor, and first published as a small book in 1859. The relevant section is Chapter II, "Of the Liberty of Thought and Discussion," in which Mill denies that people, "either by themselves or by their government" should "attempt to control the expression of opinion" (Mill [1859] 1977, 229).

"The peculiar evil of silencing the expression of an opinion is," writes Mill, in a passage frequently quoted by Feyerabend, "that it is *robbing the human race*... If the opinion is right, they are deprived of the opportunity of exchanging error for truth: if wrong, they lose, what is almost as great a benefit, the clearer perception and livelier impression of truth, produced by its collision with error" (Mill [1859] 1977, 229, my emphasis).

Mill argues for freedom of expression of opinion on four distinct grounds:

> First, if any opinion is compelled to silence, that opinion may, for aught we can certainly know, be true. To deny this is to assume our own infallibility. Secondly, though the silenced opinion be an error, it may, and very commonly does, contain a portion of truth; and since the general or prevailing opinion on any subject is rarely or never the whole truth, it is only by the collision of adverse opinions that the remainder of the truth has any chance of being supplied.
>
> Thirdly, even if the received opinion be not only true, but the whole truth; unless it is suffered to be, and actually is, vigorously and earnestly contested, it will, by most of those who receive it, be held in the manner of a prejudice, with little comprehension or feeling of its rational grounds... Fourthly, the meaning of the doctrine itself will be in danger of being lost, or enfeebled, and deprived of its vital effect on the character and conduct: the dogma becoming a mere formal profession, inefficacious for good, but cumbering the ground, and preventing the growth of any real and heartfelt conviction, from reason or personal experience." (Mill [1859] 1977, 258)

Regarding the first reason, Mill argues: "the opinion which it is attempted to suppress by authority may possibly be true. Those who desire to suppress it, of course *deny* its truth; but they are not infallible. ...To refuse a hearing to an opinion, because they are sure that it is false, is to assume that their certainty is the same thing as absolute certainty. All silencing of discussion is an assumption of infallibility" (p. 229). For Mill, such an assumption of infallibility is not merely a moral problem, it has the effect of reducing the opportunity for humanity-at-large to ascertain truth.

The second argument, in which the kernel of truth in minority opinions must be preserved—for the good of everyone –is particularly interesting. The situation is one where "conflicting doctrines, instead of being one true and the other false, share the truth between them" (p. 252). Given, then, the

> partial character of prevailing opinions, ...every opinion which embodies somewhat of the portion of truth which the common opinion omits, ought to be considered precious, with whatever amount of error and confusion that truth may be blended. No sober judge of human affairs will feel bound to be indignant be cause those who force on our notice truths which we should otherwise have overlooked, overlook some of those which we see. Rather, he will think that so long as popular truth is one-sided, it is more desirable than otherwise that unpopular truth should have one-sided asserters too; such being usually the most energetic, and the most likely to compel reluctant attention to the fragment of wisdom which they proclaim as if it were the whole." (Mill [1859] 1977, 253)

Mill's third and fourth arguments are more subtle: he insists that, even if the received opinion is *true,* those holding it need—for their own good, and the good of everyone—to be "vigorously and earnestly contested." With no discussion challenging the view, people may *believe* the truth, but they will not believe it in a reasonable manner, that is, it will simply be a prejudice. Worse yet, Mill argues, without the presence of conflicting opinions, people cannot really understand the meanings of their own beliefs. These arguments rely on Mill's vision of the nature and operation of human judgment.

The fact that human errors are *corrigible* is, Mill says, a quality of the human mind which is "the source of everything respectable in man either as an intellectual or as a moral being" (p. 231). "He is capable," Mill writes,

> of rectifying his mistakes, by discussion and experience. Not by experience alone. There must be discussion, to show how experience is to be interpreted. Wrong opinions and practices gradually yield to fact and argument: but facts and arguments, to produce any effect on the mind, must be brought before it. Very few facts are able to tell their own story, without comments to bring out their meaning. The whole strength and value... of human judgement ... [depends] on the one property, that it can be set right when it is wrong, [and] reliance can be placed on it only when the means of setting it right are kept constantly at hand. In the case of any person whose judgment is really deserving of confidence, how has it become so? Because he has kept his mind open to criticism of his opinions and conduct. Because it has been his practice to listen to all that could be said against him; to profit by as much of it as was just, and expound to himself, and upon occasion to others, the fallacy of what was fallacious. Because he has felt, that the only way in which a human being can make some approach to knowing the whole of a subject, is by hearing what can be said about it by persons of every variety of opinion, and studying all modes in which it can be looked at by every character of mind. No wise man ever acquired his wisdom in any mode but this; nor is it in the nature of human intellect to become wise in any other manner. The steady habit of correcting and completing his own opinion by collating it with those of others, so far from causing doubt and hesitation in carrying it into practice, is the only stable foundation for a just reliance on it" (p. 232).

This view of informed judgment forms the basis for Mill's conclusion that, *however true* an opinion may be, "if it is not fully, frequently, and fearlessly discussed, it will be held as a dead dogma, not a living truth" (p. 243).

Furthermore, the only way to bring the reasons and arguments supporting a true belief into genuine contact with the believer's mind, is to hear the arguments of adversaries "in their most plausible and persuasive form; he must feel the whole force of the difficulty which the true view of the subject has to encounter and dispose of; else he will never really possess himself of the portion of truth which meets and removes that difficulty" (p. 245). In fact, Mill writes, "if opponents of all important truths do not exist, it is indispensable to imagine them, and supply them with the strongest arguments which the most skillful devil's advocate can conjure up" (p. 245).

## 4.4 Feyerabend's Adopted Roles in a Millian Dynamic Moral and Intellectual World

This is where Paul Feyerabend comes in. I want to suggest that seeing Feyerabend as attempting to *enact* and *embody* these views of Mill, provides a valuable interpretive framework for his more peculiar actions and extreme views. In particular, I am suggesting that Feyerabend was compelled by Mill's account of the importance of rational discussion, and insisted, often without the consent of his interlocutors, on engaging in the *sort* of discussion which, for Mill, served as the foundation for rational opinion and conduct.

I do not mean to imply that Feyerabend kept his intentions secret actually, after reading Mill carefully, it seems that Feyerabend left stage directions all over his contributions, asides announcing to his audience: "I am now playing a believer in a minority view in order to elevate the intellectual and moral level of this discussion." Mill wrote, "I confess I should like to see the teachers of mankind endeavouring to provide a substitute for [a diversity of opinions]; some contrivance for making the difficulties of the question as present to the learner's consciousness, as if they were pressed upon him by a dissentient champion, eager for his conversion" (Mill [1859] 1977, 251). I am suggesting that Feyerabend often assigned himself the role of such a champion, and in doing so, was attempting to enact Mill's vision of the most productive form of interchange.

For instance, regarding his impassioned defenses of astrology, Feyerabend writes: "My use of examples from astrology should not be misunderstood. Astrology bores me to tears. However it was attacked by scientists, Nobel Prize winners among them, without arguments, simply by a show of authority and in this respect deserved a defence" (1991, 165).

In addition, Feyerabend carefully explained the motivation behind the form of many of his arguments: "attempts to retain well-entrenched conceptions are criticized by pointing out that the excellence of a view can be asserted only *after* alternatives have been given a chance, that the process of knowledge acquisition and knowledge improvement must be kept in motion and that even the most familiar practices and the most evident forms of thought are not strong enough to deflect it from its path" (1981a, xi).

Moreover, like Mill, Feyerabend saw the *beneficiaries* of his defenses of unpopular views as the observers of the interactions. Mill wrote, regarding the importance of the advocacy of minority opinions: "it is not on the impassioned partisan, it is on the calmer and more disinterested bystander, that this collision works its salutary effect" (Mill [1859] 1977, 257). Compare this to what might appear to be one of Feyerabend's more perverse—but typical—perspectives:

> A: Are you an anarchist?
> B: I don't know – I haven't considered the matter.
> A: But you have written a book on anarchism!
> B: And?
> A: Don't you want to be taken seriously?
> B: What has that got to do with it?
> A: I do not understand you.
> B: When a good play is performed the audience takes the action and the speeches of the actors very seriously; they identify now with the one, now with the other character and they do so even though they know that the actor playing the puritan is a rake in his private life and the bomb-throwing anarchist a frightened mouse.
> A: But they take the writer seriously!

> B: No, they don't! When the play gets hold of them they feel constrained to consider problems they never thought about no matter what additional information they may obtain when the play is over. And this additional information is not really relevant ...
> A: But assume the writer produced a clever hoax ...
> B: What do you mean – hoax? He wrote a play – didn't he? The play had some effect, didn't it? *It made people think – didn't* it?" (Feyerabend 1991, 50–51, emphasis added)

In sum, what at first (and second) glance may appear to be a scattershot and unprincipled approach to issues of scientific knowledge by Feyerabend is—plausibly—actually quite principled and unified, once we take seriously what Feyerabend says about Mill.

## 4.5 Shouldn't Sciences Be Treated as Special?

It could be objected that Feyerabend has *misapplied* Mill; after all, Mill's essay is usually remembered today for its defense of freedom of speech—especially the expression of heterodox religious views—against governmental authority. Mill's work on how *scientific* reasoning ought to proceed is presented in his *System of Logic ([1848] 1973–1974).*

I find it difficult to deny that Mill, himself, saw the sciences, especially what he called the physical sciences, as operating within a smaller, perhaps more protected, set of expectations and norms; he wrote about these norms in his *System of Logic,* the first edition of which was published in 1848, and which he continued to revise and expand through out his life. There is a sense in which Mill treated scientific topics as somewhat special cases of discussion, controversy, and resolution, for all the obvious and sensible reasons.

In contrast to Mill's discussions in his *Logic,* which concern reasoning about facts and evidence, many of Mill's points in *On Liberty* are illustrated through explicitly religious cases, in which there was no promise of factual evidence to which appeal could be made. Hence, one could argue that Mill, in *On Liberty,* is responding to extremist and sectarian religious views, cases in which there is no evidential means for settling the issues at stake. Should we follow Feyerabend in applying Mill's arguments favoring pluralism—which seem to be ad dressing primarily views that remained outside the purview of scientific inquiry—to scientific knowledge? Does Feyerabend's appeal to Mill amount to a misapplication of Mill's argument?

I do not think so. Contrary to its reputation and use today, Mill stated explicitly that *On Liberty* is *not* primarily about the limits of government interference (p. 305). He was much more concerned about *social norms,* claiming that protection is needed against "the tyranny of the prevailing opinion and feeling," and that society sometimes "practices a social tyranny more formidable than many kinds of political oppression, since, though not usually upheld by such extreme penalties, it leaves fewer means of escape, penetrating much more deeply into the details of life, and enslaving the soul itself' (p. 220).

The great bulk of the essay concerns how individual people (not governments) ought to respond when confronted with opinions and forms of life which are strange or disagreeable to them. The central point regards the inestimable value—to individuals and to society as a whole—of the existence and nurturance of a wide variety of ways of life. Given what he sees as individuals' natural, unreflective intolerance towards those who see and do things differently than they do (p. 227), Mill emphasizes the importance of reducing social and cultural sanctions against those who espouse minority views of any kind.

Perhaps even more crucial is the cultivation of attitudes and skills that, as Mill sees it, are necessary for the genuine flowering of human intelligence and creativity. These attitudes include tolerance, but Mill also demands much more; he advocates "social support for nonconformity" (p. 275). In fact, Mill argues, "In this age, the mere example of nonconformity, the mere refusal to bend the knee to custom, is itself a service. Precisely because the tyranny of opinion is such as to make eccentricity a reproach, it is desirable, in order to break through that tyranny, that people should be eccentric the amount of eccentricity in a society has generally been proportional to the amount of genius, mental vigour, and moral courage which it contained" (p. 269). Ultimately, Mill claims that "diversity of character and culture" is what has led to human progress (p. 274): "The only unfailing and permanent source of improvement is liberty, since by it there are as many possible independent centres of improvement as there are individuals" (p. 270). Most importantly for Feyerabend's adaptation of it, this value of the diversity of opinion was not, for Mill, restricted to areas such as politics and religion, for which no factual appeals were available; Mill explicitly included the sciences in his recommendation of "absolute freedom of opinion and sentiment on all subjects, practical or speculative, scientific, moral, or theological" (p. 225). Moreover, science is included not solely on the basis of a principle of liberty, but on the grounds that pluralism is needed for attaining truth: "on every subject on which difference of opinion is possible, the truth depends on a balance to be struck between two sets of conflicting reasons. Even in natural philosophy, there is always some other explanation possible of the same facts; some geocentric theory instead of heliocentric, some phlogiston instead of oxygen" (p. 244). I conclude that there is, therefore, no mis-application of Mill's views in Feyerabend's applying them to the scientific context.

## 4.6 A (Very Early) Plea for Recontextualizing Science

Thus, Mill's primary preoccupation in *On Liberty* was with positive means of cultivating human flourishing, and not with delineating the limits of legitimate government interference. There *is* something radical, though, in Feyerabend's use of Mill—a use that is consonant with Mill's own views, but which is in tension with most twentieth century philosophy of science—namely, the aim of *recontextualizing* the sciences (and philosophical discussions of the sciences) back into their larger roles in human ways of living. Mill was including modern scientific approaches in the range of ways of living that individuals in a free society may choose among. Because people differ, different ways of life will be judged valuable by them:

> If a person possess any tolerable amount of common sense and experience, his own mode of laying out his existence is the best, not because it is the best in itself, but because it is his own mode.
>
> … Such are the differences among human beings in their sources of pleasure, their susceptibilities of pain, and the operation on them of different physical and moral agencies, that unless there is a core responding diversity in their modes of life, they neither obtain their fair share of happiness, nor grow up to the mental, moral, and aesthetic stature of which their nature is capable. ([1859] 1977, 270)

Furthermore, Mill defended the notion of each individual pursuing their own path on the basis of its value to the rest of human society. Under conditions of freedom, Mill writes,

> human beings become a noble and beautiful object of contemplation; and as the works partake the character of those who do them, by the same process human life also becomes rich, diversified, and animating, furnishing more abundant aliment to high thoughts and elevating feelings, and strengthening the tie which binds every in dividual to the race, by making the race infinitely better worth belonging to.... There is a greater fulness of life about his own existence, and when there is more life in the units there is more in the mass which is composed of them." (p. 266)

What has sometimes been seen as Feyerabend's "anti-scientific" attitude is, I suggest, more appropriately interpreted as his taking this aspect of Mill's thought very seriously. The central point is not the moral one, that people should not be *made* to live in a way that they did not freely choose; it is that individuals truly can be *the best judges—better* even than any scientific experts—of which way of living is better for them, and that we could all benefit from these judgments. Consider the following passage:

> B: Ever since people were discovered who did not belong to the circle of Western culture and civilization it was assumed, al most as a moral duty, that they had to be told the truth – which means, the leading ideology of their conquerors. First this was Christianity, then came the treasures of science and technology. Now the people whose lives were disrupted in this manner had already found a way of not merely surviving, but of giving meaning to their existence. And this way, by and large, was much more beneficial than the technological wonders which were imposed upon them and created so much suffering. 'Development' in the Western sense may have done some good here and there, for example in the restriction of infectious diseases – but the blind assumption that Western ideas and technology are intrinsically good and can therefore be imposed without any consultation of local conditions was a disaster. (Feyerabend 1991, 74)

Feyerabend applies this point equally to more specific areas such as medicine, in which scientific expertise should *not—for scientific purposes as well as moral ones—be* taken as a sufficient reason to reject other forms of expertise relevant to these aspects of life. Feyerabend writes, regarding traditional Chinese medicine, for example: "alternative medical systems are often parts of entire traditions, they are connected with religious beliefs and give meaning to the lives of those who belong to the tradition. A free society is a society in which all traditions should be given equal rights *no matter what other traditions think about them.* A respect for the opinions of others, choice of the lesser evil, chance of making progress – all these things argue in favor of letting all medical systems come out into the open and freely compete with science" (1991, 75, his emphasis).

Here, Feyerabend is not appealing to a simple libertarian right of people to live as they wish; he also provides suggestions regarding promising areas of scientific research, focusing on areas in which Western scientific medicine has weaknesses that may be corrected by the introduction of Chinese techniques and understandings.

## 4.7 Feyerbend's True Roles

In summary, I have suggested following Paul Feyerabend's own advice, and interpreting Feyerabend's work in light of the principles laid out by John Stuart Mill. A review of Mill's essay, *On Liberty,* emphasizes the importance Mill placed on open and critical discussion for the vitality and progress of various aspects of human life, including the pursuit of scientific knowledge. Many of Feyerabend's more unusual stances, I suggest, are best interpreted as attempts to play certain roles especially the role of "eccentric defender of unpopular minority opinion"—that are necessary to fulfilling Mill's conditions for rational exchange and optimal human development.

I wish that I could ask Feyerabend just how explicit and extensive his intentions were, to enact Mill's context for rational judgment. I missed my chance, but this is merely my painful loss. I have attempted here to characterize *our* loss—as philosophers of science—in terms that I believe cast a rather different light on Feyerabend's obstreperous and, to many, slightly insane defenses of unpopular viewpoints. He can be seen as trying to provoke us into protecting us from ourselves, and to highlight and actually *enact* the most fundamental principles by which human life is actually improved. For this, I would like to thank him.

## References

Feyerabend, Paul K. 1975. *Against Method: Outline of an Anarchistic Theory of Knowledge.* London: Verso.

———. 1978. *Science in a Free Society*. London: Verso.

———. 1981a. *Realism, Rationalism and Scientific Method: Philosophical Papers*. Vol. 1. Cambridge: Cambridge University Press.

———. 1981b. *Problems of Empiricism: Philosophical Papers*. Vol. 2. Cambridge: Cambridge University Press.

———. 1991. *Three Dialogues on Knowledge*. Oxford: Basil Blackwell.

Lloyd, Elisabeth A. 1996. The Anachronistic Anarchist. *Philosophical Studies* 81: 247–261.

Mill, John S. [1848] 1973–74. In *A System of Logic*, ed. J.M. Robson. Toronto: University of Toronto Press.

———. [1859] 1977. On Liberty, Essays on Politics and Society. In *Collected Works of John Stuart Mill*, ed. J.M. Robson, vol. 18. Toronto: University of Toronto Press.

# Chapter 5
# Feyerabend and Mach

**John Preston**

**Abstract** Of all the influences on the work of Paul Feyerabend, Ernst Mach's was probably the most long-standing, and undoubtedly among the most important. I first show that Feyerabend's earliest mentions of Mach are heavily under the influence of Karl Popper. Early Feyerabend characterises Mach in traditional terms as a positivist whose philosophy is flawed in comparison with critical rationalism. Next, in the papers Feyerabend published during the early and mid-1960's, Mach appears in two main guises: not only as an anti-realist, but also as an anti-pluralist. I show that while there are some flashes of insight about Mach here, by and large Feyerabend sticks to the traditional way of reading him as aligned with the logical empiricists. Feyerabend did, however, come to change his mind about Mach. He eventually came to be at the forefront of those who initiated a re-evaluation of Mach, thereby beginning to move opinion away from the 'received view'. Feyerabend's published struggle on Mach's behalf began in earnest around 1970. From this point onwards, his attitude to Mach was relatively constant. While endorsing several of the ways in which Feyerabend came to characterise Mach's thought, I take issue with some other themes in these publications. I suggest that we should not follow Feyerabend's mature reading of Mach in its entirety, since he had turned away from certain sensible aspects of Mach's ideas. Most notably, Feyerabend misinterpreted what Mach said about world views, he wrongly imagined that Mach endorsed the 'constructivist' ideas that human cognition transforms (non-cognitive) facts, and that facts themselves get adapted, and he fancifully ascribed to Mach his own view that *all* concepts are 'theoretical' concepts.

For helpful comments on a draft of this paper, I'm indebted to Luca Guzzardi, Michael Stöltzner, Friedrich Stadler, Pietro Gori, María de Paz, and an anonymous referee.

J. Preston (✉)
Department of Philosophy, University of Reading, Reading, UK
e-mail: j.m.preston@reading.ac.uk

© The Author(s), under exclusive license to Springer Nature Switzerland AG 2024

S. Gattei, R. Corvi (eds.), *Feyerabend in Dialogue*, Boston Studies in the Philosophy and History of Science 346,
https://doi.org/10.1007/978-3-031-71938-7_5

**Keywords** Paul Feyerabend · Ernst Mach · Positivism · Anti-realism · World views · Theoretical concepts

## 5.1 Introduction

Of all the influences upon Paul Feyerabend, the work of Ernst Mach is perhaps the most long-standing, and undoubtedly among the most important. In Feyerabend's more mature estimations, only the work of John Stuart Mill ranked as high, and Mill's influence on Feyerabend was nowhere near as wide-ranging. (Ludwig Boltzmann, who Feyerabend also greatly admired, he discussed far less often, and in much less depth).

In his autobiography, *Killing Time*, Feyerabend recounted something of his early acquaintance with Mach's works, as follows:

> I was interested in both the technical and the more general aspect of physics and astronomy but I drew no distinction between them. For me, Eddington, Mach (his *Mechanics* and *Theory of Heat*), and Hugo Dingler (*Foundations of Geometry*) were scientists who moved freely from one end of their subject to another. I read Mach very carefully and made many notes. Then I forgot about him. Decades later I heard Mach described as a short-sighted and narrow-minded positivist. 'That is not the Mach I know', I said, and re-read the *Mechanics*. And, indeed, it turned out that much of the Mach scholarship (so called) is ideology, not fact, and that it can be exploded by a little relaxed reading of Mach's basic works (Feyerabend 1995a, p. 30).

Feyerabend continued to refer very occasionally to Eddington and Dingler in his published work, but to nothing like the same extent as he referred to Mach. His work contains important claims about Mach's philosophy, and he was at the forefront of those who initiated a re-evaluation of Mach in the English-speaking context during the late twentieth century, thereby beginning to move opinion away from the 'received view' of Mach as a simple pre-logical positivist whose doctrines were paradigm but also extreme versions of sensationalism, psychological atomism, phenomenalism, instrumentalism, classical epistemological foundationalism, epistemically conservative empiricism, inductivism, and opposition to theorising in science. In this he did us a great service. Nothing that follows here is intended to detract from this. Along the way, Feyerabend engaged in disputes about Mach with what we might call more 'traditional' readers of Mach such as Albert Einstein, Gerald Holton, Elie Zahar and Klaus Hentschel (see Zahar 1977; Feyerabend 1980/1981, 1984/1987; Zahar 1981; Hentschel 1985).

This paper will not be a full comparison of the philosophies of Mach and Feyerabend. Neither will it be a defence of modern readings of Mach, such as Feyerabend's, against the received view—the untenability of the received view will mostly be *presupposed* here. Rather, what will be attempted is an account and critical evaluation of the most significant things Feyerabend said about Mach. I will first trace the ways in which Mach appears in Feyerabend's works, from the mid-1950s right up until the mid-1990s. As we will soon see, Feyerabend's own account, above, is part of the story, but by no means the whole.

## 5.2 Dingler

Let us dwell for a moment, though, on the lesser-known figure of Hugo Dingler. In a curriculum vitae he wrote in 1951, recently published in the first volume of *Feyerabend's Formative Years*, Feyerabend stated 'My interest in philosophy followed a lively interest in the natural sciences. I read Duhem, Mach and (less understandably to me today) Dingler' (Collodel and Oberheim 2020, p. 47). Feyerabend's interest in physics dated from age 12 (ibid., p. 45), i.e., from 1936 or 1937 (he also says, ibid., p. 51). Through reading books by Mach, Duhem, Poincaré and Dingler, he was then led to his interest in philosophy (ibid.). This trajectory is confirmed in certain autobiographical remarks he made many years later, in 1976, where he began by saying:

> Hugo Dingler was one of my first loves in the philosophy of science. When I was 16 I picked up his *Grundlagen der Geometrie* (1st edition) in a second-hand bookstore. I must have read all his books (though I never found his book on the Jewish question) and I was a determined Dinglerian for years afterwards. I think that his book *Das Experiment*, his *Zusammenbruch der Wissenschaft* and, of course, his *Grundlagen der Geometrie* are well worth reading. I also learned a lot from his short essay on Mach (with excerpts from Mach's diaries) (Collodel and Oberheim 2020, p. 19).

Dingler is usually thought of as an unorthodox kind of conventionalist, and often also as a kind of Machian. (For a sketch of his life and views, see Wolters 1988). He was certainly interested in and influenced by Mach—the 'short essay on Mach' that Feyerabend mentions is his book *Die Grundgedanken der Machschen Philosophie* (Dingler 1924), a positive appreciation of Mach's philosophy.

Perhaps unsurprisingly, given Dingler's rejection of Einstein's theory of relativity, but also the fact that he joined the Nazi party, Feyerabend does not mention his having been a 'determined Dinglerian' in his correspondence with Karl Popper. In fact, by October 1965, he comments that he has met Paul Lorenzen, with whom he 'got on splendidly, despite the sometimes quite violent debates we had (he wants to revive Hugo Dingler)' (Collodel and Oberheim 2020, p. 404. See also p. 20).

## 5.3 An Early Machian Phase?

That Feyerabend had an early Machian phase, probably in his early to mid-20s, is suggested by something he said in a *curriculum vitae* he produced in 1951, which was submitted to the University of Vienna along with his dissertation on basic statements. In this document he credits Walter Hollitscher with 'repeated encouragement to verify more precisely my philosophical views and to substantially correct them (from the Machian positivism to the position I took here)' (quoted in Stadler 2008, p. 206). This tallies with the version of a 1951 curriculum vitae recently published in *Feyerabend's Formative Years*, where he says 'Thanks mainly to [Hollitscher], my obscure and therefore very dogmatic theory of elements (following Mach) came to waver and I defend [in my PhD dissertation] a view that is diametrically opposed to the one I had held about five years earlier and displays somewhat more differentiation' (Collodel and Oberheim 2020, p. 48).

It is notable that Feyerabend here characterises his early stance as 'Machian positivism'. The only published document that survives from the pre-doctoral and 'positivist' phase of Feyerabend's work is his 1948 article 'The Concept of Intelligibility in Modern Physics' (Feyerabend 1948/2016). It certainly explains and seems to recommend a perspective that Feyerabend calls 'today's positivism' (p. 7), but although it does mention a 'principle of continuity' (p. 5), it is generally too vague to be said to present any *Machian* positivism. Since this article is from 1948, presumably it is too late to feature the Machian view he took himself to have held in 1946. As we will now see, though, this way of characterising *Mach*, as a positivist, remained in place for quite some time before Feyerabend eventually repudiated it.

## 5.4 The Negative Phase: Mach the Anti-realist, and Anti-pluralist

The earliest mention of Mach in any of Feyerabend's published works comes in his essay 'Physics and Ontology', published in 1954, where Feyerabend is keen to characterise some very different kinds of philosophies of science as 'positivistic'. The first kind in question comprises philosophies from the Platonic-Aristotelian tradition, deemed positivistic because they have an instrumentalist conception of laws. In addition, though, and more familiarly, Feyerabend argues that a second kind of positivistic philosophy of science can be traced back to Bishop George Berkeley. His account of Berkeley and the instrumentalist tradition owes a great deal to Karl Popper's then very recently-published 'Note on Berkeley as a Precursor of Mach and Einstein' (Popper 1953/1963). Feyerabend proceeds to criticise this version of positivism, which, he says, 'found an excellent advocate, above all, in E. Mach' (Feyerabend 1954/2016, p. 12), and he does so by deploying the objection (common among anti-positivists, including Popperians),[1] that attempts to reject and avoid metaphysics always themselves involve *another*, rival metaphysics:

> [T]he positivistic version… proceeds from a definite theory of 'that which is', that is, from a definite ontology, from an ontology of things (sense data, objects of the everyday world, classical situations), about whose existence predictions should be made. In order to keep the description as general as possible from the beginning, I name these things *elements* (ibid.).

[1] The objection is not original to Popperians, though: Charles Sanders Peirce had already levelled it around 1898 (see Erwin Hiebert's introduction to Mach 1905/1976, p. xvii).

It is perhaps significant that Feyerabend takes over Mach's term 'elements' here, explaining it by remarking that 'Those who are more comfortable with illustrations may thereby easily think of sense data, simple feelings – in short, of all phenomena which are subjectively present in observations' (ibid.). Perhaps this is what he then understood Mach to be referring to by that same term. Later we shall see his more mature understandings of this term.

Here, Feyerabend takes this 'ontology of elements' to be associated with a specific theory of meaning, according to which 'an expression is meaningful if, and only if, it refers to an element, or else it belongs to the class of logical expressions' (ibid., p. 13). He takes this to imply that the meaning of non-logical terms must be fixed in advance of the activity of the scientist. And his objections to this supposed Machian positivism are (a), that there can be no impenetrable 'border' between physics and philosophy and (b), that what is meaningful changes as science progresses. These are the first appearances of what will become two constant *leitmotivs* of Feyerabend's philosophy, already in place before he had published his review of the work of Wittgenstein (which never succeeded in dislodging the first of them).

The next time Feyerabend mentions Mach is in his 1958 encyclopaedia entry on '*Naturphilosophie*', where he elaborates on another Popperian theme, portraying Mach as the positivist behind Niels Bohr's philosophy of quantum-mechanics (Feyerabend 1958/2016, p. 339). He also has a curious take on Mach's ambitions there:

> According to Ernst Mach (1838–1916), who proceeds more radically [than Immanuel Kant], the concept of nature has no longer any basis in experience, and must be removed. For science (besides which no independent natural philosophy is allowed) can and should do nothing else but systematically order experience; a concept, or a theory, is allowed as long as it contributes to this ordering. Such an epistemological principle must now dispose not only of the general concept of nature, but also of the special concepts of substance, absolute space, absolute time, and atoms; but this means that it must also rid itself of the very heart of the mechanical concept of the world, that is, mechanics (ibid., p. 335).

This characterisation of Mach is mostly fanciful. Mach was interested neither in 'removing' the concept of nature,[2] nor in eliminating mechanics from science. Feyerabend here has inflated something which really can be found in Mach, a critique of the 'mechanical world view', into two theses he would not have accepted. Mach did want to cut mechanics down to size, and he certainly explored alternatives to the mechanical world view. He even occasionally suggested that some other branch of physics might prove to be more fundamental than mechanics. But his ambition did not extend to abolishing mechanics. Neither did he have any animus against the concept of nature. In his book *History and Root of the Principle of Conservation of Energy*, for example, Mach was happy to talk liberally about 'the investigation of nature', about those who investigate nature, and about laws of nature (Mach 1872/1911, pp. 49, 50, 56, 57, 61, 63, 73, 74, 81). In fact, references to nature occur throughout his works.

[2] This idea is repeated in Feyerabend's book *Philosophy of Nature* (Feyerabend 2016b, p. 196), material for which was written around 1975.

From this point onwards, in a curious development, the papers Feyerabend published during the early 1960's feature Mach in two main guises: as an anti-pluralist, as well as an anti-realist. In 'Professor Bohm's Philosophy of Nature', for example, Feyerabend's review of David Bohm's book *Causality and Chance in Modern Physics*, Mach's opposition to the mechanical theory of heat is deemed to be a case of 'using a theory (the second law of thermodynamics) to exclude alternatives' (Feyerabend 1960/1981, p. 223n), and the quantum theory is said to be 'the first theory of some importance that satisfies to some extent the programme of Berkeley and Mach' (ibid., p. 224). Around the same time, in an article on methodology, Feyerabend takes Mach to have had 'a veritable phobia of hypotheses' (Feyerabend 1961, p. 884), and to be part of 'a current hostile to theories' (ibid., p. 895). These ways of characterising Mach are familiar components of the received view.

Further, in Feyerabend's *major* papers from the early 1960s, the ones that made his name in philosophy of science, Mach continues to appear as one of those who accepts what is identified there as the perniciously conservative (anti-pluralistic) assumptions of contemporary empiricism. In 'Explanation, Reduction and Empiricism', for example, he identifies what he calls the 'consistency condition': 'Only such theories are admissible (for explanation and prediction) in a given domain which either contain the theories already used in this domain, or are at least consistent with them' (Feyerabend 1962/1981, p. 56). He then claims that this was 'adopted, at least implicitly, by all those who defend the logical empiricist theory of explanation and reduction, and by many others', and he takes Mach to be an example, quoting a remark from the *Wärmelehre* which he then takes to 'insinuate' that mechanical hypotheses should not be taken too seriously (ibid.).

This is all reiterated in one of Feyerabend's best-known and most-reprinted papers, 'How to be a Good Empiricist' (Feyerabend 1963/1999, pp. 83–4). There, in addition, he quotes Mach as follows, from the same book:

> I would like to emphasize that *every and any* idea is admissible as a means for research, provided it is helpful; still, it must be pointed out, on the other side, that it is very necessary from time to time to free the presentation of the *results* of research from all inessential additions (this is Feyerabend's own translation. For another, see Mach 1896/1986, p. 333).

Mach's idea is that although metaphysical considerations may be important when the task is to invent a new physical theory, as soon as the theory has been developed in a formally satisfactory fashion and has received sufficient confirmation to be regarded as empirically successful it can and must forget its metaphysical past. Feyerabend complains that this advice does not allow metaphysical considerations to survive beyond the initial phase of theory-invention, and illegitimately gives empirical considerations the upper hand over metaphysical ones (Feyerabend ibid., p. 100). These points are then reiterated again in his long paper 'Problems of Empiricism' (Feyerabend 1965, pp. 161, 166). Mach still appears here as an anti-pluralist villain, aligned with the logical empiricists.

Within these same papers, though, another, more prescient note is sounded. In a footnote to 'How to be a Good Empiricist', Feyerabend says:

> In about 1925 philosophers of science were bold enough to stick to their theses even in those cases where they were inconsistent with actual science. They meant to be *reformers* of science, and not *imitators*. (This point was explicitly made by Mach in his controversy with Planck (Cf. his *Zwei Aufsätze*). In the meantime they have become rather tame (or beat). (Feyerabend 1963/1999, p. 89 note).

This footnote, together with a similar one from 'Problems of Empiricism' (Feyerabend 1965, pp. 245–6, note 114), is the first clear published ancestor of Feyerabend's later take on Mach. Another occurs in his 'Reply to Criticism' (also from 1965), where he describes modern science and its methodology as a rebellion of a few people (like René Descartes and Galileo Galilei) against established modes of thought, and says 'Ernst Mach, who initiated the movement of modern positivism, criticized the science of his time in the very same spirit. This is the attitude which has led to the sciences as we know them today' (Feyerabend 1965/1981, p. 105 note). Along the same lines, 'Dialectical Materialism and the Quantum Theory', from 1966, includes a passage in which Feyerabend contrasts modern philosophers of science with their illustrious predecessors:

> So great is the awe in which science is held by philosophers, and so small is the confidence in the power of philosophy to advance the state of our knowledge, that all a philosopher is now prepared to do is either to 'analyze' science, that is, to present its results and methods in a language of his own, or to make it the basis of philosophical generalizations. But all this overlooks the fact that science is full of ideology, although its ideological commitments are usually masqueraded as 'obvious methodological rules' or as 'well-established experimental facts'. Now obviously such commitments can be examined only by a philosophy that is bold enough to oppose the sciences. Mach provided such a philosophy (Feyerabend 1966/2016, p. 223).

However, since these three footnotes and one passage praise Mach (and the early logical positivists) for being *critical of* certain scientific developments, and for wanting to *eliminate* metaphysical components from science, they sit uncomfortably alongside Feyerabend's usual and insistent proselytization for the kind of pluralism which wants *all* theories to bloom. One may well feel that this tension is not effectively resolved in Feyerabend's work (and I will return to this issue below).

The dominant themes in Feyerabend's early mentions of Mach, then, are Popperian. The characterisation of Mach as an anti-realist is solidly Popperian, and the related portrayal of him as an anti-pluralist is, we might say, Feyerabend's explicit development of an idea, implicit in Popper's philosophy, which became his own particular hobby-horse. However, we should note that the general nature of this characterisation of Mach as a classic positivist cannot entirely be put down to the fact that Feyerabend was going through a Popperian phase (as one might well imagine Feyerabend later wanting to do!). As we saw, he took his own former stance to have been a version of Machian positivism in 1951, by which time he had known Popper for more than 2 years.

Alongside these characterisations of Mach in positivist terms, though, we ought to harken to another, much softer note. In his fairly well-known review of Wittgenstein's *Philosophical Investigations*, Feyerabend lists Mach (among others) as someone who anticipated what he supposes to be Wittgenstein's achievements in that book, namely the critique of a certain theory of meaning, and its replacement by another such theory (Feyerabend 1955/1981, p. 128n). Although this hint is never really developed, Feyerabend was right to recognise at least some kind of kinship between Mach and Wittgenstein (Preston 2019 relates some of its aspects, but much more remains to be said). Friedrich Stadler has insightfully described Feyerabend's piece as 'One of the first reviews of Wittgenstein's *Philosophical Investigations* in the context of Austrian philosophy (Ernst Mach, Heinrich Gomperz)' (Stadler 2008, p. 209).

## 5.5 'Philosophy of Science: A Subject with a Great Past' (1970)

Feyerabend's struggle on Mach's behalf really began in earnest with his paper 'Philosophy of Science: A Subject with a Great Past', published in 1970. The great past in question comprised what Feyerabend called 'the heroic time of the scientific philosophy' (Feyerabend 1970/1999, p. 128), that is, the scientific revolution of the sixteenth and seventeenth centuries, and most notably the work of Galileo (ibid., p. 127). According to Feyerabend, the new philosophy of this time is 'not content just to mirror a science that develops independently of it; nor is it so distant as to deal just with alternative philosophies. It plays an essential role in building up the new science that was to replace the earlier doctrines' (ibid., p. 128). Isaac Newton, though, was essentially responsible for replacing this 'active and critical philosophy' by 'a more conservative creed' which generates 'technical problems of its own which are in no way related to specific scientific problems' (e.g., Hume's problem of induction), and which gives rise to 'a special subject that codifies science without acting back on it (Kant)' (ibid., p. 128). Feyerabend also contrasted the 'great past' of the scientific revolution with what he considered to be the impoverished *present*, which he understandably took to be represented by the project of 'logical reconstruction', featuring formal puzzles concerning confirmation, prediction, counterfactuals, predicates like 'grue', etc. (ibid., pp. 127, 134–5).

His paper then introduces, with a wonderful flourish, the one philosopher of the nineteenth century who 'was not prepared to accept the status quo, who was not content to criticise science from the safe distance of a special subject either, but who proceeded to suggest concrete means for its change. The nineteenth century produced *Ernst Mach*' (ibid., p. 129).

According to Feyerabend, Mach's 'philosophy' (and Feyerabend is well aware that Mach disavowed that title), 'contains a general criticism of the science of his time, [...] and a philosophy of science that completely abandons the idea of a foundation of knowledge' (ibid., p. 129). Against more 'traditional' readers, Feyerabend insists that Mach's philosophy has been fruitful (although the case he makes for this idea has to wait until his next article).

Feyerabend is at his most fast-and-loose here, though, when it comes to the idea of a *general methodology*. He first makes the following eminently reasonable demand:

> While it should be possible, in a free society, to introduce, to expound, to make propaganda for any subject, [...] it must also be possible to *examine* what is being expounded by reference, not to the *internal* standards of the subject (which may be but the method according to which a particular madness is being pursued), but to standards which have the advantage of being simple, commonsensical, and accepted by all (ibid., p. 127).

Mach might have approved of this, for he allotted an important role to such standards. However, it soon transpires that what Feyerabend has in mind is that a 'general methodology' must be 'independent of any particular assertion about the world, however trivial, and however obvious. It is supposed to provide a point of view from which *all* such assertions can be judged and examined' (ibid., pp. 129–130). This means that a methodology must make *no* assumptions which may later need to be revised, even such 'assumptions' as the banal one Feyerabend identifies here, that scientists are perceiving subjects who investigate an objective world and gradually increase their knowledge of it (ibid., p. 130).

It also means that methodology must be *normative*, understood as meaning that philosophers 'doing' (studying?) methodology ought to be *appraisers* of science. This is an idea that Feyerabend may well have inherited from Karl Popper. Popper's account of methodology was problematic (I have argued in Preston 1994), but both he and, at this point, Feyerabend, took it be part of the responsibility of philosophers not merely to criticise scientists for not sticking to the agreed methodological rules governing science, but also to appraise and if necessary criticise those methodological rules themselves.

This, in my view, goes way beyond Mach. While Mach knew that assumptions sometimes have to be examined, he would rightly have regarded the idea of a methodology which makes absolutely *no* assumptions as a chimera. In fact, what Feyerabend has done here is to set up an illusory philosophical ideal, and one which bears little relation to Mach's philosophy. When Feyerabend rightly notes that 'According to Mach the task of science is to find simple and regular connections between *elements*' (ibid., p. 130), this does not mean that Mach imagined that this task could be undertaken in a way that involves *no* presuppositions. On the contrary, as we will soon see, Mach was explicitly a fallibilist about his own method and its results—this is part of the import of his declarations that his 'elements' were *provisional*.

Notice, too, that Feyerabend's derisive treatment of Hume and Kant implies that *philosophical* problems, or at least philosophical problems that do not have their roots and life in science, are not worth addressing. Mach would have agreed with this only in a sense in which Feyerabend did not intend it, since a crucial step in certain of his major works is to show that at least some such problems are *pseudo-problems*, which cannot be solved, but must be dissolved and thereby avoided (see Preston 2023). There is an important connection here with the work of Heinrich Hertz, Boltzmann, Wittgenstein *and* the Vienna Circle, a connection which

Feyerabend misses (or ignores). *His* attitude to philosophical problems that do not have their roots in science, that they are simply not worth addressing, was more dismissive (and, in a way, more fervently anti-philosophical) than Mach's.

A further misconception of Feyerabend's emerges when we consider his idea that Mach's 'elements' 'are *not* sensations' (ibid., p. 131, emphasis added), indeed that they 'are *carefully distinguished from* sensations' (ibid., emphasis added). Noticing Mach's declaration that 'Speaking of sensations, already entails acceptance of a one-sided theory' (Mach 1886/1959, p. 22), Feyerabend takes his explanation of the term to mean that neither sensations nor perceptions figure among Mach's 'elements'. But this is not correct. What Mach means is that we associate the *word* 'sensation' (or rather '*Empfindung*') exclusively with the psychological, the mental, whereas the 'elements' he speaks of are equally physical *and* psychological.[3] This is why he *prefers* to speak of 'elements' (although, unfortunately, he did not always stick with that decision). But it does *not* mean that *Empfindungen* are not among those elements. The idea of Mach's 'monistic standpoint' (which others later call his 'neutral monism') is that all 'elements' (including 'sensations') can be investigated in *two* ways: the physical way, and the psychological way (see Preston 2021). As we shall see, Feyerabend did eventually get this crucial point right.

Feyerabend, then, even after his Machian turn, was fully capable of misrepresenting Mach *within* his generally insightful reading. However, from the early 1970s onwards his *attitude* to Mach remained relatively constant. Some new themes emerge, others pass into desuetude, and emphases change, but Mach appears as a constant role-model for Feyerabend.

## 5.6 Major Themes in Feyerabend's Later Remarks on Mach

Feyerabend's most important paper on Mach, 'Mach's Theory of Research and its Relation to Einstein' (Feyerabend 1984/1987) is now well-known, and indeed fêted, among those interested in Mach. His other later remarks on Mach are not as well-known, but should certainly feature in a full picture of his later reading of Mach. This is what will be attempted here. Looking at all the many articles and books in question, it is possible to identify the themes which most often occur in them, or which occur with the greatest emphasis. Not all of these themes, understandably, pertain to Mach as directly as is relevant for my purposes here.[4] I will focus, first, on some of the things that Feyerabend gets most right about Mach, and about which

[3] Mach sometimes describes them thus, but sometimes describes them as *neither* physical *nor* psychological.

[4] So, for example, one of the ideas most prevalent in these publications is the way and extent to which the philosophy of science degenerated after Mach (see., e.g., Feyerabend 1970/1999, 1974/1981, 1978, 1980/1981, 1981a, b, 1984/1987, 1995a). Although this was clearly one of Feyerabend's major concerns, it will not be treated here.

his reading is a salutary and necessary corrective to more traditional readings. After that, I will give a more detailed treatment of some of Feyerabend's more implausible claims about Mach. We will start, though, with the points on which Feyerabend is, in my view, at his best, and I begin with the least problematic points among these. It will be possible to proceed fairly briskly here.

## 5.7 Feyerabend's Positive Contributions

### *5.7.1 Mach as a Fallibillist*

Mach's fallibilism, and the fact that he was a fallibilist about *various* different matters, should stand out from his writings. Surprisingly, though, Feyerabend's first explicit published recognition that Mach was an early and important fallibilist occurs in *Against Method*, where he says:

> Scientific laws were once thought to be well-established and irrevocable. The scientist discovers facts and laws and constantly increases the amount of *safe* and *indubitable* knowledge. Today we have recognized, mainly as a result of the work of Mill, Mach, Boltzmann, Duhem and others, that science cannot give any such guarantees (Feyerabend 1975, p. 188).

There is more than one *slide* here, from the idea that no individual scientific *law* is irrevocable, to the idea that science cannot increase our '*safe* and *indubitable* knowledge', as well as from the idea of laws being well-established to their being *indubitable*, but we must let these pass.

In his reply to Zahar, Feyerabend also urged that Mach's methodology and psychology of knowledge were tentative:

> 'Above all there is no Machian philosophy' he writes in the introduction to *Erkenntnis und Irrtum* – 'there is, at most, a methodology of science and a psychology of knowledge *and like all scientific theories these things must be regarded as preliminary and incomplete attempts'* (Feyerabend 1980/1981, p. 94, note 28, quoting Mach 1905/1976, p. xxxiii, note 2, but with emphasis added by Feyerabend).

In the Introduction to the second volume of his own *Philosophical Papers*, he noted that Boltzmann and Mach 'treat the history of science as a special instance of the history of species and regard the ideas of their time as late but by no means final results' (Feyerabend 1981b, p. 20). And in his 1986 review of the multi-volume *Postscript* to Popper's *The Logic of Scientific Discovery*, he again noted Mach's fallibilism:

> The pluralism implicit in [Duhem, Helmholtz and Boltzmann] found support in Darwin's theory. Before Darwin it had been customary to view organisms as divinely created and therefore perfect solutions to the problem of survival. Darwin drew attention to numerous 'mistakes': life is not a carefully-planned and meticulously performed realisation of clear and stable aims; it is unreasonable, wasteful, it produces an immense variety of forms and leaves it to the particular stage it has reached... to define and eliminate the failures.

> Similarly, so Mach, Boltzmann, and other followers of Darwin inferred, the development of knowledge is not a well-planned and smoothly running process; it, too, is wasteful and full of mistakes; it, too, needs many ideas and procedures to keep it going. Laws, theories, basic patterns of thinking, facts, even the most elementary logical principles, are transitory results, not defining properties of this process (Feyerabend 1986/1987a, p. 188).

Fallibilist flavours, it should probably be said, are easy to spot in Mach, and it remains surprising that they were *not* more generally spotted by advocates of the received view. It is true that they do sit alongside certain remarks which suggest their contrary, some form of classical foundationalism. A full treatment would of course have to look at all these remarks. But on Feyerabend's behalf I would suggest that the fallibilist notes have a certain priority, since they occur in places where Mach is discussing the *general* character of his views. Mach treated science as a whole as fallible, and scientific theories and laws, as its fallible products. Feyerabend was one of the first to recognise this.

### 5.7.2 *Mach as Evolutionary Epistemologist*

That Mach's epistemological thinking was an early example of an *evolutionary* epistemology is now a commonplace (it has perhaps been most urged by Milič Čapek (1968), by Dario Antiseri (2000/2006), and by Isabel Serra and Elisa Maia (2018)). Feyerabend did recognise this, for he remarked in his review of Popper's book *Objective Knowledge* that Mach was a pioneer evolutionary epistemologist (Feyerabend 1974/1981, p. 192). (It is also implicit in a remark of Feyerabend's which will be quoted below in connection with the issue of pluralism). Feyerabend also claimed that Mach had gone *further* than Popper:

> Evolutionary ideas were introduced into epistemology by Spencer, Mach, and Boltzmann and [...] these thinkers went much further than Popper: if knowledge is part of evolution, then even its most basic constituents such as the laws of logic, or the laws of arithmetic, or the use of argument as an instrument of progress must be regarded as temporary phases which are in need of improvement and which will be overcome at a later stage (some of us believe that this 'later stage' has already arrived) (ibid.).

Boltzmann, it is true, saw the laws of logic (in their nineteenth-century guise as the supposed 'laws of thought') as subject to evolution. Whether Mach had the same conception is not so obvious, and Feyerabend cites no evidence.

Feyerabend, though, never laid the kind of stress upon this feature of Mach's thought, his being an evolutionary thinker, that more recent authors do, perhaps because, as we shall see below, it is in tension with his own take on theoretical pluralism.

However, Feyerabend was importantly correct, I believe, when he claimed that what some think of as Mach's epistemology just *wasn't* his epistemology. Einstein is the particular target of Feyerabend's remark here (Feyerabend 1984/1987, p. 193), but the point applies to most traditional readers, as well as to some more recent ones. To put matters in a nutshell: the first chapter of *The Analysis of Sensations* is *not* the location of Mach's epistemology. (Indeed Mach distinguishes that book from his 'epistemologico-physical writings' in the Preface to its second edition (Mach 1886/1959, p. xxxvii)). 'Mach's epistemology' is really presented in certain chapters of *Knowledge and Error* (plus his 1883 inaugural address 'On Transformation and Adaptation in Scientific Thought'). What traditional readers identify as 'Mach's epistemology' is in fact, says Feyerabend, 'a general scientific theory (or theory-sketch), comparable in form (though not in content) to atomism, and different from any positivistic ontology' (ibid.). While this isn't exactly how I would put the matter, I agree that it gets a lot closer to the truth than the traditional reading.

### *5.7.3 Mach as Historically-Oriented*

In the late 1960s, Stephen Toulmin found it necessary to draw attention to the 'historical-critical' side of Mach's work, since the received view of Mach had often ignored it (Toulmin 1969, p. 50). But the fact that Mach was an historically-oriented thinker is now well-recognised. Three of his major books are explicitly presented as 'historical-critical' or historical-epistemological studies (Mach 1883/1960, 1896/1986, 1921/1926). More recently, this has been emphasised in a variety of works (see, for example, Fuller 1994; de Paz 2018; Guzzardi 2021). Feyerabend took explicit notice of this character of Mach's thinking surprisingly late, in *Against Method* (Feyerabend 1975, p. 146, note), and then again in the introduction to the first volume of his own *Philosophical Papers*, where he says:

> Mach differs from the positivists [...] in two ways: he does not assume a two-layer model of knowledge (except locally [...]) and he examines the historical (physiological, psychological) determinants of scientific change. He gives an account of this change and of the difficulties facing the individual scientist that is much more realistic than the accounts of the philosophers (Feyerabend 1981a, p. 13).

Mach's emphasis on history also gets mentioned in Feyerabend's 1984 paper, 'Mach's Theory of Research and its Relation to Einstein' (Feyerabend 1984/1987, p. 210), but we have to wait until *Farewell to Reason* for this to be fleshed out. There, Feyerabend emphasises that for Mach science is an historical tradition (Feyerabend 1987a, p. 15), and he claims, notably, that Mach's *most important* achievement was giving an account of theory-construction that combined historical, theoretical and psychological considerations, thereby providing a model for Einstein's methods of research (ibid.).

### 5.7.4 Mach's Philosophy as a Critical Philosophy

Feyerabend regularly located Mach in the critical tradition (not the Kantian tradition, but simply the broader philosophical tradition of criticism rather than system-building, or any school-philosophy). Most often, we find him portraying Mach as a critic of *particular* scientific theories:

> Mach criticized the theory of relativity because it did not pay attention to psychological phenomena. Cf. the introduction to his *Physical Optics* (Feyerabend 1975, p. 259 note).

Feyerabend's most detailed presentation of this critique of relativity, as well as of Mach's critique of atomism, appear in Section 6 of 'Mach's Theory of Research and its Relation to Einstein'.

At other times Feyerabend portrays Mach as a critic of *science as a whole*, or of certain supposedly fundamental *parts* of science:

> Mach's 'philosophy' contains a general criticism of the science of his time, including the house philosophy of the contemporary Newtonians (Feyerabend 1970/1999, p. 129. See also pp. 130, 131, 133–4).
>
> Mach was not satisfied with the science of his time. As he saw it science had become partially petrified. It had used entities such as space and time and objective existence but without examining them. Moreover philosophers had tried to show and scientists had started believing that these entities could not be examined by science because they were 'presupposed' by it. This Mach was not prepared to accept. For him every part of science, 'presuppositions' included was a possible topic of research and subjected to correction (Feyerabend 1978, p. 196).
>
> Mach was critical towards science *as a whole* (Cf. his debate with Planck). Modern philosophers occasionally make a big show of their independence and their expert knowledge by criticizing particular scientific theories and suggesting minor changes. But they would never dare to criticize science as a whole. They are its most obedient servants. Secondly, Mach criticized scientific ideas not by comparing them with external standards (criteria of meaning or demarcation) but by showing how *scientific research itself* suggested a change (ibid., pp. 196–7. See also pp. 198, 202).
>
> Mach… wanted to examine even the most general and the most firmly entrenched ingredients of science (Feyerabend 1984/1987, pp. 202–3).
>
> Mach emerges as a critical scientist trying to make some very fundamental parts of science conform to principles already practised in other regions of scientific research (ibid., p. 216).

But we also find Feyerabend portraying Mach as criticizing matters that are not straightforwardly or merely scientific. Mach, he says, 'criticized simpleminded conceptions of reality and indicated how scientific research can be used to change them' (Feyerabend 1981b, p. 2, note 2). Also:

> Scientists themselves have started criticizing the separatist view of human beings, the view, that is, that there exists an 'objective world' and a 'subjective realm' and that it is imperative to keep them apart. Thus Ernst Mach pointed out, more than a century ago, that the separation cannot be justified by research, that the simplest sensation is a far reaching abstraction and that any act of perceiving is inextricably tied to physiological processes (Feyerabend 1987b, p. 141).

I think Feyerabend was quite right to present Mach in these ways, not just as a comprehensively critical thinker but also as within a broad tradition of critical philosophy. This follows, after all, not only from his 'historical-critical' approach to philosophical problems, but also from his being an evolutionary epistemologist.

### *5.7.5 Mach No Positivist*

We saw that Feyerabend initially took his own stance to be a version of Machian positivism, and that although he subsequently came no longer to think of himself as a positivist, he continued to characterise Mach in these terms, in his published works, until the mid-1960s. However, we also saw that, by the time of his autobiography, Feyerabend noted that he had challenged the received view of Mach as a positivist (Feyerabend 1995a, p. 30). It was in opposition to the more traditional reading of Zahar (Zahar 1977) that he initiated this challenge (Feyerabend 1980/1981, esp. p. 93).

He repeated this idea in the Introduction to his *Philosophical Papers*, there ascribing to Mach the idea that 'all concepts are theoretical concepts', and associating Mach's view with the viewpoint of his *own* earlier papers on scientific realism (Feyerabend 1981a, p. 13). I will discuss this below, since it is important but wrong.

That Mach was no positivist is also one of the central contentions of 'Mach's Theory of Research and its Relation to Einstein'. There, Feyerabend argued that Einstein's criticism of Mach (to the effect that he was a positivist) can be refuted (Feyerabend 1984/1987, pp. 198–199), that Arthur Miller and Gerald Holton, two other rather traditional readers, also got Mach wrong in this respect (ibid., pp. 196–1977), and that Einstein was much closer to positivism than Mach (ibid., pp. 204–210). The first of these conclusions was reiterated in Feyerabend's review of Popper's *Postscript*, where he declared that Mach's positivism 'does not exist' (Feyerabend 1986/1987a, p. 177).

Deciding the vexed issue of whether we should really think of Mach as a positivist would take a much larger and more involved publication than this. There is evidence for thinking of him thus, and much of it was highlighted by the received view. Some of that evidence has been ignored or downplayed by more recent readers, Feyerabend included. But Feyerabend excelled at presenting the contrary evidence, the reasons for *not* thinking of Mach as a positivist.[5]

Although I am going to criticise aspects of Feyerabend's reading of Mach in this paper below, I want to stress that most of its *major* points are unaffected. The most major of these is that 'Mach's physical arguments [...] constitute a philosophy of science which differs from positivism, is in line with Einstein's research practice [...], and entails entirely reasonable objections against nineteenth century atoms

[5] When he remarked that the 'battle' over whether Mach was a positivist had been a net of confusions (Feyerabend 1984/1987, p. 208), he seems to have meant only that those on the *other* side of the battle were confused.

and the special theory of relativity' (ibid., p. 193). These are important correctives to the received view.[6]

Strangely, though, Feyerabend was not always very good at identifying those who thought of Mach as a positivist. In his review of Larry Laudan's book *Progress and its Problems*, he berated Laudan for buying into the idea that 'Boltzmann was an old-fashioned realist fighting valiantly against positivist deadbeats such as Mach' (Feyerabend 1981c, p. 240). This represented something of a turnaround on Feyerabend's part, for in his crucial earlier paper 'Philosophy of Science: A Subject with a Great Past' he himself had drawn favourable attention to Laudan's work on Mach (Feyerabend 1970/1999, adverting to material which ultimately went into Laudan 1976/1982, presumably).[7] It was also unfair, since none of Laudan's mentions of Boltzmann or Mach in that book (Laudan 1977, pp. 50, 58, 85, 100) or in his 1976 paper portray either of them in the way Feyerabend suggests. In fact Laudan's 1976 paper, one of the best things ever written about Mach and atomism, is informed by a wider historical understanding than Feyerabend's remarks on Mach. When it comes to reacting against the received view of Mach, Laudan was on the same side as Feyerabend.

### 5.7.6 Mach as Pluralist

Although, as we have seen, Feyerabend did not initially think of Mach as a theoretical or methodological pluralist, he certainly *came* to think of him thus. However, it took quite a while for Feyerabend to change his mind about Mach, and to start portraying him not as an anti-pluralist (because of his critique of the mechanical world view), but as the exact opposite. Even in papers from the early 1970s, such as 'On the Limited Validity of Methodological Rules', we find him complaining that Mach's objection to the kinetic theory of matter, on the ground that it contradicted (the supposedly well-confirmed) phenomenological thermodynamics, 'almost succeeded in pushing the kinetic theory out of science' (Feyerabend 1972/1999, p. 140n).

This is not incompatible with recognising that Mach was a pluralist, which Feyerabend had already done in 'Consolations for the Specialist' (published in 1970, but presented to Popper's LSE seminar back in 1967). There we are told that 'The idea that a struggle of alternatives is decisive for science, too, was introduced by Mach (*Erkenntnis und Irrtum*) and Boltzmann (*Populäre Schriften*), mainly under the impact of Darwin' (Feyerabend 1970/1981, p. 144).

[6] I just think Feyerabend over-egged it, and that the received view of Mach may not have been as comprehensively wrong as Feyerabend (or modern Mach fans) suppose—should that come as a surprise?

[7] As a referee pointed out, there is an irony here, for Feyerabend was the external examiner of Laudan's doctoral dissertation, which was on the history of positivism.

It has to be said, though, that Feyerabend persistently mixed up two sorts of issues. On the one hand there are questions about whether Mach was a pluralist. When thinking about science (rather than philosophy, for example), this question can be asked with respect to a variety of things, but the two most salient here are structures such as theories, and methodology. Feyerabend made his own name advocating *theoretical* pluralism (from the mid-1950s onwards), and can also be seen as having later advocated *methodological* pluralism (from the late 1960s onwards), notably in *Against Method*.

Unfortunately, Feyerabend mixed up issues of this kind with the question: did Mach feel free to *criticise* existing science, or existing scientific theories? Feyerabend rightly noted that Mach did. But his early perspective on Mach confused this with the idea that Mach was *not* a theoretical pluralist. The reason for the confusion is that Mach's pluralism was restrained, whereas Feyerabend's was not. Feyerabend, as someone once said, was not merely in favour of letting a thousand flowers bloom; he also sought to cast doubt on attempts to distinguish between flowers and weeds. That Mach would not have endorsed the *unrestrained* theoretical pluralism of the kind Feyerabend favoured is clear, and should have been clear to Feyerabend from the comparison he himself uses to describe Mach's views. Darwinian struggle, after all, constitutively involves some lineages *failing to survive*, and Mach would not have been as squeamish as Feyerabend about the idea that some theories and world views really are inadequate, or about their demise.

## 5.8 More Problematic Themes—Feyerabend as a Follower of Mach?

We now come to the ideas in Feyerabend's remarks on Mach which are, as I see it, wrong. They can all be associated with the rough-and-ready notion that Feyerabend took himself to be a follower of Mach. Although perhaps not naturally a follower (or a leader, for that matter), there is evidence that Feyerabend came to think of himself thus. In a letter to Friedrich Stadler, apparently sent while 'Mach's Theory of Research and its Relation to Einstein' was still in press (between 1982 and 1984), and recounting some of his intellectual influences, Feyerabend confided:

> A further teacher, unfortunately no longer alive when I was a student, was Mach and as a convinced follower of Mach I am presenting a further exegesis which more or less contradicts everything which erudite gentlemen such as Holton and others presently have to say about Mach (quoted in Stadler 2008, p. 220).

Feyerabend superfan Steve Fuller, too, has noted that Feyerabend can be thought of thus. He says 'Feyerabend identified the role of Mach's views in the education of scientists as the inspiration for his own project', referring to the material on Mach in Feyerabend's *Science in a Free Society*, (i.e., Feyerabend 1978, pp. 195–205) (Fuller 1994, p. 203). While Feyerabend doesn't really *say* (there) that Mach inspired his own project, his sympathies obviously did lie with Mach, so this isn't too much of an exaggeration.

However, claims that Feyerabend came to follow Mach's philosophy, whether made by Feyerabend or by others on his behalf, should not go unchallenged. The truth is that although there is significant affinity, their views differ not only because there are aspects of Feyerabend's views that Mach would not have endorsed, but also because Feyerabend mis-characterised aspects of Mach's position.[8] Let us first look at an example of the latter which is probably not of the first importance.

### *5.8.1 Mach as a Critic of World Views*

Connected with his early mistaken belief that Mach wanted the concept of nature 'removed', Feyerabend also took Mach to disavow the idea that science should strive towards a world view or world picture. *This* error, though, persisted throughout Feyerabend's career. (It has also been reproduced in a fainter form by Fuller, who contrasts Max Planck's advocacy of science (physics, really) as a world view with Mach's advocacy of science as an instrument, or a collection of instruments (Fuller 1994, p. 205)).

Although Feyerabend did use the term 'world view', he did so in a vaguer way than Mach, and he also blew hot and cold about world views. In *Farewell to Reason*, for example, he denies that contemporary scientists have a unified theory of reality, or a unified view of the physical world (Feyerabend 1987a, pp. 60–61, 100), arguing instead that our physical theories cannot be 'readily united' (p. 169). In another representative text, his 1992 lectures later published as *The Tyranny of Science*, he asks whether there is a single scientific world view (Feyerabend 2011, pp. 35, 36), and seems inclined to respond in the negative, urging that the sciences do not 'speak with a single voice' (pp. 55–6). But there are also several mentions of world views (scientific and other) which clearly presuppose or imply that such things *do* exist, such as 'World views are not only incomplete, they also deceive and, to use a somewhat bloated phrase, they diminish our humanity' (ibid., p. 11. See also pp. 37, 38, 40). Certainly Feyerabend worried about the ways in which a supposed scientific world view threatens, or would threaten, *other* traditions. Perhaps he also took himself to be a critic of the very idea of world views, or perhaps simply of the idea that there is a single *scientific* world view, even though neither of these really fits with everything he said.

In any event, this particular erroneous Feyerabendian view of Mach started very early on in his oeuvre, in 'Physics and Ontology', where he argued that however hostile positivists and traditional philosophers may otherwise be to each other, they agree on one point, 'that from science, especially from physics, we cannot expect

[8] Luca Guzzardi has suggested to me that Feyerabend was ungenerous in his treatment of Gereon Wolters' work on Mach. I suspect he is right, but omit this topic here for lack of space.

any picture of the world [*Weltbild*]' (Feyerabend 1954/2016, p. 12). That this view somehow persisted can be seen from Feyerabend's 1994 paper 'Quantum Theory and Our View of the World':

> When I was a student in Vienna, in the late 1940s, we had three physicists who were known to a wider public: Karl Przibram, Felix Ehrenhaft and Hans Thirring. They differed in many ways. But they were united in their belief that world views are dangerous things and that physics should do without them. The belief had been held before – by Mach, Boltzmann, Franz Exner and his group, and by the members of the Vienna Circle. Only a few of these thinkers realised that they were guided by a rival world view which, though humanitarian in intention, disregarded important features of human existence (Feyerabend 1994, pp. 149–150. Repeated in Feyerabend 1999a, p. 162).[9]

Mach would never have agreed with this. He disavowed only certain *specific* world views, not the very idea of a world view or the idea that science should aim at such a thing. He most clearly distances himself from the *mechanical* world view, as I mentioned earlier. But Feyerabend and Fuller are wrong to take that move as the disavowal of world views *per se*.

A *full* treatment of this issue would have to investigate all Mach's uses of the expressions *Weltanschauung*, *Weltbild*, *Weltaufassung*, and *Weltansicht*. Lacking space for that here, I merely note a few of Mach's positive mentions of these things, from across his career. In *The Science of Mechanics*, for example, he says:

> Physical science does not pretend to be a *complete* view of the world [*Weltanschauung*]; it simply claims that it is working toward such a complete view in the future. The highest philosophy of the scientific investigator is precisely this *toleration* of an incomplete conception of the world and the preference for it rather than an apparently perfect, but inadequate conception' (Mach 1883/1960, p. 559).

In *Principles of the Theory of Heat* we find that:

> The aim of scientific economy is to provide us with a picture of the world as complete as possible – connected, unitary, calm and not materially disturbed by new occurrences: in short a world-picture of the greatest possible stability (Mach 1896/1986, p. 337).

Mach is even prepared to disagree with one of his closest followers on this issue:

> I cannot agree with Petzoldt when he says that thought does not wish to mirror the world. In order to place ourselves in any relation to our surroundings we need a picture of the world; and to obtain this in an economical way, we cultivate science (ibid., p. 361).[10]

Finally, in a lecture on projectile flight we find:

> Modern science strives to construct its picture of the world not from speculations but so far as possible from facts. It verifies its constructs by recourse to observation. Every newly observed fact completes its world-picture, and every divergence of a construct from observation points to some imperfection, some lacuna in it (Mach 1895/1896, p. 310).

[9] It is worth noting the irony that this last remark echoes his old (1954) Popperian (and strongly anti-Machian) accusation that anti-metaphysicians are merely rival metaphysicians.

[10] Feyerabend really should have re-read Mach's *Wärmelehre* when he re-read his *Mechanik*, for one simply could not read the former carefully and still conclude that Mach disavowed world views!

(Is it any wonder that Feyerabend's 'relaxed reading' does not acknowledge *this* passage!). Feyerabend's idea that Mach opposed all world views is just false.

It's fairly clear what has gone wrong here. In his 1910 reply to Planck's critique, Mach rejected Planck's view that the 'the colourless kinetic world picture' of contemporary physics would be 'valid for all times and peoples, even for the inhabitants of Mars' (Mach 1910/1970, p. 36). He did so in dramatic terms, characterising this view as 'highly premature, and indeed almost ludicrous' (ibid.), and worrying that Planck was on his way to founding a physicists' church. From such a church Mach famously agreed to be ex-communicated. But Feyerabend mistook Mach's opposition to Planck's idea that the scientific world view was the world view *of contemporary physics* for opposition not only to the idea of a scientific world view but to the very idea of world views. None of that is warranted by what Mach said.

Whether there are such things as world views, and whether world views are desirable, are issues that did not explicitly play a major role in mid-twentieth century academic philosophy of science. This is not true of the issues we now move on to. Perhaps the most egregious aspect of Feyerabend's 'relaxed reading' of Mach is that he attributes to Mach various views characteristic of the mid-twentieth century historical turn in the philosophy of science (more specifically, of *his own* such turn). Notably, Feyerabend credits Mach with the beliefs that science involves *transforming* facts, or *adapting* facts, and that all concepts are theoretical concepts. Let us take these one at a time.

### *5.8.2 Transforming Facts?*

In the first case, Feyerabend works up to his conclusion gradually, by making a reasonable point against the received view of Mach. Noting the role of abstraction in Mach's work, he rightly observes that its 'negative' effects (i.e., the omission of real physical properties such as temperature, friction, air resistance) are 'the side-effect of *positive* and *constructive* work that is "added" (*Erkenntnis und Irrtum*, S.316) by the scientist' (Feyerabend 1984/1987, p. 201).[11] So far so good.

Feyerabend, though, immediately takes this to mean that the scientist 'rebuilds' facts (Feyerabend ibid., p. 201). But when Mach there distinguishes between what is *obvious* to Kepler, to Galileo, and to Newton, that is, what they 'can see', and those things which they have to add 'by their own efforts from their own store of thoughts', he is making the point that scientists often need to use hypotheses that go beyond the data. This, as Feyerabend knows, does speak against the received view of Mach, but it does not implicate Mach in supposing that *facts* are transformed.

Again, from the idea that the scientific process is 'a truly creative enterprise', with which Mach would certainly have agreed (although Feyerabend, ironically,

---

[11] Feyerabend says (ibid., note 4) that he is quoting from an edition of *Erkenntnis und Irrtum* published in Leipzig during 1917. A searchable copy of that edition is available online here: http://www.zeno.org/Philosophie/M/Mach,+Ernst/Erkenntnis+und+Irrtum

should not! See Feyerabend 1987b, plus 1987a, ch.4), Feyerabend immediately draws the conclusion that it 'connects facts and ideas *by changing and rebuilding both*' (Feyerabend 1984/1987, p. 201, emphasis added). He then quotes, in a footnote, a passage from *Knowledge and Error* which he regards as supporting this idea:

> It is *logically* possible that someone produces a purely phoronomic [kinematic] analysis of Keplerian motion and hits upon the idea of describing it by accelerations which are in the inverse square to the radius from the sun and directed towards it. But this process, in my opinion, is *psychologically* unthinkable. Why should somebody without a physical conception to guide him, hit upon *accelerations*? Why not use the first or the third differential quotient? Why use, out of the infinitely many possible analyses of motion into two components the one which yields such a simple result? For me already the analysis of the parabolic trajectory of thrown objects is very difficult without the guiding idea of gravitational acceleration (Feyerabend ibid., p. 201, note 23, all in Feyerabend's own translation).

Here, though, Mach is criticising the suggestion of psychologist Franz Hillebrand that hypotheses played no role in Newton's theory of gravitation (Mach 1905/1976, pp. 180–181). Just as above, he is arguing that the use of 'guiding ideas' (i.e., hypotheses), by Newton, was psychologically indispensable. Doubtless this involved the changing and rebuilding of previous ideas, but Mach makes no suggestion that any *facts* are being changed or rebuilt.

Slightly earlier in the same paper, purporting to sum up Mach's view, Feyerabend says:

> Productive scientists [...] do not enumerate facts and arrange them in lists, they either 'reconstruct' them (*Mechanik*, S.30) or engage in 'constructive efforts', building 'ideal cases' (*Erkenntnis und Irrtum*, SS.190ff.) from their own 'reservoir of ideas' (ibid., S.316 (this is the same '*Gedankenvorrat*' mentioned above)) (Feyerabend 1984/1987, p. 199).

Here, though, Feyerabend is working with a false trichotomy. Mach *does*, of course, allow that scientists engage in 'constructive efforts' and that they build 'ideal cases'. In the places Feyerabend has in mind, Mach talks of 'thought construction' (Mach 1905/1976, p. 144), and says 'Kepler's laws are only ideal cases, neglecting perturbations' (ibid., p. 139). These are *his* alternatives to 'enumerat[ing] facts and arrang[ing] them in lists' and Mach is not in a position to object to the idea that they involve the use of hypotheses. But '*reconstructing*' facts, whatever that would be, is not an alternative that Mach suggests or, as far as I can see, countenances.

It is noteworthy that Feyerabend, who didn't usually indulge in this habit, was here reduced to quoting only *individual words* ('rebuilds', 'reconstructs') or tiny phrases ('constructive efforts') from Mach's texts. This is because Mach simply did not say, or mean, what Feyerabend wanted him to mean.

Mach did indeed say that we can, 'by operating with the exact concepts reconstruct the facts and acquire a scientific and logical mastery of them' (Mach 1883/1960, p. 40). His original German words are: 'Nun erst können wir die Tatsachen mit exakten Begriffen operierend selbsttätig rekonstruieren, wissenschaftlich, logisch, beherrschen' (Mach 1883/2012, S.45). Facts are being said to be reconstructed here, but what does Mach *mean* by what he says? Only by looking at the context can we really say. He seems to mean something like 'form the correct interpretation of' the facts, in other words 'reconstrue' them. (And perhaps by his own term '*rekonstruiren*' Mach meant 'reconstrue', rather than 'reconstruct'). Alternatively, he might have meant 'reconstruct *what the facts were*', i.e., reconstruct what was the case. This fits well with Mach's mention of *mastering* the facts using exact concepts. After all, to 'reconstruct' something, such as a crime scene, for example, need not be to *construct* anything of that *same* kind (in the way in which a crime scene can be physically reconstructed). It might only mean '*mentally* reconstruct', that is, mentally model the construction of a (non-mental) complex. In neither of these readings is Mach committed to the idea that scientists reconstruct facts *in the philosophically weighty sense in which Feyerabend wants*, i.e., the 'rebuilds' sense.

Feyerabend simply moves too fast here. He extrapolates from one of Mach's mentions of facts being 'reconstructed', deploying it twice (Feyerabend ibid., pp. 194, 199), and taking it to indicate that Mach countenanced the idea that facts can be 'rebuilt' (ibid., p. 201), apparently using 'rebuilt' as a synonym for 'reconstructed'. But he provides no reason to think that by 'reconstructed' Mach really *meant* 'rebuilt' and, needless to say, no defence of the idea that it even makes sense to say of *facts* that they can be rebuilt.

There are a few other places in which Mach talked of facts (or even nature) being constructed or reconstructed by science. In *The Science of Mechanics*, for example, he says:

> The so-called descriptive sciences must chiefly remain content with reconstructing nature, individual facts. Where it is possible, the common features of many facts are once for all placed in relief. But in sciences that are more highly developed, rules for the reconstruction of great numbers of facts maybe embodied in a *single* expression. [...] In nature there is no *law* of refraction, only different cases of refraction. The law of refraction is a concise compendious rule, devised by us for the mental reconstruction of a fact, and only for its reconstruction in part, that is, on its geometrical side (Mach 1883/1960, p. 582).

But the 'reconstruction' in question here is clearly 'mental'. (In each case, the word translated by 'reconstructing' or 'reconstruction' is a cognate of '*nachbilden*' (SS.507-8), meaning to reproduce or copy). This is even more obviously true of the 'construction' of facts in the following passage from *Principles of the Theory of Heat*:

> Description is a construction of facts in thought [*ein Aufbau der Tatsachen in Gedanken* (S452)], and such a construction in the experimental sciences often makes possible an actual production. For the physicist the units of measurement are the building-stones, conceptions are instructions for building, and facts are the results of building. [...] Our thought construction [*Unser Gedankengebilde*] is for us an almost complete equivalent of the fact, all of whose properties we can ascertain from the construction (Mach 1896/1986, p. 370).

The same is true of mentions of facts being reconstructed in *Knowledge and Error* (e.g., Mach 1905/1976, p. 213).

Mach's mentions of facts being 'constructed' or 'reconstructed', therefore, do not make him a 'constructivist' about facts. They would in any case have to be set against the enormous number of times he mentions facts and talks about them in the *other* way, the non-philosophical way, the way in which to talk of facts being 'rebuilt' is out of the question.

### *5.8.3 Facts as Adapted?*

Trying to play the same game when it comes to Mach's discussions of adaptation, Feyerabend supposes that

> Mach's famous slogan 'science means adapting ideas to facts and to each other' (*Mechanik*, S.478, and many other places)... does not mean *repeating* the *unchanged* facts in the medium of thought, it is a dialectical process that transforms both ingredients (Feyerabend 1984/1987, p. 201).

Although Mach talked a lot about adapting ideas to facts, and adapting ideas to one another (e.g., Mach 1895/1896, ch.10, Mach 1905/1976, pp. 2, 15, 65, 81, plus the whole of ch.X), he never talked about adapting facts (to anything). Consider, for example, his 1883 essay 'On Transformation and Adaptation in Scientific Thought', a *locus classicus* of his ideas on adaptation. There Mach talks about the need for thought to correspond to observed facts (Mach 1895/1896, p. 219), of things being conceived *as they are* when thoughts are adapted to observation and activity (p. 220), of thoughts being adapted to events (p. 221), of ideas being adapted to the field of observation (p. 222), of thought as adapting to what is observed (p. 222), and of adapting notions to facts (p. 232). At no point is there any suggestion that facts (or how things are, or events, or even what is observed) must be adapted to thoughts (or ideas, or notions, or anything 'mental') or transformed.

And in the *other locus classicus*, *Knowledge and Error*'s chapter on 'Adaptation of Thoughts to Facts and to Each Other', the story is exactly the same. We hear of ideas adapting to facts (p. 120), of thoughts adapting to facts (pp. 120, 122, 128), of ideas adapting to one another (pp. 120, 121, 122, 123, 124), of ideas adapting to thoughts (p. 125), of the mutual adaptation of thoughts (pp. 120, 126, 127), and of assertions being adapted to experience (p. 126). At no point is there any suggestion that any facts do any adapting or transforming.

The process of adaptation involved in science is thus *not* (what Feyerabend calls) a dialectical process.[12] For Mach, abstraction is indeed positive and constructive work, but it is not work that involves transforming (or rebuilding or reconstructing) *facts*. Of course, Mach would have agreed that facts can change in the ordinary sense in which people say 'when the facts change, I change my mind'. But what Feyerabend needs to find him saying is 'when we change our minds, the facts change', and such 'constructivism' (or 'constructionism') about facts is not a

[12] Feyerabend's earlier suggestion that Mach 'by his opposition to the supremacy of mechanics opened the way to a more dialectical point of view' (Feyerabend 1966/2016, pp. 223 note 11) is either no better (if it refers to Hegelian dialectic) or hopelessly vague. And for Mach, after all, 'dialectics is often the art of leading astray others and sometimes oneself' (Mach 1905/1976, p. 132, note 3).

feature of Mach's outlook.[13] As someone with a robust sense of the biologicality and the limitations of human beings, Mach would not have had the hubris (or the philosophical pretentiousness) that this view implies. Neither would he have sympathised with the relativist and transcendental academic philosophies which generate such views of facts. Feyerabend was right to insist, against Holton, that Mach was happy to use the term 'fact' to refer to not only to particular facts but also to *general* facts (Feyerabend 1984/1987, p. 197). But although Mach used the term 'fact' very widely, he did so in the everyday sense,[14] not in some special philosophical sense, and his uses of the term respect a rule built into that everyday concept, according to which human cognitive changes do not alter facts (facts that have nothing to do with human cognition, that is). Feyerabend's case for Mach's having endorsed the same constructivism about facts as he did, therefore, is flimsy.

It is true that Mach's subscribing to an evolutionary epistemology might well lead one to expect that he *would* allow that facts change in response to human cognitive changes. Such epistemologies rest on an analogy between organisms and their environments, on the one hand, and ideas (or thoughts) and their environments, on the other. If the environment of ideas is thought of as what is the case (i.e., facts), this analogy might tempt one into asserting that facts can change in response to ideas, just as environments can and do change somewhat in response to the organisms that inhabit them. Unfortunately for this argument, Mach's being an evolutionary epistemologist did not mean that he accepted the analogy as flat-footedly and wholeheartedly as this. In fact, in 'On Transformation and Adaptation in Scientific Thought', he explicitly warned that 'ideas, as such, do not comport themselves in all respects like independent organic individuals' (Mach 1895/1896, p. 218).

### 5.8.4 All Concepts as Theoretical Concepts?

Finally, Feyerabend goes even further in his attempt to saddle Mach with views from his own philosophy of science. In the Introduction to the first volume of his *Philosophical Papers*, he remarked that according to Mach

> *science explores all aspects of knowledge, 'phenomena' as well as theories, 'foundations' as well as standards; it is an autonomous enterprise not dependent on principles taken from other fields*. This idea according to which all concepts are theoretical concepts, at least in principle, is definitely in conflict with the positivistic version of scientific realism and it is very close to the point of view of ch.2.6, thesis I. It was this idea of Mach's which led to the conceptual revolutions of the twentieth century. And it is this idea which also explains

[13] Whether Mach was a 'constructivist' about *concepts*, or anticipated constructivism about *mathematics*, as Feyerabend also claims (Feyerabend 1980/1981, p. 93, note 17; Feyerabend 1984/1987, p. 208; Feyerabend 1995b, p. 516), are completely different issues. Feyerabend was right about both, I believe.

[14] Although it is notable that he also counted (what others would think of as) misleading appearances as facts (e.g., Mach 1905/1976, p. 122), just as he counted as *Empfindungen* anything we can be said to even *think* we sense.

> Mach's opposition to nineteenth-century atomism and the more dogmatic versions of Einstein's theory of relativity (Feyerabend 1981a, p. 13).

He went on to say 'For Mach all concepts are theoretical, as we have seen; even sensation talk involves a "one sided theory". A distinction is recognized – but it is regarded as temporary and as being subjected to further research' (ibid.).

The idea that 'all concepts are theoretical concepts' is Feyerabend's extreme version of the claim that that scientific observation is theory-laden. The 'Thesis I' he refers to in the quotation above, from 'An Attempt at a Realistic Interpretation of Experience', one of his most important early papers, was that '[t]he interpretation of an observation language is determined by the theories which we use to explain what we observe, and it changes as soon as those theories change' (Feyerabend 1958/1981, p. 31). This idea, which he ascribed to Galileo and other scientists, was also the core of Feyerabend's version of scientific realism (see Preston 1997, ch.2). Note that this is *not*, though, the relatively modest idea that scientific observation and observation-statements are 'theory-laden', but rather the idea that they are '*fully* theoretical; they have *no* "observational core"' (Feyerabend 1981a, p.x, emphasis added. See Preston ibid., ch.3). It is this extreme version of the 'theory-ladenness' view (and not merely the relatively modest one that most philosophers of science now accept) that he ascribed to Mach.

But the attribution of such ideas to Mach is completely false, a travesty of Mach. In fact there are two wilful misunderstandings of Mach here. The first is the idea that for Mach sensation talk is 'theoretical' because it involves a 'one-sided theory'. What Mach says is:

> [P]erceptions, presentations, volitions, and emotions, in short the whole inner and outer world, are put together, in combinations of varying evanescence and permanence, out of a small number of homogeneous elements. Usually, these elements are called sensations. But as vestiges of a one-sided theory inhere in that term, we prefer to speak simply of elements, as we have already done. The aim of all research is to ascertain the mode of connexion of these elements (Mach 1886/1959, p. 22).

We have already seen Feyerabend misinterpret this passage in one way, his understanding in 1970 of what Mach meant by 'elements' being problematic, but here he misinterprets it in a different way. Consider, alongside the passage just quoted, this similar passage from Mach's 1882 address 'The Economical Nature of Physical Inquiry':

> The world consists of colors, sounds, temperatures, pressures, spaces, times, and so forth, which now we shall not call sensations, nor phenomena, because in either term an arbitrary, one-sided theory is embodied, but simply elements (Mach 1895/1896, pp. 208–9).

Both passages have the same import. Mach preferred the term '*Element*' to the term 'sensation' (or rather, '*Empfindung*') simply because he worried that the *term* '*Empfindung*' involved the 'one-sided theory' which he considered idealism to be, not because he thought 'sensation talk' (that is, *talk about sensations*) is theoretical. Mach continually wanted to talk about what he thought of (misguidedly, I would argue) as sensations, but he had a good (anti-metaphysical) reason for wanting to call those sensations 'elements', rather than 'sensations'.

This misinterpretation has to be set alongside the fact that, in *other* parts of this 1984 paper, Feyerabend does get Mach's understanding of 'elements' right. Section 4 of 'Mach's Theory of Research and its Relation to Einstein' contains the following explanation:

> According to Mach [...] the world consists of elements which can be classified and related to each other in many different ways (*Erkenntnis und Irrtum*, p. 7ff. [Mach 1905/1976, p. 13, perhaps (JP)]). Elements are sensations, 'but only insofar' as we consider their dependence on a particular complex of elements, the human body; 'they are at the same time physical objects, namely insofar as we consider other fundamental dependencies' (*Analyse der Empfindungen*, p. 13 [Mach 1886/1959, p. 16 (JP)]) – 'the elements, therefore, are both *physical* and *psychological* facts' (*Erkenntnis und Irrtum*, p. 136 [see Mach 1905/1976, pp. 7, 17, 99, 203 (JP)]). The elements are not final – 'they are just as tentative and preliminary as were the elements of alchemy, and as are the elements of chemistry today' (*Erkenntnis und Irrtum*, p. 12 [Mach 1905/1976, p. 12 (JP)]). Nor is it necessary to refer every piece of research back to them. Formally, Mach's monism and the atomic hypothesis have much in common. Both assume that the world consists of certain basic entities, both use scientific research to discover their true nature, both agree that the assumption is not necessary and has to be tested by experience (Feyerabend 1984/1987, pp. 203–4).

Although this is brief, and relies heavily on quotes from Mach, it is as lucid a short account of Mach's term as appears anywhere in the literature, fully prefiguring recent sophisticated understandings of Mach as something like a 'neutral monist' (notably Banks 2003).

Feyerabend's second misattribution, though, which goes well beyond his thinking that for Mach the distinction between observational and theoretical concepts is 'temporary, and subject to further research', is that Mach thought that all concepts are theoretical concepts. Feyerabend cites no text of Mach's which says or implies this, and I cannot see that Mach would have endorsed it. Mach's discussions of observation, or concepts, and of theories, some of which we will see in a moment, all tell against it.

A full treatment of Mach's views on observation and theory, experience and hypothesis, would require a paper even longer than this one, so here I can only pick out highlights of his views. Although he did not devote attention to the subject of observation *statements*, as those who later took the 'linguistic turn' would do, Mach used the term '*Beobachtung*' and its derivatives very liberally. But, as Elske de Wahl and Sjang ten Hagen have recently argued, Mach inherited the fact/theory distinction from early and mid-nineteenth-century German physicists, taking for granted both that distinction and the prevalent understanding of facts as observable (de Waal and ten Hagen 2020, p. 60). At no point did he mount the sort of systematic challenge to this distinction which would be necessary for him to count as a precursor to Feyerabend's view. So although his epistemological magnum opus, *Knowledge and Error*, for instance, contains many occurrences of 'observe' and its cognates, it features no sustained discussion of the distinction that would come to obsess his successors.

Mach's attitude was consistently that hypotheses and theories, while temporarily useful and often necessary, are ultimately dispensable. In *History and Root of the Principle of Conservation of Energy* we find him declaring that: 'The object of natural science is the connexion of phenomena; but the theories are like dry leaves which fall away when they have long ceased to be the lungs of the tree of science' (Mach 1872/1911, p. 74). He says the same more than two decades later in his 1894 essay 'On the Principle of Comparison in Physics' (Mach 1895/1896, pp. 248, 257).

In his article 'Facts and Mental Symbols', Mach tells us that 'In the further progress of my physical work I soon discovered that it was very necessary sharply to distinguish between what we see and what we mentally supply' (Mach 1892, p. 200). Admittedly this is a comment only about his *physical* work. But it does show that he can't have thought there was no such distinction. He immediately goes on to explain that the danger of theories is that they prevent us from seeing what lies before us, i.e., *facts*. And, then, returning to the distinction between observation and theory which Feyerabend scouts, Mach says 'As a general rule we have every reason to distinguish sharply between our theoretical conceptions of phenomena and that which we observe. The former must be regarded merely as auxiliary instruments that have been created for a *definite* purpose and which possess permanent value only with respect to that purpose' (ibid., p. 201). This establishes not only that he did not regard the distinction as useful only in his physical work, but also that he thought the distinction could be drawn *sharply*. Empiricists do not *have* to think that, but Mach did so.

For later confirmation of his attitude to the relationships between facts, observations, hypotheses and theories, we can turn to Chapter VII of *Principles of the Theory of Heat*, 'The Development of the Theory of the Conduction of Heat'. In the first section of that chapter, Mach's favourite example of a physical theory, Fourier's theory of heat conduction,[15] the one he calls 'an ideal physical theory' (Mach 1896/1986, p. 113), is said to be 'founded, not upon a hypothesis but upon an observable fact', Mach contrasting the *security* of this foundation in fact with the insecurity of foundation upon an hypothesis. In the second section of the same chapter, Mach portrays this same theory as a mere inventory of facts:

> The entire theory of Fourier really consists only in a consistent, quantitatively exact, abstract conception of the facts of conduction of heat – in an easily surveyed and systematically arranged *inventory* of facts, or rather in an introduction to the developing of this inventory (ibid.).

Finally, in the third section of this same chapter, Mach tells us that the result of Fourier's theory can be expressed so that 'it appears almost self-evident, and very close to our instinctive perception. It lies as close as the observation that all heavy bodies left to themselves sink', which he then calls 'an obvious fact' (ibid., p. 114).

A great deal more material from this text could be assembled here, but life is short. Suffice it to say that Mach would never have agreed with Feyerabend that 'all concepts are theoretical concepts'.

What should we conclude, then, about Feyerabend's claim to have become a follower of Mach? Giving a 'relaxed reading' is one thing, but Feyerabend's 'reading' of Mach is, in *this* respect, so relaxed as to be comatose. Mach ought not to be

[15] Zeljko Loparić called it 'Mach's favourite case of theory construction' (Loparić 1984, p. 41).

lumbered with the wilder suppositions of mid-twentieth century philosophy of science. And Feyerabend's claim to have become a Machian should be taken about as seriously as his somewhat later claim to have 'become a Wittgensteinian' (Feyerabend 1995a, p. 94).

In the end, it's just *not necessary* to attribute these views to Mach in order to explain why he opposed atomism and the rather dogmatic reading of Einstein's theory of relativity that was arguably promulgated by, for example, Max Planck. In fact, Feyerabend does a good job of explaining that, in his 1984 article, without having to suppose that Mach was a constructivist about facts, or thought of all concepts as theoretical.

## 5.9 When, and Why, Did Feyerabend 'Return to Mach'?

As we have seen, Feyerabend had a major change of mind about Mach. Why did he do so? Intriguingly, it seems possible to say both when and, perhaps, why. In the essay 'Farewell to Reason', as translated in his book of the same name, Feyerabend tells us:

> Having absorbed Ernst Mach while about fifteen [which would make that 1939] and having been a student of Hans Thirring and Felix Ehrenhaft in physics, I took it for granted that the work of scientists was self-supporting and did not need any outside legitimation. I got impatient with people who though lacking any experience of the complexity of scientific research still claimed to know what it was all about and how it could be improved. I guess I was a kind of scientific libertarian and my battle-cry could have been 'leave science to the scientists!'. Of course I had once been a rationalist myself – but it needed only a simple practical example, it needed only Prof. von Weizsäcker's concrete arguments, in Hamburg, back in 1965 (I believe), to reveal the shallowness of rationalistic orations and to make me return to Mach (Feyerabend 1986/1987b, pp. 316–7).

Carl Friedrich von Weizsäcker, physicist and philosopher, was a Professor of Philosophy at the University of Hamburg from 1957 to 1969. Feyerabend mentions this interaction with him several times, and even credits von Weizsäcker with unwittingly initiating his turn to 'anarchism' in the revised edition of *Against Method* (Feyerabend 1988, p. 283). Marij van Strien has persuasively argued that Feyerabend's change towards 'anarchism' took place more gradually, from the early 1960s (van Strien 2020, p. 14). But this does not rule out his rediscovery of Mach being dateable to 1965, even though that is a related development.

Why would this interaction have initiated a return to *Mach*, though? Perhaps only work on the relationship (e.g., any correspondence) between Feyerabend and von Weizsäcker will yield a full answer (see Collodel and Oberheim 2020, p. 33, note 43, pp. 391–2 note 4, and pp. 402–3, note 3). But it is possible to say a little about why Mach might have seemed to fit the bill. Mach, after all, did not present himself as a 'rationalist', and did not present science as autonomous, that is, as a matter to be left to scientists alone. He stressed that the roots of science lay in everyday life, that science is a development of our biology, and that science should be used to improve people's lives. He also made the connection between science and practice more obvious than rationalist thinkers typically do. Perhaps this is enough to explain why it was Mach, in particular, who began to appeal to Feyerabend again after von Weizsäcker's intervention.

As well as von Weizsäcker, though, I venture to suggest that Feyerabend's take on Mach might have been influenced by the Belgian physicist Léon Rosenfeld. In a 1957 discussion of papers by Bohm and Rosenfeld, in which Feyerabend was involved, Rosenfeld is recorded as having said:

> There's a great difference between the early positivists for whom I have much respect – even for Mach, in spite of his aversion against atomism – and the later positivists. I have not the same respect for the later positivists because I think that where positivism went wrong was when it became a system in which those nice points which I think were made already by Mach, were fossilized into principles of absolute validity; it was maintained that science was only a language and that one could decide about statements only by reducing them to sense data and so on. Mach himself was not very consistent and he made many mistakes which encouraged that direction (Feyerabend 1957/2016, p. 279).[16]

Rosenfeld (sometimes referred to as 'Bohr's bulldog') was a Marxist sympathiser who defended Bohr's version of the quantum theory (against Bohm and others) partly because he thought Bohr's principle of complementarity was 'dialectical' (see van Strien ibid.). Feyerabend would *not* have liked Rosenfeld's dogmatic defence of Bohr. But he might conceivably have learnt something about Mach from him. What Rosenfeld said in that conversation, as recorded above, clearly anticipates Feyerabend's later reading of Mach, and Feyerabend might well also have approved of Rosenfeld's sympathetic take on Mach's opposition to atomism (Feyerabend 1957/2016, p. 280). However, these ideas must have taken some time to sink in since, as we saw, in the 1950s Feyerabend was taking a line less sympathetic to Mach.

## 5.10 Conclusion

We saw Feyerabend say that he had studied Mach's *Mechanik* and *Wärmelehre*, and that he re-read the former. But we have also seen him quote from *The Analysis of Sensations*, from some of Mach's *Popular Scientific Lectures*, from *Knowledge and Error*, from *The Principles of Physical Optics*, and from his debate with Max Planck. In fact, Feyerabend not only mentions but discusses almost all of Mach's major works (except *The History and Root of the Principle of Conservation of Energy*), and he was clearly aware of Mach's early works in psychophysics (he refers to Floyd Ratliff's volume *Mach Bands* (Ratliff 1965), and to 'Mach's investigations on the physiology of perception' (Feyerabend 1987a, p. 9), although it's not

[16]Although this discussion is reproduced in the fourth volume of Feyerabend's *Philosophical Papers*, the words are Rosenfeld's not Feyerabend's.

clear whether by this he meant those same papers on Mach bands, or rather Mach's early papers on the perception of movement (Mach 1875/2001)).

Feyerabend's first mentions of Mach, through the 1950s and early to mid-1960s, exhibit something very much like the 'received view' that Mach can be counted a positivist. His conversation with von Weizsäcker seems to have initiated a change of mind in the mid-1960s, though, and from that point onwards, up until his last article on Mach, which was a small entry in Ted Honderich's *Oxford Companion to Philosophy* (Feyerabend 1995b), Feyerabend consistently portrayed Mach as a liberating critic of existing science. From the late 1960s onwards, Mach was a major influence on Feyerabend. Not as obvious or as large or as trumpeted an influence as John Stuart Mill, perhaps, but nevertheless one of Feyerabend's most important heroes and influences, someone he repeatedly turned to after he had become disillusioned with the way philosophy of science was pursued in the second half of the twentieth century.

Stadler says that 'It was only at the end of his life that [Feyerabend] returned to the Vienna roots of his intellectual development, i.e., the programmatic unity of philosophy and history of science, with Mach, Boltzmann, Alois Riegl and the historical tradition of Logical Empiricism' (Stadler 2008, p. 216), and that 'By the time Feyerabend received a call to the ETH in Zürich in the 1980s, he had returned both physically and mentally to his early intellectual socialization by rereading Mach and elaborating his relativism and pluralism' (ibid., p. 217). I would say that Feyerabend's return to this particular Viennese root, at least, started earlier, building in intensity from the late 1960s onwards.

## References


Antiseri, Dario. 2000/2006. *La Vienna di Popper*, (Rubbetino: Soveria Mannelli, 2000). Translated as *Popper's Vienna*. Aurora, CO: The Davies Group, 2006.

Banks, Erik C. 2003. *Ernst Mach's World Elements: A Study in Natural Philosophy*. Dordrecht: Kluwer Academic Publishers.

Čapek, Milič. 1968. Ernst Mach's Biological Theory of Knowledge. *Synthese* 18: 171–191.

Collodel, Matteo, and Eric Oberheim, eds. 2020. *Feyerabend's Formative Years, volume 1: Feyerabend and Popper*. Switzerland: Springer.

de Paz, María. 2018. Reconsidering Mach in the Light of the Interplay of Practices. *Revista Portuguesa de Filosofia* 74: 219–246.

de Waal, Elske, and Sjang L. ten Hagen. 2020. The Concept of Fact in German Physics around 1900: A Comparison between Mach and Einstein. *Physics in Perspective* 22: 55–80.

Dingler, Hugo. 1924. *Die Grundgedanken der Machschen Philosophie*. Leipzig: J.A. Barth.

Feyerabend, Paul K. 1948/2016. *The Concept of Intelligibility in Modern Physics*, as translated in Feyerabend 2016a, 3–8.

———. 1954/2016. *Physics and Ontology*, as translated in Feyerabend 2016a, 9–24.

———. 1955/1981. Wittgenstein's *Philosophical Investigations*. *The Philosophical Review* 64: 99–130, as reprinted in Feyerabend 1981b.

———. 1957/2016. *Excerpts from Discussions with Léon Rosenfeld and David Bohm (and Others)*, as reprinted in Feyerabend 2016a, 275–293.

———. 1958/1981. An Attempt at a Realistic Interpretation of Experience. *Proceedings of the Aristotelian Society* 58: 17–36, as reprinted in Feyerabend 1981a.
———. 1958/2016. Natural Philosophy, as translated in Feyerabend 2016a, 323–345.
———. 1960/1981. Professor Bohm's Philosophy of Nature. *British Journal for the Philosophy of Science* 10: 219–235, as reprinted in Feyerabend 1981a.
———. 1961. Methodologie. In *Les Grands Courants de la Pensee Mondiale Contemporaine*, ed. M.F. Sciacca, vol. I, 871–927. Milan: C. Marzorati.
———. 1962/1981. Explanation, Reduction and Empiricism. In *Scientific Explanation, Space, and Time: Minnesota Studies in the Philosophy of Science*, vol. 3, ed. H. Feigl and G. Maxwell, 44–96. Minneapolis: University of Minnesota Press, as reprinted in Feyerabend 1981a.
———. 1963/1999. How to be a Good Empiricist: A Plea for Tolerance in Matters Epistemological. In *Philosophy of Science: The Delaware Seminar*, vol. 2, ed. B. Baumrin, 78–103 New York: Interscience Press. As reprinted in Feyerabend 1999b
———. 1965. Problems of Empiricism. In *Beyond the Edge of Certainty, University of Pittsburgh Studies in the Philosophy of Science*, ed. R.G. Colodny, 145–260. New Jersey: Prentice-Hall.
———. 1965/1981. Reply to Criticism. In *Boston Studies in the Philosophy of Science, volume 2: In Honour of Philipp Frank*, ed. R.S. Cohen and M.W. Wartofsky, 104–131. Dordrecht: D. Reidel, as reprinted in Feyerabend 1981a.
———. 1966/2016. Dialectical Materialism and the Quantum Theory. *Slavic Review* 25: 219–224, as reprinted in Feyerabend 2016a.
———. 1970/1981. Consolations for the Specialist. In Criticism and the Growth of Knowledge, ed. I. Lakatos and A. Musgrave, 131–161. Cambridge: Cambridge University Press, as reprinted in Feyerabend 1981b.
———. 1970/1999. Philosophy of Science: A Subject with a Great Past. In *Minnesota Studies in the Philosophy of Science*, vol. 5, ed. R. Stuewer, 127–137. Minneapolis: University of Minnesota Press, as reprinted in Feyerabend 1999b.
———. 1972/1999. *On the Limited Validity of Methodological Rules*, as translated in Feyerabend 1999b, 138–180.
———. 1974/1981 Popper's *Objective Knowledge*. *Inquiry* 17: 168–201, as reprinted in Feyerabend 1981b.
———. 1975. *Against Method*. London: New Left Books.
———. 1978. *Science in a Free Society*. London: New Left Books.
———. 1980/1981. Zahar on Mach, Einstein, and Modern Science. *British Journal for the Philosophy of Science* 31: 89–98. 1980, as reprinted in Feyerabend 1981b.
———. 1981a. *Realism, Rationalism & Scientific Method: Philosophical Papers*. Vol. 1. Cambridge: Cambridge University Press.
———. 1981b. *Problems of Empiricism: Philosophical Papers*. Vol. 2. Cambridge: Cambridge University Press.
———. 1981c. More Clothes from the Emperor's Bargain Basement: A Review of Laudan's *Progress and its Problems*. *British Journal for the Philosophy of Science* 32: 231–246. as reprinted in Feyerabend 1981b.
———. 1984/1987. Mach's Theory of Research and its Relation to Einstein. *Studies in History and Philosophy of Science* 15: 192–218. 1984, as reprinted in Feyerabend 1987a.
———. 1986/1987a. Trivializing Knowledge: A Review of Popper's *Postscript*. *Inquiry* 29: 162–191. as reprinted in Feyerabend 1987a.
———. 1986/1987b. *Farewell to Reason*, as translated and reprinted in Feyerabend 1987a, 280–319.
———. 1987a. *Farewell to Reason*. London & New York: Verso.
———. 1987b. Creativity: A Dangerous Myth. *Critical Inquiry* 13: 700–711. (A revised version, 'Creativity', is included in Feyerabend 1987a, 128–142).
———. 1988. *Against Method*. Revised Edition. London & New York: Verso.
———. 1994. Quantum Theory and Our View of the World. In *Physics and Our View of the World*, ed. J. Hilgevoord, 149–168. Cambridge: Cambridge University Press.

———. 1995a. *Killing Time: The Autobiography of Paul Feyerabend*. Chicago: University of Chicago Press.
———. 1995b. Mach, Ernst. In *The Oxford Companion to Philosophy*, ed. T. Honderich, 516. Oxford: Oxford University Press.
———. 1999a. *Conquest of Abundance: A Tale of Abstraction Versus the Richness of Being*, ed. B. Terpstra. Chicago: University of Chicago Press.
———. 1999b. *Knowledge, Science and Relativism: Philosophical Papers*, ed. J.M. Preston, vol. 3. Cambridge: Cambridge University Press.
———. 2011. *The Tyranny of Science*, ed. E. Oberheim. Cambridge: Polity Press.
———. 2016a. *Physics and Philosophy: Philosophical Papers*, Vol.4, ed. S. Gattei and J. Agassi. Cambridge: Cambridge University Press.
———. 2016b. *Philosophy of Nature*, ed. H. Heit and E. Oberheim. Cambridge: Polity Press.
Fuller, Steve. 1994. *Retrieving the Point of the Realism-Instrumentalism Debate: Mach vs. Planck on Science Education Policy*, PSA 1994, Volume 1, ed. D. Hull, M. Forbes, and R.M. Burian, 200–208. East Lansing, MI: Philosophy of Science Association.
Guzzardi, Luca. 2021. Holding the Hand of History: Mach on the History of Science, the Analysis of Sensations, and the Economy of Thought. In *Interpreting Mach: Critical Essays*, ed. J.M. Preston, 164–183. Cambridge: Cambridge University Press.
Hentschel, Klaus. 1985. On Feyerabend's Version of "Mach's Theory of Research and its Relation to Einstein". *Studies in History and Philosophy of Science* 16: 387–394.
Laudan, Larry. 1976/1982. The Methodological Foundations of Mach's Anti-Atomism and their Historical Roots. In *Motion and Time, Space and Matter*, ed. P.K. Machamer and R.G. Turnbull, 390–417. Columbus, OH: Ohio State University Press. Reprinted as 'Ernst Mach's Opposition to Atomism', in his *Science and Hypothesis: Historical Essays on Scientific Methodology*, (Dordrecht: D.Reidel, 1981), 202–225.
———. 1977. *Progress and its Problems: Towards a Theory of Scientific Growth*. Berkeley: University of California Press.
Loparić, Zeljko. 1984. Problem-Solving and Theory Structure in Mach. *Studies in History and Philosophy of Science* 15: 23–49.
Mach, Ernst. 1872/1911. *Die Geschichte und die Wurzel des Satzes von der Erhaltung der Arbeit*, 2nd ed. Prag: J.G. Calve'sche, 1909. Translated by P.E.B. Jourdain as *History and Root of the Principle of the Conservation of Energy*. Chicago: Open Court, 1911.
———. 1875/2001. *Grundlinien der Lehre von den Bewegungsempfindungen*. Leipzig: W. Engelmann. Translated as L.R. Young, V. Henn and H. Scherberger eds., *Fundamentals of the Theory of Movement Perception*. New York: Kluwer Academic/Plenum Publishers, 2001.
———. 1883/1960. *Die Mechanik in ihrer Entwickelung historisch-kritisch dargestellt*. Leipzig: F.A. Brockhaus, 1883. 9th German edition translated as *The Science of Mechanics, A Critical and Historical Exposition of its Principles*, 6th American edition, trans. T.J. McCormack. LaSalle, IL: Open Court.
———. 1883/2012. *Die Mechanik in ihrer Entwicklung, historisch-kritisch dargestellt (Ernst Mach Studienausgabe, Band 3)*, ed. G. Wolters and G. Hon. Berlin: xenomoi.
———. 1886/1959. *Beiträge zur Analyse der Empfindungen*. Jena: Gustav Fischer, 1886. Further German editions in 1900, 1902, 1903, 1906, 1911, 1918, 1922. Translated by C.M. Williams and S. Waterlow as *The Analysis of Sensations and the Relation of the Physical to the Psychical*. Chicago & London: The Open Court Publishing Company, 1914, reprinted New York: Dover Publications, 1959.
———. 1892. Facts and Mental Symbols. *The Monist* 2: 198–208.
———. 1895/1896. *Popular Scientific Lectures*, translated by T.J. McCormack. Chicago: Open Court, 1895. (German edition, *Populär-wissenschaftliche Vorlesungen*, (Leipzig, 1896. Further German editions, 1897, 1903, 1910, 1923)).
———. 1896/1986. *Die Principien der Wärmelehre, historisch-kritisch entwickelt*. Leipzig: J.A. Barth, 1896. Further German editions, 1900, 1919. Translated by T.J. McCormack, P.E.B. Jourdain & A.E. Heath as *Principles of the Theory of Heat, Historically and Critically Elucidated*, ed. B.F. McGuinness, (Dordrecht: D.Reidel, 1986).

———. 1905/1976. *Erkenntnis und Irrtum. Skizzen zur Psychologie der Forschung*. Leipzig: J.A. Barth, 1905. Further German editions, 1906, 1917, 1920, 1926. Translated by T.J. McCormack & P. Foulkes as *Knowledge and Error: Sketches on the Psychology of Enquiry*, ed. B.F. McGuinness (Dordrecht: D.Reidel, 1976).

———. 1910/1970. Die Leitgedanken meiner naturwissenschaftlichen Erkenntnislehre und ihre Aufnahme durch die Zeitgenossen. *Scientia*, 7, 1910, 2225–240, and *Physikalische Zeitschrift*, 11, 1910, 599–606. Translated by Ann Toulmin as 'The Guiding Principles of My Scientific Theory of Knowledge and its Reception by My Contemporaries', in S.E. Toulmin (ed.), *Physical Reality: Philosophical Essays on Twentieth-Century Physics*, (New York: Harper Torchbooks, 1970), 28–43.

———. 1921/1926. *Die Principien der physikalischen Optik. Historisch und erkenntnispsychologisch entwickelt* (Leipzig), translated by J.S. Anderson & A.F.A. Young as *The Principles of Physical Optics: An Historical and Philosophical Treatment* (London: Methuen & Co., 1926. Reprinted, New York: Dover Publications, 1953).

Popper, Karl R. 1953/1963. A Note on Berkeley as a Precursor of Mach and Einstein. *British Journal for the Philosophy of Science* 4: 166–174. 1953, as reprinted in his *Conjectures and Refutations: The Growth of Scientific Knowledge*, (London: Routledge & Kegan Paul, 1963).

Preston, John M. 1994. Methodology, Epistemology and Conventions: Popper's Bad Start. In *PSA 1994, Proceedings of the 1994 Biennial Meeting of the Philosophy of Science Association, Volume 1: Symposia and Invited Papers*, ed. D. Hull, M. Forbes, and R.M. Burian, 314–322. East Lansing, MI: Philosophy of Science Association.

———. 1997. *Feyerabend: Philosophy, Science and Society*. Cambridge: Polity Press.

———. 2019. Mach, Wittgenstein, Science and Logic. In *Ernst Mach – Life, Work, Influence, (Vienna Circle Institute Yearbook)*, ed. F. Stadler, 63–90. Cham, Switzerland: Springer.

———. 2021. Phenomenalism, or Neutral Monism, in Mach's *Analysis of Sensations*? In *Interpreting Mach: Critical Essays*, ed. J.M. Preston, 235–257. Cambridge: Cambridge University Press.

———. 2023. The Idea of a Pseudo-problem in Mach, Hertz, and Boltzmann. *Journal for General Philosophy of Science* 54: 55–77.

Ratliff, Floyd. 1965. *Mach Bands: Quantitative Studies on Neural Networks in the Retina*. San Francisco: Holden Day Inc.

Serra, Isabel, and Elisa Maia. 2018. Looking for Routes in Mach's Epistemology. *Revista Portuguesa de Filosofia* 74: 173–196.

Stadler, Friedrich. 2008. Paul Feyerabend and the Forgotten "Third Vienna Circle". In *The Migration of Ideas*, ed. R. Scazzieri and R. Simili, 203–224. Sagamore Beach: Science History Publications.

Toulmin, Stephen E. 1969. From Logical Analysis to Conceptual History. In *The Legacy of Logical Positivism: Studies in the Philosophy of Science*, ed. P. Achinstein and S.E. Barker, 25–53. Baltimore: The Johns Hopkins Press.

Van Strien, Marij. 2020. Pluralism and Anarchism in Quantum Physics: Paul Feyerabend's Writings on Quantum Physics in Relation to his General Philosophy of Science. *Studies in History and Philosophy of Science* 80: 72–81.

Wolters, Gereon. 1988. Hugo Dingler. *Science in Context* 2: 359–367.

Zahar, Elie G. 1977. Mach, Einstein, and the Rise of Modern Science. *British Journal for the Philosophy of Science* 28: 195–213.

———. 1981. Second Thoughts about Machian Positivism: A Reply to Feyerabend. *British Journal for the Philosophy of Science* 32: 267–276.

# Part II
# Feyerabend and the History of Science

# Chapter 6
# Feyerabend and Galileo

**Stefano Gattei**

**Abstract** In *Against Method* (1975), Feyerabend's main argument for epistemological anarchism is drawn from a long case-study of Galileo's defence of the heliocentric cosmology. Feyerabend shows that Galileo broke all sorts of methodological rules to argue for the Copernican hypothesis against the received Aristotelian view, often introducing (or pushing) new interpretations whenever empirical evidence seemed to be unmistakably in contrast with the earth's rotation. Given the widely-accepted understanding of Galileo as the "father of modern science" and one of the "heroes" of the Scientific Revolution, upholding freedom of research against any dogmatism, Feyerabend claims that, if the Copernican worldview is currently accepted, it is not because it is in agreement with reason, but, rather, because "reason was overruled at some time in their past". Feyerabend's challenging portrayal of Galileo (in *Against Method* and elsewhere, including *Farewell to Reason* and "Problems of Empiricism") is not supposed to be an attack on Galileo, or on science, though, but aims at offering a critique of the views of contemporary philosophers of science (chiefly Popper and Kuhn), who made Galileo a hero of method and rationality. Rather than a historical reconstruction of the Galileo case, Feyerabend aims, on the one hand, to show the intrinsic limits of any rational reconstructions of a historical event; and, on the other, to call attention to the all too frequent risk of turning science into dogma, thereby neutralizing it.

**Keywords** Paul Feyerabend · Galileo Galilei · Rationality · Dogmatism · Tower experiment · Scientific method · Historical reconstructions

---

S. Gattei (✉)
Department of Sociology and Social Research, University of Trento, Trento, Italy
e-mail: stefano.gattei@unitn.it

© The Author(s), under exclusive license to Springer Nature Switzerland AG 2024

S. Gattei, R. Corvi (eds.), *Feyerabend in Dialogue*, Boston Studies in the Philosophy and History of Science 346,
https://doi.org/10.1007/978-3-031-71938-7_6

GALILEO: Look, I proved to you…
ANDREA: But last night I realized that if the earth turned that way I'd be hanging head downwards every night. And that's a fact.
GALILEO (*takes an apple from the table*): Right, now this is the earth.
ANDREA: Don't keep on taking that sort of example, Signor Galilei. They always work.
GALILEO (*putting the apple back*): All right.
ANDREA: Examples always work, if you are clever.

Bertolt Brecht, *Life of Galileo* (1945)

## 6.1 The Sapienza Affair

In the autumn of 2007, the rector of the University of Rome "La Sapienza" invited Pope Benedict XVI to inaugurate the new academic year. This caused a bitter polemic, triggered by a letter of protest addressed to the rector by Marcello Cini, which was published in the left-wing newspaper *Il manifesto* on 14 November. Cini, who was then Emeritus Professor of theoretical physics, contested the decision to invite a pontiff, the sovereign of a confessional State, to a solemn event at a secular institution. On November 23, a second letter, signed by a few dozen university professors, and subsequently by several other professors and scientists (Italian and otherwise), further inflamed the debate on the appropriateness of such an invitation, also leading to various demonstrations by students. In the end, on 15 January 2008, two days before the scheduled inauguration ceremony, the Holy See decided to decline the invitation, and the *lectio* the pope was to deliver was read by the pro-rector.[1]

Much has been said and written on the matter, on either sides, and I do not intend to return to it.[2] The reason why I recall the incident here is that the reasons offered by Cini against the invitation of the pontiff—Pope Benedict XVI had already been the target of heated criticism for a lecture he had given at the University of Regensburg on 12 September 2006[3]—were supplemented (in the second letter) by a

[1] The two letters are available online, at http://appellolaico.wordpress.com/documenti/; Pope Benedict XVI's speech may be accessed at https://www.vatican.va/content/benedict-xvi/en/speeches/2008/january/documents/hf_ben-xvi_spe_20080117_la-sapienza.html. For the pope's eventual decision to decline the invitation, see Pani 2008, pp. 75–76; the *lectio* he was to deliver is found in Guarini 2014, pp. 143–155.

[2] Cf. Pani 2008 and Guarini 2014.

[3] The lecture was delivered in German and titled "Glaube, Vernunft und Universität. Erinnerungen und Reflexionen" ("Faith, Reason and the University: Memories and Reflections"). It focuses on faith and religion, with references ranging from ancient Jewish and Greek thinking to Protestant theology and modern secularity, and was meant to call attention to the tendency to "exclude the question of God, making it appear an unscientific or pre-scientific question"; in the pope's eyes, this equalled to reducing the scope of rational discourse. In the part devoted to Islam, the pope quoted a strong criticism of Islam, which he deemed "unacceptable", and which was advanced by Byzantine emperor Manuel II Palaiologos (or Palaeologus) in 1391. The reference was widely

quote from Paul Feyerabend on Galileo, which Cardinal Joseph Ratzinger had made in 1990. According to Ernst Bloch, he said,

> The two methodological spheres [that is, science and divine revelation] are clearly distinguished from one another [...], and the rights, as well as the limitations, of each are acknowledged. But the summary of the skeptical agnostic philosopher P. Feyerabend sounds much more aggressive when he writes: "The Church at the time of Galileo kept much more closely to reason than did Galileo himself, and she took into consideration the ethical and social consequences of Galileo's teaching too. Her verdict against Galileo was rational and just, and the revision of this verdict can be justified only on grounds of what is politically opportune".[4] C. F. von Weizsäcker (to take one example) goes even one step farther in considering the practical effects when he sees a "perfectly straight path" leading from Galileo to the atomic bomb.[5]

Ratzinger's quote from Feyerabend, although taken out of context—indeed, it appears in a book programmatically titled *Against Method*, which praises those (and Galileo in particular) who were brave enough to break the rules of method and the canons of rationality received at their times, thereby allowing science to progress[6]—

---

perceived as an insulting characterization of Islam, and sparked an international controversy, during which mass street protests were mounted in several Islamic countries. In the published version of the lecture, Pope Benedict XVI further clarified that the quote did not express his personal views, but merely intended "to draw out the essential relationship between faith and reason. On this point I am in agreement with Manuel II, but without endorsing his polemic". See https://www.vatican.va/content/benedict-xvi/en/speeches/2006/september/documents/hf_ben-xvi_spe_20060912_university-regensburg.html

[4] Ratzinger quotes from the second edition of the German translation of *Against Method*, published in 1983 (p. 206). The quote is found at the beginning of Ch. 14, as a sort of summary of the chapter contents. Feyerabend added such summaries to the second edition of the German translation of the work, and does not appear in other editions of the book—such as, for instance, the Italian edition (1979), which is a translation from the first English edition (1975); the summaries were then added to subsequent English editions, published in 1988 (p. 129) and 1993 (p. 125); the quote appears, however, at the opening of Ch. 13. Feyerabend used to revise his works in view of a translation, reprint or new edition, and often introduced significant variants. This is how the original quote from Feyerabend reads in German: "Die Kirche zur Zeit Galileis hielt sich viel enger an die Vernunft als Galilei selber, und sie zog auch die ethischen und sozialen Folgen der Galileischen Lehren in Betracht. Ihr Urteil gegen Galilei war rational und gerecht, und seine Revision läßt sich nur politisch-opportunistisch rechtfertigen"(p. 8). And this is how the same passage reads in the English edition: "'The Church at the time of Galileo not only kept closer to reason as defined then and, in part, even now; it also considered the ethical and social consequences of Galileo's views. Its indictment of Galileo was rational and only opportunism and a lack of perspective can demand a revision".

[5] The passage is from a lecture Cardinal Ratzinger first delivered in Rieti, on 16 December 1989, and then at the University of Rome "La Sapienza", on 15 February 1990: cf. Pani (2008), pp. 21–24. The text of the lecture, first published in pamphlet form in Parma, in 1990, later appeared as Ratzinger 1990 (p. 15). Here I am quoting from the English translation (p. 104) of the slightly revised German version, Ratzinger 1991 (p. 71).

[6] As Giorgio Parisi (one of the signatories of the second letter to the rector of the University of Rome "La Sapienza", and winner of the Nobel Prize for Physics in 2021) pointed out on 17 January 2008, in his reconstruction of the events, "Cardinal Ratzinger does not distance himself from Feyerabend's statement, and indeed he uses it to argue that Galileo was not a vicim of the Church's obscurantism. He concludes by saying that he does not wish to use the text for 'hasty apologetics',

was reiterated in an interview with Marcello Frediani on 12 May 1990, upon the publication of the publication of the Italian translation of Feyerabend's *Farewell to Reason* (a collection of essays that includes a chapter on "Galileo and the Tyranny of Truth"):

> My thesis was presented correctly. The Church was right in stating that scientists are not the final authority in scientific matters.[7]

It would be a grave mistake to say, based on these words, that Feyerabend was an advocate of Cardinal Bellarmino or a supporter of the Inquisition, or—as has repeatedly been said[8]—an "enemy" of Galileo's, or of science more generally. In fact, Feyerabend was a defender of science, as was his master, Karl R. Popper—Feyerabend's provocative views in *Against Method*, his portrayal of the mixed results in "How to Defend Society Against Science" (1975) and his attacks on scientific colonialism in *Farewell to Reason* notwithstanding. From Popper, Feyerabend learned that science is characterized chiefly by its critical spirit, which makes it a body of knowledge that is constantly evolving. Whenever Feyerabend seemed to be distancing himself from this perspective, even with very harsh tones, he did it with a precise, polemic aim, namely, calling attention to the limits of the scientific enterprise. Not in order to attack science, destroy or diminish its achievements—rather, in order to avoid turning scientific knowledge into an absolute form of knowledge, one that sees science as a corpus of dogmas controlled by few 'experts'. In a metaphor, Feyerabend's attitude is that of a tight-rope walker: whenever balance is lost, he must move the pole to maintain stability.[9] This, I believe, the required premise to understand the larger part of Feyerabend's rich and varied philosophical output, which is to be read against the background of the epistemological debate in the second half of the twentieth century (as is well known, in his works we do find a large number of explicit or implicit critical references to the normative prescriptions advanced by Thomas S. Kuhn, Imre Lakatos, Norwood R. Hanson and the logical positivists, besides the aforementioned Popper) as well as Feyerebend's peculiar writing style, dense with polemic and controversial statements, often written with the piercing and sparking irony that characterizes a brilliant philosopher.

---

but does not deny its validity. Moreover, the quotation from Feyerabend, made withou reference to Feyerabend's philosophical views, completely distorts the thought of this philosopher of science. As opposed to what it seems from this isolated quotation, Feyerabend has always extolled Galileo's creativity and intellectual audacity—but he expresses himself through paradoxes, and his entire view is a critique of 'reason'. Accordingly, in saying that the Church was on the side of reason, Feyerabend is not blaming Galileo, but the Church": http://appellolaico.wordpress.com/documenti/#parisi. For a different reading of the affair, cf. Pani 2008.

[7] Frediani (1990), p. 55. Besides the interview with the pope, Frediani's article includes an excerpt from *Farewell to Reason* (1987), titled "Ma la Chiesa fu più razionale" ("But the Church was more rational [than Galileo]", pp. 56–57). Cf. also Feyerabend (1985) and (1995), p. 178. On the incident at La Sapienza, and on Feyerabend's role in it, see also John Heilbron's contribution to this volume.

[8] See, for example, the emblematic title of Horgan (1993).

[9] The image is Robert Farrell's: see his (2003), especially Chs. 1 and 8.

## 6.2 Anything Goes

As famously stated at the very beginning of the book, the central claim in *Against Method*—Feyerabend's most iconic work, the one that better presents his attack on traditional philosophy of science[10]—is

> The idea of a method that contains firm, unchanging, and absolutely binding principles for conducting the business of science meets considerable difficulty when confronted with the results of historical research. We find, then, that there is not a single rule, however plausible, and however firmly grounded in epistemology, that is not violated at some time or other. It becomes evident that such violations are not accidental events, they are not results of insufficient knowledge or of inattention which might have been avoided. On the contrary, we see that they are necessary for progress.[11]

The case of Galileo, in this sense, is emblematic:

> The development of the Copernican point of view from Galileo to the 20th century is a perfect example of the situation I want to describe. We start with a strong belief that runs counter to contemporary reason and contemporary experience. The belief spreads and finds support in other beliefs which are equally unreasonable, if not more so (law of inertia; the telescope). Research now gets deflected in new directions, new kinds of instruments are built, "evidence" is related to theories in new ways until there arises an ideology that is rich enough to provide independent arguments for any particular part of it and mobile enough to find such arguments whenever they seem to be required. We can say today that Galileo was on the right track, for his persistent pursuit of what once seemed to be a silly cosmology has by now created the material needed to defend it against all those who will accept a view only if it is told in a certain way and who will trust it only if it contains certain magical phrases, called "observational reports". And this is not an exception – it is the normal case: theories become clear and "reasonable" only *after* incoherent parts of them have been used for a long time. Such unreasonable, nonsensical, unmethodical foreplay thus turns out to be an unavoidable precondition of clarity and of empirical success.[12]

To conclude,

> the idea of a fixed method, or of a fixed theory of rationality, rests on too naive a view of man and his social surroundings. To those who look at the rich material provided by history, and who are not intent on impoverishing it in order to please their lower instincts, their craving for intellectual security in the form of clarity, precision, 'objectivity', 'truth', it will

---

[10] The core of *Against Method* was originally published in 1970, as a contribution to Vol. 4 of the "Minnesota Studies in the Philosophy of Science". The work was later reworked and enlarged in dialogue with Feyerabend's close friend Imre Lakatos. They were planning to publish a joint work, provisionally titled *For and Against Method*, in which Lakatos was to play the role of the 'rationalist', advocating the need for scientific method, whilst Feyerabend that of the 'anarchist', contesting any norms that might constrain the liberty and creativity of the scientists. Lakatos's sudden death, on 2 February 1974, left the project unfinished, and in 1975 Feyerabend decided to publish his own part alone. The Lakatos-Feyerabend correspondence, together with the preparatory material for the joint work, is now available in Lakatos and Feyerabend (1999).

[11] *Against Method* (henceforth, *AM*), p. 14; unless otherwise stated, all references to *AM* are to the third edition (1993).

[12] *AM*, pp. 17–18.

> become clear that there is only one principle that can be defended under all circumstances and in all stages of human development. It is the principle: *anything goes*.[13]

No method can guarantee continuous and linear progress. Far from it: as the history of science shows with abundance of examples—and the example of Galileo is certainly one of the most significant—great progress has been made by those who challenged the rules in place at their times, going beyond pre-established boundaries. If Feyerabend claims that the Church kept more closely to reason, that is, abode by the canons in place at the time of the trial of Galileo, he does so in a work programmatically titled *Against Method*—a work in which the Pisan scientist, as opposed to the scientists of the Roman College, is praised for his ability to break the rigid rules of a pre-established method, thus allowing science to progress. Abiding by the rules, complying with instructions, behaving in what appears to be the rational way may turn out to be an obstacle: science, in Feyerabend's eyes, is an essentially anarchic enterprise, that is, an activity free from any sorts of constraints. It is the only way it may progress.

The possible agreement with 'experimental results' is a measure of the success of a theory: it is the very essence of empiricism, that is, of any theories of confirmation. Feyerabend, by contrast, suggests we complement standard methodological rules with necessary 'counter-rules': we should advance and elaborate on hypotheses that openly contradict established theories or well-known facts. Not only because, on the one hand, there is no theory that agrees perfectly with established facts, but also—and above all—because the proliferation of competing hypotheses (as opposed to Kuhn's theoretical monism) is not merely a catalyst for progress, as Lakatos would put it, but its very engine.

The first step of Feyerabend's criticism of concepts that were regarded as standard in the philosophy of science consists in breaking away from the accepted tradition and devise a new conceptual system that open clashes with established observational results and with received theoretical principles, or else import such system from other sciences, or fields that are not related to science. Exactly as Galileo did, Feyerabend claims.

There are no unalterable rules in the development of scientific knowledge: a science based on Law and Order (that is, on methodological rules) is a science that tends to conservation, not to progress. Against any (fictional) verificationist or falsificationist image of science, Galileo—Feyerabend's anarchic scientist *ante litteram*—achieved all his major results by systematically violating any norms and principles, and by systematically appealing, in their stead, to propaganda, conceptual fraud, theoretical deception as well as cultural and scientific trickery. In Feyerabend's eyes, he even opted for defending unacceptable observations by way of falsified theories. Without the appeal to such means, his new ideas would have run the risk of being nipped in the bud by close adherence to tradition, which is able to exert on scientists an actual "tyranny of truth":

[13] *AM*, pp. 18–19.

> allegiance to the new ideas will have to be brought about by means other than arguments. It will have to be brought about by *irrational means* such as propaganda, emotion, *ad hoc* hypotheses, and appeal to prejudices of all kinds. We need these 'irrational means' in order to uphold what is nothing but a blind faith until we have found the auxiliary sciences, the facts, the arguments that tum the faith into sound 'knowledge'.[14]

The historical case of Galileo clearly shows that

> there are situations when our most liberal judgements and our most liberal rules would have eliminated a point of view which we regard today as essential for science, and would not have permitted it to prevail – and such situations occur quite frequently. The ideas survived and they *now* are said to be in agreement with reason. They survived because prejudice, passion, conceit, errors, sheer pigheadedness, in short because all the elements that characterize the context of discovery, *opposed* the dictates of reason *and because these irrational elements were permitted to have their way*. To express it differently: *Copernicanism and other 'rational' views exist today only because reason was overruled at some time in their past*.[15]

Counter-inductive appeal to hypotheses that are clearly at odds with well-established theories, or openly contradict accepted experimental results, perfectly illustrates the creative growth of scientific knowledge. And that is how Feyerabend's Galileo proceeded, introducing and advocating a falsified hypothesis such as the Copernican hypothesis, which not only clashes with previously accepted theories, but does not come with a plausible and reliable empirical foundation. And, faced with difficulties, does not hesitate to appeal to a whole series of clever propagandistic tricks and subtle psychological devices, eventually managing to achieve a resounding victory.[16]

## 6.3 Galileo's Tower Experiment

Facts may support or disprove a theory, but our reading of facts involves a whole series of elements that we take for granted: within a theory's observational terms there are ideas that implicitly play a major constitutive role in the construction of experience itself. In the history of philosophy and of science, the tacit presence of

---

[14] *AM*, p. 114.

[15] *AM*, p. 116.

[16] See, for example, *AM*, pp. 55 (Galileo's "attempts at concealment [of reality]"), 65 ("Galileo uses *propaganda*. He uses *psychological tricks* in addition to whatever intellectual reasons he has to offer. These tricks are very successful: they lead him to victory. [But they] obscure the fact that the experience on which Galileo wants to base the Copernican view is nothing but the result of his own fertile imagination, that it has been *invented*"), 121 ("whenever possible [Galileo] replaced old facts by a new type of experience he simply *invented* for the purpose of supporting Copernicus"), 127 ("[Galileo's] lies and attempts at deception"). Thanks to a whole set of intellectual tricks, Galileo skilfully defused the counterarguments of his adversaries, attempting to pass off as traditional—hence, innocuous—new ideas and new phenomena that were, in fact, revolutionary innovations: "Galileo's first step, in his joint examination of the Copernican doctrine and of a familiar but hidden natural interpretation", Feyerabend concludes, "consists therefore in *replacing the latter by a different interpretation*. In other words, *he introduces a new observation language*" (p. 63).

such ideas has given rise to various interpretations: from the Kantian one, which assimilated them as inescapable *a priori* presuppositions, to the Baconian one, which qualified them as prejudices from which we must free ourselves. Galileo, argues Feyerabend, was one of the few thinkers who chose a path halfway, deciding not to preserve natural interpretations forever, nor to eliminate them completely. Instead, he opted for subjecting them constantly to a rigorous and systematic critical analysis, to decide which to preserve and which to replace with new theoretical points of view.

One of the chief anti-Copernican arguments discussed in Galileo's *Dialogue* (1632) is the famous tower experiment,[17] described by Salviati, Galileo's spokeperson, in The Second Day:

> As the strongest reason of all is adduced that of heavy bodies, which, falling down from on high, go by a straight and vertical line to the surface of the earth. This is considered an irrefutable argument for the earth being motionless. For if it made the diurnal rotation, a tower from whose top a rock was let fall, being carried by the whirling of the earth, would travel many hundreds of yards to the east in the time the rock would consume in its fall, and the rock ought to strike the earth that distance away from the base of the tower.[18]

Faced with the apparently undeniable contradiction between the Copernican theory and experimental evidence, Galileo does not question the correctness of the observation ("I [...] have never seen nor ever expect to see the rock fall any way but perpendicularly, just so do I believe that it appears to the eyes of everyone else"), but urges that we learn "to be more circumspect and less confident about that which the senses represent to us at first impression, for they may easily deceive us". Confronted with what by all means seems an obvious refutation of the rotational motion of the earth, the best move for Feyerabend's Galileo is not to give up the Copernican theory, but rather use it to identify the hidden principles responsible for the refutation itself:

> It is therefore better to put aside the appearance, on which we all agree, and to use the power of reason either to confirm its reality or to reveal its fallacy.[19]

Galileo invites his readers to set aside the "appearance", on which everyone agrees, because the senses can be deceptive: they are not a secure point of reference, they do not constitute the solid foundation upon which we can build a stable theoretical edifice. Of course, we must start from experience, but we must never stop uncritically at the immediacy of the senses. We must, on the contrary, force them by appealing to reason, that is, correct experience in the light of rational reasoning:

> Nor can I ever sufficiently admire the outstanding acumen of those who have taken hold of this opinion and accepted it as true; they have through sheer force of intellect done such violence to their own senses as to prefer what reason told them over that which sensible experience plainly showed them to the contrary.[20]

---

[17] Cf. Aristotle, *De caelo*, II 14, 296 b 21–25. A similar argument is found in Ptolemy's *Almagest*, I 7, as well as at the end of Copernicus, *De revolutionibus*, I 7. For a discussion, see also Andersson 1991.

[18] Galilei (1632), p. 126 (henceforth, *Dialogue*).

[19] *Dialogue*, p. 256.

[20] *Dialogue*, p. 328.

Galileo's experience is no longer an immediate, "pure" experience: it is sense experience (*sensata esperienza*) as we perceive it through our senses and our experiments—an experience, that is, structured and shaped in the light of a particular theory or a specific practical action, constructed according to procedures capable of producing reliable results.[21] Whereas the Peripatetics believed that experience coincides with what appears to the immediacy of the senses, Galileo is fully aware that the empirical level (possibly supplemented by technical instruments such as the telescope) and the theoretical level must complement each other.[22] In other words: without *discorso* (reason), there is no possibility of authentic experience.[23] Unlike the Peripatetics, however, Galileo is aware of the fact that immediate experience alone cannot provide us with correct information on the true physical constitution of the world: the natural philosopher's task is not simply to passively record the phenomena that are placed before his eyes, but to critically investigate nature, pressing it with exact questions, forcing it to provide precise answers to a given question at any given time. Galileo returns to this point in The Third Day of the *Dialogue*, in possibly one of the work's most beautiful pages:

> You [Sagredo] wonder that there are so few followers of the Pythagorean opinion, whereas I am astonished that there have been any up to this day who have embraced and followed it. Nor can I ever sufficiently admire the outstanding acumen of those who have taken hold of this opinion and accepted it as true; they have through sheer force of intellect done such violence to their own senses as to prefer what reason told them over that which sensible experience plainly showed them to the contrary. For the arguments against the whirling of the earth which we have already examined are very plausible, as we have seen; and the fact that the Ptolemaics and Aristotelians and all their disciples took them to be conclusive is indeed a strong argument of their effectiveness. But the experiences which overtly contradict the annual movement are indeed so much greater in their apparent force that, I repeat, there is no limit to my astonishment when I reflect that Aristarchus and Copernicus were able to make reason so conquer sense that, in defiance of the latter, the former became mistress of their belief.[24]

It is the natural (Aristotelian) interpretation associated with the individual observational data, not the empirical facts themselves, that must be questioned and replaced with a new interpretation: the same observation, in other words, must be placed within two different theoretical and interpretative frameworks. According to the Aristotelian perspective, every motion is operative; hence the fall of a rock at the base of the tower constitutes the irrefutable proof of the earth's immobility. Conversely, the failure to observe a motion along a slating line constitutes proof

---

[21] Cf. Baroncini (1984).

[22] Cf. *Dialogue*, p. 328: "[…] were it not for the existence of a superior and better sense than natural and common sense to join forces with reason, I much question whether I, too, should not have been much more recalcitrant toward the Copernican system than I have been since a clearer light than usual has illuminated me".

[23] The converse is true, too: without direct reference to experience we would be unable to come up with theories provided with genuine cognitive value. We turn to reality led by our theoretical expectations, but it is reality itself that selects such expectations and sets precise constraints to our speculations: "one will philosophize better by giving assent to those conclusions depending on manifest observations than by persisting in opinions contrary to sense perception itself, and confirmed only by probable or apparent reasons" (Galilei 1613, p. 129).

[24] *Dialogue*, pp. 327–328.

with high empirical content, capable of refuting the Earth's alleged mobility. In Galileo's eyes, by contrast, the natural interpretation coincides with the assumption that relative motion only is truly operative; hence, a falling rock striking the ground at the base of the tower, and not to the west of it, simply demonstrates that there is no relative motion between the point at which the rock was let go and earth itself:

> Thus it is likewise true that the earth being moved, the motion of the stone in descending is actually a long stretch of many hundred yards, or even many thousand; and had it been able to mark its course in motionless air or upon some other surface, it would have left a very long slanting line. But that part of all this motion which is common to the rock, the tower, and ourselves remains insensible and as if it did not exist. There remains observable only that part in which neither the tower nor we are participants; in a word, that with which the stone in falling measures the tower.[25]

For these reasons, at the beginning of The Fourth Day of the *Dialogue*—before turning to the explanation of the tides, the last and (at least in his eyes) most compelling argument in favour of the Copernican hypothesis—Galileo can conclude that "all terrestrial events from which it is ordinarily held that the earth stands still and the sun and the fixed stars are moving would necessarily appear just the same to us if the earth moved and the others stood still".[26] With the shift from the Aristotelian paradigm (according to which all motion is operative, and therefore detectable through the senses) to the Galilean paradigm (according to which only relative motion is operative, and therefore detectable through the senses) a profound conceptual change takes place.[27] In doing so, however, Galileo does not merely defuse one of the strongest arguments against the Copernican hypothesis, but surreptitiously goes so far as to replace the Aristotelian paradigm with a new paradigm, which applies to all cases, celestial and terrestrial. Here lie the "propagandistic machinations",[28] the unscrupulous recourse to theoretical subterfuges and psychological tricks. In Feyerabend's words:

> Galileo's proposal amounts to a partial revision of our observation language or of our experience. An experience which partly *contradicts* the idea of the motion of the earth is turned into an experience that *confirms* it, at least as far as "terrestrial things" are concerned. This is what *actually happens*. But Galileo wants to persuade us that no change has taken place,

[25] *Dialogue*, pp. 172–173.

[26] *Dialogue*, p. 416.

[27] For an Aristotelian, the motion of a body is inherent to its nature and does not at all depend on the physical situation within which the body finds itself: the nature of a body, in other words, completely determines its motion. It is this substantial and ontological conception of motion that Galileo rejects, contrasting it with a profoundly different view. The two conceptions, the Aristotelian and the Copernican, diverge on a theoretical level, involving two different conceptual horizons and two different ways of investigating the very nature of motion. Both points of view contemplate the possibility of perceiving relative motions, but they play radically different roles in the two paradigms: whereas for Aristotle relative motion is secondary, insofar as motion is defined by the nature of the body, for Galileo it is of decisive importance; whereas for the Aristotelian paradigm the observability (or otherwise) of a motion is a more or less irrelevant detail, in the Galilean paradigm it comes to play an epistemic role capable of shattering the Aristotelian-Ptolemaic model.

[28] *AM*, p. 72.

> that the second conceptual system is already universally *known*, even though it is not universally *used*. Salviati, his representative in the *Dialogue*, his opponent Simplicio and Sagredo the intelligent layman all connect Galileo's method of argumentation with Plato's theory of *anamnesis* – a clever tactical move, typically Galilean one is inclined to say.[29]

The psychological 'leverage' of anamnesis crowns the operation, facilitating the tacit process of extending the principle of relativity from "terrestrial things" (rocks, birds, ships, carriages and whatnot) to the earth as a whole. In so doing, Galileo masks the revolutionary substitution of the new operative concept of relative motion for the traditional Aristotelian concept of the operative character of all motion. His is a clear example—a textbook example, one might say—of the counter-inductive method recommended by Feyerabend:

> an initially inadequate view, the Copernican theory, is supported by another inadequate view, the idea of the non-operative character of shared motion, and both theories gain strength and give support to each other in the process. It is this change which underlies the transition from the Aristotelian point of view to the epistemology of modern science.[30]

## 6.4 The Target of Feyerabend's Criticism

While the tower experiment is a particularly interesting (and qualifying) point in the argumentative structure of Galileo's *Dialogue*, Feyerbend's analysis is not limited to it. In the following chapters of *Against Method*, other key aspects of Galileo's work are addressed in detail, from the use of the telescope to his fundamental contributions to mechanics, the principle of inertia and the principle of relativity of motion. Feyerabend's procedure is quite similar to that adopted for the tower experiment, and its discussion would exceed the limits of the present contribution.

The provocative picture of Galileo as a charlatan and swindler, who succeeded in undermining the received tradition (or anyway contributing decisively to this process) only by appealing to rhetorical and propagandistic weapons, as well as psychological tricks, is usually considered an exaggeration. Historians claim it is unsupported by available documents, or misleading at best. Indeed, several objections may be raised against Feyerabend's reading of the texts and the events, especially in light of the enormous progress made by Galileo studies in recent decades. There is no doubt, however, that in his own cultural battle—in his writings, first and

[29] *AM*, p. 71.

[30] *AM* (1988), pp. 74–75; the passage is omitted in the third edition of *AM*. Cf. also *AM* (1993), p. 105: "while the pre-Copernican astronomy *was in trouble* (was confronted by a series of refuting instances and implausibilities), the Copernican theory *was in even greater trouble* (was confronted by even more drastic refuting instances and implausibilities); but that being in harmony *with still further inadequate theories* it gained strength, and was retained, the refutations being made ineffective by *ad hoc* hypotheses and clever techniques of persuasion. This would seem to be a much more adequate description of the developments at the time of Galileo than is offered by almost all alternative accounts".

foremost, but also in his long chess match against the Aristotelian tradition and the Catholic Church, in favour of Copernicanism and a new physics—Galileo repeatedly appealed to clever rhetorical devices and psychological stratagems for propagandistic purposes. The *Dialogue*, after all, as has been repeatedly emphasised, is more of a manifesto aimed at renewing period culture, rather than a scientific treatise operating within an already renewed cultural horizon. The "propagandistic" component of Galileo's work is therefore not only undeniable, but fundamental, and indeed functional for the purpose that Galileo set himself. In this sense, Feyerabend's analysis—the rivers of ink poured to refute it (and to portray its author as a supporter of the Church's view, or as an enemy of science[31]) notwithstanding—retains its validity, contributing to focusing on aspects previously overlooked.

When reading Feyerabend, it is crucial to keep in mind the main target of Feyerabend's philosophical attitude:

> Always remember that the demonstrations and the rhetorics used do not express any "deep convictions" of mine. They merely show how easy it is to lead people by the nose in a rational way. An anarchist is like an undercover agent who plays the game of Reason in order to undercut the authority of Reason (Truth, Honesty, Justice, and so on).[32]

In fact, Feyerabend's aim was not so much a historical reconstruction of Galileo's case—let alone a re-evaluation of the trial of 1633, or of Cardinal Bellarmino's work. His targets were the views of contemporary philosophers of science (chiefly Popper, of whom Feyerabend was a great admirer and the most recalcitrant pupil, and Kuhn, whose theoretical monism he could not but contrast with his pluralistic approach to science), who had made Galileo a hero of method and rationality. In conclusion, *Against Method* contains both a challenge and a provocation: with clever irony and brilliant insight, Feyerabend suggests, on the one hand, to show the intrinsic limits of any rational reconstructions of a historical event; on the other, to call attention to the all too frequent risk of turning science into dogma, thereby neutralizing it: one of the most powerful and far-reaching enterprises of the human mind, one whose very nature identifies with relentless and uncompromising criticism.

Even the most disruptive criticism, if absolutized, may lose its innovative and revolutionary character. The smile with which (I like to think) Feyerabend took the interviewer's question as to whether his remarks had been misunderstood or used instrumentally by Cardinal Ratzinger—who was then Bellarmino's successor at the Congregation for the Doctrine of the Faith, formerly the Holy Office—had precisely this aim.

---

[31] Cf. the title of Horgan (1993); for a different view, see Preston, Munévar and Lamb, eds. (2000).

[32] *AM*, p. 23.

## References


Andersson, Gunnar. 1991. The Tower Experiment and the Copernican Revolution. *International Studies in the Philosophy of Science* 5 (2): 143–152.

Baroncini, Gabriele. 1984. Sulla galileiana «esperienza sensata». *Studi secenteschi* 25: 147–172.

Farrell, Robert P. 2003. *Feyerabend and Scientific Values: Tightrope-Walking Rationality*. Dordrecht-Boston-London: Kluwer.

Feyerabend, Paul K. 1970. Against Method: Outline of an Anarchistic Theory of Knowledge. In *Analysis of Theories and Methods of Physics and Psychology*, ed. Michael Radner and Stephen Winokur, 17–130. Minneapolis: University of Minnesota Press.

———. 1975. *Against Method: Outline of an Anarchistic Theory of Knowledge*. London: New Left Books and Atlantic Heights, NJ: Humanities Press. Second edition, London-New York: Verso, 1988. Third edition, Verso: London-New York, 1993.

———. 1976. *Wider der Methodenzwang: Skizze einer anarchistischen Erkenntnistheorie*. 2nd ed, 1983. Frankfurt: Suhrkamp.

———. 1985. Galileo and the Tyranny of Truth. In *The Galileo Affair: A Meeting of Faith and Science*, ed. George V. Coyne, Michael Heller, and Jożef M. Życiński, 155–165. Vatican City: Specola Vaticana. reprinted in Feyerabend (1987), pp. 247–264.

———. 1987. *Farewell to reason*. London: Verso.

———. 1995. *Killing Time*. Chicago/London: The University of Chicago Press.

Frediani, Marcello. 1990. Un giudice per Galileo. *Il Sabato* 19: 54–57.

Galilei, Galileo. 1613. *Istoria e dimostrazioni matematiche intorno alle macchie solari e loro accidenti*, Rome: Giacomo Mascardi; in *Le Opere di Galileo Galilei*, National Edition, Antonio Favaro editor-in-chief, Vol. 5, Florence: G. Barbèra, 1895, pp. 72–149. Engl. ed. in Galileo Galilei and Christoph Scheiner, *On Sunspots*, translated and introduced by Eileen Reeves and Albert Van Helden, Chicago-London: The University of Chicago Press, 2010, pp. 87–105, 107–168 and 251–304.

———. 1632. Galileo Galilei, *Dialogo sopra i due massimi sistemi del mondo tolemaico e copernicano*, Florence: Giovanni Battista Landini; in *Le Opere di Galileo Galilei*, National Edition, Antonio Favaro editor-in-chief, Vol. 7, Florence: G. Barbèra, 1897, pp. 21–520. Engl. ed., Galileo Galilei, *Dialogue Concerning the Two Chief World Systems—Ptolemaic & Copernican*, translation by Stillman Drake, Berkeley-Los Angeles: University of California Press, 1962, 1967[2].

Guarini, Renato. 2014. *Sapienza e libertà: Come e perché papa Ratzinger non parlò all'Università di Roma*, with contributions by Carlo Cosmelli, Vincenzo D'Adamo and Gianluca Senatori, edited by Pier Luigi De Lauro, Rome: Donzelli.

Horgan, Paul. 1993. The Worst Enemy of Science. *Scientific American* 268 (5): 36–37.

Lakatos, Imre, and Paul K. Feyerabend. 1999. *For and Against Method*, edited with an introduction by Matteo Motterlini: Chicago/London: The University of Chicago Press.

Pani, Giancarlo. 2008. In *Ratzinger-Galileo alla Sapienza: Il caso Galileo, il metodo scientifico e la Bibbia*, ed. Vincenzo D'Adamo. Palermo: Sigma.

Preston, John M., Gonzalo Munévar, and David Lamb, eds. 2000. *The Worst Enemy of Science? Essays in Memory of Paul Feyerabend*. Oxford: Oxford University Press.

Ratzinger, Joseph [Pope Benedict XVI]. 1990. Prospettive e compiti del Cattolicesimo per il futuro dell'Europa. *Il Nuovo Areopago*, IX, 2, pp. 7–24.

——— [Pope Benedict XVI]. 1991. Wege des Glaubens im Umbruch der Gegenwart. In Joseph Ratzinger, *Wendezeit für Europa? Diagnosen und Prognosen zur Lage von Kirche und Welt*, Freiburg im Breisgau: Johannes Einsiedeln, pp. 59–81; English translation, "Path of Change in the Revolutionary Change of the Present Day", in Joseph Ratzinger, *A Turning Point for Europe? The Church in the Modern World: Assessment and Forecast*, San Francisco. Ignatius Press, 1994, 2010[2], pp. 87–117.

## References

[illegible]

# Chapter 7
# Feyerabend, Bohr and Quantum Physics

**John L. Heilbron**

**Abstract** Between 1952 and 1975, between his arrival in England and the publication of *Against method*, Feyerabend transformed himself from a critic of quantum mechanics into an anti-philosopher of science. When he moved to Berkeley, in 1958, he distanced himself from Karl Popper, argued with Thomas Kuhn, and came to grips with the issues raised by the "free speech" movement. The exigencies of his journey drove him to intellectual history, particularly to the Greeks and Galileo, in a search for plausible arguments for his anti-philosophy. In the first part of this paper, Heilbron presents Feyerabend the critic of quantum mechanics: starting from Popper's arguments against Bohr's complementarity and the diffuse "Copenhagen interpretation," Feyerabend supported David Bohm's challenge to Copenhagen in the early 1950s; after moving to Berkeley, Feyerabend defended Bohr's views against Popper's criticism. In the second part, Heilbron approaches Feyerabend the historian through his apt objections to Kuhn and his unpersuasive account of Galileo's science and sufferings.

**Keywords** Paul Feyerabend · Niels Bohr · Karl Popper · Thomas Kuhn · David Bohm · Quantum mechanics · Copenhagen interpretation · Galileo

Between 1952 and 1975, between his arrival in England to work with Popper and the publication of *Against method*, Feyerabend transformed himself from a responsible and authoritative critic of quantum mechanics into a wreckage-wreaking anti-philosopher of science. During the quarter century in which he advanced or regressed from the serious work he did under Popper's influence to the repudiations and

John L. Heilbron died before publication of this work was completed.

J. L. Heilbron (✉)
University of California at Berkeley, Berkeley, CA, USA

California Institute of Technology, Pasadena, CA, USA

© The Author(s), under exclusive license to Springer Nature Switzerland AG 2024

S. Gattei, R. Corvi (eds.), *Feyerabend in Dialogue*, Boston Studies in the Philosophy and History of Science 346,
https://doi.org/10.1007/978-3-031-71938-7_7

exaggerations of his destructive period, he lived a Bohemian life as a professor of philosophy at the University of California at Berkeley. There he distanced himself from Popper, argued with Thomas Kuhn, and came to grips with the issues raised by the "free speech" movement. The exigencies of his journey drove him to intellectual history, particularly to the Greeks and Galileo, in a search for plausible arguments for his anti-philosophy. I should report that I met him when I was Kuhn's graduate assistant and later became his colleague on the Berkeley faculty.

I have divided my description of Feyerabend's life and work during the period into two parts. The first presents Feyerabend the critic of quantum mechanics. Starting from Popper's arguments against Bohr's complementarity and the diffuse "Copenhagen interpretation," Feyerabend supported David Bohm's challenge to Copenhagen in the early 1950s. The challenge exhibited a form of quantum theory that the great mathematician John von Neumann had "proved" impossible; Feyerabend sought to discover von Neumann's errors or gratuitous assumptions and to provide an alternative solution to the stubborn problem of measurement at the heart of quantum mechanics. After moving to Berkeley, Feyerabend joined the enemy and defended Bohr's views against Popper's criticism.

In the second part of my paper, I approach Feyerabend the historian through his apt objections to Kuhn and his unpersuasive account of Galileo's science and sufferings. In exploiting this account to establish his case against method, Feyerabend encouraged the resistance to alternative views for which he faulted Kuhn's *Structure of scientific revolutions*.

## 7.1 The Quantum Mechanic

### *7.1.1 Neutralizing Neumann*

In the fall of 1952 Feyerabend joined Popper's group at the London School of Economics. When he presented himself for duty, he was surprised to be asked what he planned to do. In answer he pitched on David Bohm's textbook, *Quantum mechanics* (1952), which he had been studying. Attracted by the book's unusual emphasis on physical interpretation and foundational questions, Feyerabend proposed to review it.[1] Even more attractive to his contrarian streak were two papers Bohm published in the spring of 1952 demonstrating that the rules of quantum mechanics left open the possibility of a deterministic theory of the microworld. This demonstration upset the physicists in Copenhagen, according to whom the microworld was radically indeterministic and, consequently, could at best be described by statistical statements about the outcomes of experiments made on it.

Feyerabend knew at firsthand how shocking Bohm's bombshell was to the keepers of the faith. During the summer of 1952 he had heard Bohr himself on the subject. Perhaps more puzzled than disturbed, Bohr was "neither dismissive nor

[1] *KT*, 86–7.

shaken...[just] amazed" by Bohm's counter example. Not so his dedicated followers like Heisenberg and Pauli. They knew that Bohm had to be wrong because, in addition to their master's teachings, they had the firm rock of von Neumann's mathematical proof, which was widely believed, even by people who had read it, to rule out the possibility of a deterministic theory that could deliver the confirmed results of quantum mechanics.[2]

Discussions with Feyerabend and his fellow student Joseph Agassi reenergized Popper's interest in the interpretation of quantum mechanics, which he had attempted to enrich almost 20 years earlier by exposing flaws in Heisenberg's elucidation of the uncertainty relations. Popper mistakenly believed that he had designed a decisive thought experiment and he abandoned the subject from chagrin at being corrected.[3] The self-inflicted injury had lost some of its affect by the time he took on von Neumann with his bright and pushy students. They spent some time trying to master it, at first with little success; according to Popper, it was he who first recognized that the proof failed, which, of course, it had to do to allow Bohm's counter example. Popper thought he had priority and construed Feyerabend's continuing hunt for weak spots in von Neumann's proof as plagiarism. He worked himself into a nonsensical fit over it: "Feyerabend has no original ideas, but he poses as an original thinker: he is a compulsive plagiarist."[4]

Feyerabend worked compulsively at breaking the chains with which von Neumann held physicists in thrall to Copenhagen. The famous proof was then attracting enough belated interest that Princeton University Press made von Neumann's *Mathematische Grundlagen der Quantenmechanik* (1932) accessible to Anglophone theorists by publishing a translation in 1955. In his review of the book, Feyerabend compliments von Neumann for a presentation that might rescue philosophers from basing their analyses of quantum physics on wave mechanics and that clarified without settling the nature of measurement or the relationship between quantum mechanics and the Copenhagen interpretation. Avoiding plagiarism, Feyerabend credited Popper with perceiving that complementarity does not follow from von Neumann's formalism.[5]

Von Neumann had become (please pardon my French) the *bête noire* of Feyerabend the *enfant terrible*. He thought it shameful that physicists should have allowed themselves to fall for a faulty proof of an unlikely thesis directed at the foundations of their discipline. Their uncritical acceptance of the proof was an exemplar of the hamstringing of science by the premature foreclosing of thought. Or, even more cowardly, it showed them sheltering under mathematics they had not bothered to understand, giving or withholding their assent as their mathematical

[2] *KT*, 77–8, and PF to Peat, 9 Jul and 30 Aug 1993, in Peat, *Infinite potential* (1996), 129.

[3] Popper, *Logik* (1935), 178–81, 220–2, in *Logic* (1959), 236–46, esp. 243; he nursed and rehearsed his embarrassment for years, e.g., in *Postscript 3* (1982), 15, and *Quest* (2002), 94.

[4] Agassi (1991a), in Munévar, *Beyond reason* (1991), 384, and in Sachs, *Einstein versus Bohr* (1991b), xx; Del Santo, *SHPMP*, 67 (2019), 79–80, 86 (quote, from a letter to Henry Margenau, 21 Feb 1969).

[5] PF, *BJPS*, 8:32 (1958), 343–7; Jammer, *Philosophy* (1974), 272.

authority required. The phenomenon was widespread. As will appear, Feyerabend contrived to discover in Galileo, of all people, the sort of slavishness that quantum physicists had shown to von Neumann.

While Feyerabend was beginning to identify von Neumann's proof as an exemplary obstacle to the pursuit of science, Popper went lecturing to the United States. In Princeton Bohr and Einstein attended his seminar. Einstein said a few words in agreement; Bohr ("the most wonderful person") went on until only he and Popper remained on the field of battle. Einstein was happy to talk at length one-on-one with Popper about Bohr's special version of the Copenhagen interpretation, complementarity, although, he said, "in spite of the greatest efforts, he could not understand what Bohr meant by [it]."[6] Popper thought that he understood it well enough to criticize it as a sort of positivism, the most distasteful to him of all philosophies. Or, to use bigger words, Popper diagnosed Bohr as an instrumentalist and instrumentalism as "a misleading and even a vicious doctrine."[7]

Complementarity is not an ordinary form of positivism since it admits the possibility of knowledge about the world interpreted in the refined common-sense language of classical physics. To be sure, it did not permit knowledge of entities as they are, but as they behave in experiments devised to expose their properties. To Bohr's insistence that no more could be known about micro-entities than probabilities of their behavior under defined circumstances, Popper could oppose not only his philosophical convictions, but also Einstein's endorsement of them. Feyerabend accepted Popper's depreciation of complementarity and his misleading assurance that the difficulties of quantum mechanics had an easy solution within the grasp of the Popperian school. For Popper, the solution began with a proper understanding of statistics.[8] For Feyerabend, it began by exorcising "Neumannian nightmares."[9]

Since Bohm had exhibited a deterministic version of quantum mechanics, artificial to be sure but logically sustainable, the nightmare must contain an unwarranted fright. Bohm thought that it could be found in the morass of measurement. Von Neumann had assumed tacitly that the "hidden variables," which on Bohm's theory allowed a deterministic account of quantum processes, would obey the general rules of quantum-mechanical systems. But the processes of primary interest in quantum mechanics concerned interactions between microsystems and measuring instruments and von Neumann had not shown that such processes had to obey the general rules. Why should hidden variables obey rules that they are introduced to subvert? And, reversing the direction of argument, why should hidden variables not give results indistinguishable from those of standard quantum mechanics? Feyerabend accepted this argument in his first published analysis of von Neumann's proof.[10]

---

[6] Popper, *Quest* (2002), 148, 151, and in *Postscript 3*, 9 (quote).

[7] Popper, *Postscript 3*, 42 (text of 1966, quote), 99 (1956/7).

[8] Popper, *Logik* (1935), 161–72, = *Logic* (1961), 223–36; continued in *Postscript 3*, 35–95, passim.

[9] *AM*, 50n25 (quote); *PP4*, 19, 22 (text of 1954).

[10] Bohm, *PR, 85* (1952), 187–8; PF, in *PP4*, 40–42 (1954).

Looking more deeply, he noted that the proof rested on the claim that every ensemble or collection of quantum mechanical elements consisted of sub-elements whose statistical properties did not differ from those of the ensemble. This definition has a considerable bite. Assume that the ensemble is a mixture of elements each of which has a sharp value, buckshot, let us say, of different sizes. The different sizes could be separated by a series of sieves, producing sub-ensembles unlike the master ensemble: the separated fractions would be "dispersion [or scatter] free," homogeneous, not further analyzable or reducible. Von Neumann proved that in quantum mechanics there can be no dispersion-free sub-ensembles; scatter is inevitable, no matter how deeply we probe; scatter is an indelible property of a quantum mechanical ensemble; the theory is incurably probabilistic.[11]

So what? Popper had pointed out that von Neumann's conditions—an aggregate and its sub-aggregates have the same statistical properties and there are no dispersion-free aggregates—also hold for dice games, and yet no one denied that a clever and determined physicist could give a deterministic account of the throw of a single die. Consequently, the poison pill must lie somewhere else in von Neumann's nightmare. The Popperians thought they had found it as far down as assumption "S" in the itemized proof: a *deterministic* description must always lead to scatter-free aggregates. Von Neumann thought that his "S" held for classical but not for quantum physics, but in fact it always fails when statistics are invoked. "[Therefore] hidden parameters, which were introduced for a more exact description of statistical processes, will likewise scatter." The shake of a die admits a deterministic account, which, owing to the complexities of the process, we cannot give. We retreat to statistics and estimate the scatter, but do not in consequence surrender determinism.[12]

In his review of Princeton's translation of von Neumann's *Grundlagen*, Feyerabend identified the poison pill in a deduction from the assumption of the identity of the statistics of an aggregate and its sub-aggregates; von Neumann had claimed that the assumption of identity "allows for the application to individual systems of probabilities in the sense of relative frequencies;" universal determinism contradicts quantum mechanics. Perhaps von Neumann's proof falters here; other analysts have identified other weak spots; but the bottom line for Feyerabend was that the proof must fail and with it the myth that the Copenhagen Interpretation had firm mathematical backing.[13] Or, to speak in his straightforward language, the claim put forward by the Copenhagen school that theirs was the only tenable interpretation of quantum mechanics was just plain "dishonest."[14]

Feyerabend's study of von Neumann culminated in a paper on the theory of measurement he offered to a high-powered audience of physicists and philosophers in 1957. The venue was the University of Bristol, where Feyerabend had just begun a

[11] *PP4*, 36–7, 48 (1954), 48 (2956).

[12] *PP4*, 48.

[13] Cf. for myth, *PP4*, 109, 221, and, for more weak spots in von Neumann's proof, Pinch, in Mendelsohn, *Production* (1977), 187–92, 206–8, Seevinck, in Crull and Bacciagaluppi, *Hermann* (2017), 110–15, and Diecks, *SHPMP*, 60 (2017), 139–47.

[14] Feyerabend, cited by Gattei and Agassi, in *PP4*, xxvii, xxv, resp.

three-year contract as a lecturer in the department of philosophy. Despite insulting the committee that interviewed him, he got the job, primarily, he thought, through a recommendation from Schrödinger and perhaps also from Bohm, who had joined Bristol's physics department in 1957. Bohm and Feyerabend became such good friends that Bohm could enter Feyerabend's apartment without an invitation, talking so enthusiastically about the latest twist in physics that once he carried on for half an hour without noticing that Feyerabend was not there.[15]

Feyerabend helped to organize the conference of 1957 and edited its proceedings. That brought him into heady company: Léon Rosenfeld, Bohr's close colleague and first-line defender of the Copenhagen view; Eugene Wigner, von Neumann's boyhood friend and Princeton colleague; Bohm and his powerful collaborators Jean-Pierre Vigier and Yakir Aharonov; Boris Podolsky and Nathan Rosen, who had collaborated with Einstein in 1935 in a famous and eventually fruitful challenge to Bohr; Wigner's brother-in-law and sparring partner, Paul Dirac; and, though in absentia, the fearless philosopher Popper. Apart from Rosenfeld, these men agreed that the Copenhagen Interpretation could not be right, but not about the nature of its replacement.

### *7.1.2 Applying Popper's Propensities*

Popper's contribution at a distance concerned his easy resolution of "the great quantum muddle." What he had in mind and condensed in his lecture is more readily assimilated from the "postscript" to his earlier work that he had then just finished revising for the press.[16] He reasserted that physicists could clear up their muddle if they harkened to his understanding of probabilities. Aristotle had the right word and concept for the confused quantum physicist, *potentia*; the probability that $\alpha$ should occur in the circumstances $\beta$ is an objective "propensity." In evaluating the possible outcomes of measuring $\alpha$, the circumstances $\beta$ of the experiment play an essential and non-separable part; the probability that figures in quantum physics is not an expression of the experimenter's subjective ignorance of the details of the object under investigation but is owing solely to the fact that the propensities of the situation determine a range of possible outcomes. The main business of quantum mechanics is to calculate these propensities, which should be ranked as real as fields and forces, for they can add and subtract and interfere with one another.[17]

Heisenberg's uncertainty relations are measures of the dispersion or scatter in the outcomes and do not limit the precision to which the position and momentum of a particle can be obtained. Or, rather, obtained in retrodiction: measure exactly where

---

[15] *KT*, 102; PF to Peat, 9 Jul 1993, in Peat, *Infinite potential* (1996), 188.

[16] Popper's reworking of his *Logik der Forschung* was finally published in three volumes in 1982 as a *Postscript* to *Logic of scientific discovery* (1959).

[17] *Postscript*, *3*, 126–7.

an electron scattered from a very small hole strikes a photographic plate and you have its straight-line trajectory with sharply defined $p$ and $q$ at all points, just as Einstein, Podolsky, and Rosen (EPR) had concluded in another way.[18] Moreover, there is no problem of measurement, of wave-function collapse, or of understanding why, as Bohr had insisted, the range of possible results from a given experiment depend sensitively on all its circumstances and conditions.

Propensities give the famous double-slit experiment its ultimate explanation. A beam of photons cannot build up an interference pattern by shining it through one hole (the other being blocked) and then the other, because changing the experiment changes the propensities. The pattern builds up with both holes open even if the beam delivers only one photon an hour. How does that happen? How does a photon know where its predecessors and successors will go? Answer: the propensity of each photon to travel a given path depends on the entire experimental arrangement. The reasoning applies to classical cases too. A ball entering a pinball machine has a range of propensities that, when the game is repeated many times, gives rise to a distribution of the balls in pockets at the bottom of the machine. If a pin is removed, the propensities change and so does the distribution, even though some or many balls do not go near the position of the missing pin.[19]

In his Bristol lecture, Popper illustrated the difference between a subjective and objective understanding of probability with a lucid account of one of his dice games. The subjectivist and objectivist agree that the probability of a particular outcome on a single throw is the relative frequency with which that throw would occur in a long sequence of throws. But subjectivists, having gathered all the information they can from the frequencies and the manner of throwing, ascribe their recourse to probabilities in placing their bets to their inadequate knowledge of the situation. Objectivists who understand probabilities as propensities of the sequence, though no more successful at dice than subjectivists, are content to know that whatever they know has no effect on the probability they assign to outcomes. Popper gives the example of a die loaded so that the six comes up one time in four. If a true die were substituted occasionally for the bad one, the subjectivist would say that the probability of an individual throw would be affected, whereas the objectivist would stay with 1/4, even for the true die, because the rare substitutions do not change the frequency distribution for the sequence.[20]

The propensities of a sequence depend on many circumstances. For example, the asymmetry of the loaded die becomes less effective in a weak and more effective in a strong gravitational field. Therefore, it is better to conceive of probabilities as likelihoods of the realization of propensities that are joint properties of the system under investigation (the die) and the experimental setup (the manner of throwing the die, the surface receiving it, the gravitational field acting on it). Schrödinger's $\psi$ function is just an inventory of weighted propensities, a vector in the "space of

[18] Ibid., 146–9.

[19] Ibid., 131, 151–3.

[20] Popper, in Koerner, *Observation* (1957), 65–6.

possibilities." Here there is no room for "the intrusion of the subject into physics." Bohr should be content with propensities, Popper supposed, because they express the essential role of the experimental arrangement; Schrödinger and Einstein should also be content, because propensities are real, objective, and relational. And Aristotle would be joyful that his concept of *potentia*, suitably extended, resolved the pressing problems of a microphysics of which he had had no inkling.[21] Really, it is all very simple, if we do not surrender to imposing formulae: "that is the road to perdition." The right road is waymarked by "clear thinking, simplicity, and intellectual responsibility." Several important quantum physicists, including Bohm, accepted Popper's propensities.[22]

Propensities and simplicity might have mitigated the muddle, but they did not indicate how it happens that when one possibility is realized in an experiment all the others simultaneously vanish. Feyerabend undertook to make this difficulty vanish too. He began his Bristol lecture with a brief clear statement of von Neumann's assumptions and proof. The assumptions include the formalism of quantum mechanics and Born's probability interpretation of ψ; the proof includes the argument that every ensemble of quantum-mechanical systems is either a pure case or a mixture of pure cases and no pure case is free from statistical scatter. From these propositions the applicability of quantum mechanics to individual systems followed. Feyerabend here observed, gratuitously, that propensities do not help us to obtain information about the actual state of a given system and, "therefore," alas, can have "no influence upon the problems of quantum mechanics."[23]

What will help? Von Neumann required a special process to squeeze information from the ψ inventory because ψ with all its possibilities evolves regularly in time as if it were an ordinary wave and would keep on evolving after the system it describes hits the measuring apparatus. The inventory of possibilities would alter but the system would remain suspended in them. Von Neumann had escaped from this awkwardness by assuming that ψ collapsed to a single realized possibility on being measured (underwent a quantum jump from nowhere in particular to the observed state) and, finding no defensible place for the collapse within the apparatus whether or not it included the observer's retina and optic nerve as well as the instrument, did what came naturally to a mathematician and located it in the only unique entity in the chain, the observer's consciousness. This was anathema to the Popperian principle of objectivity. Feyerabend saved it by taking the confrontation of the system with the instrument as the place of effective collapse.[24]

The question now is how the confrontation puts the instrument in a single identifiable state indicated by, say, the position of a needle on a dial. For it too is a quantum system and should have its own ψ of possibilities before the confrontation, a joint ψ with the system during their interaction, and another ψ after separation.

---

[21] Ibid., 68–70.

[22] Popper, *Postscript 3* (1982), text of 1956/7; Del Santo, *SHPMP*, 67 (2019), 83–4.

[23] PF, in Koerner, *Observation* (1957), 122 and 122n8.

[24] Ibid., 126.

How does it finally settle on one possibility and so allow an inference about the $\psi$ of the separated microsystem? Feyerabend argued that because the instrument and the observer are macroscopic, the state of the instrument cannot be read finely enough to distinguish between closely associated states. In practice, the apparently distinct state read from the instrument differs negligibly from the relevant set of entangled propensities created by the confrontation of system and apparatus. With this realization, the Schrödinger equation for $\psi$ supplies all that is needed, quantum jumps disappear, the observer's consciousness is not wanted, and quantum mechanics rises from a predictive tool to a descriptive theory. The explanation of this last mental jump is that, since classical theory is taken as descriptive, its necessary incorporation in the process of measurement gives quantum theory a similar status.[25]

During the discussion of Feyerabend's paper, the paradoxical status of Schrödinger's cat came up. By taking the cat as the instrument and the radioactive atom whose decay would activate the mechanism that would kill it as the system, Feyerabend declared that the animal would never be in the paradoxical situation in which Schrödinger had left it. In the paradox the cat is neither dead nor alive before being observed, when it would choose one state or the other, although, if dead, might find the quantum jump a challenge. In Feyerabend's scheme, the cat as a macroscopic measurer would have assumed a definite state before the observation. A doctoral student at the University of Munich, Georg Süssmann, observed that the paradox arose from forgetting that the $\psi$ inventory did not represent the state of the cat, but the investigator's knowledge of its state, and that Feyerabend's classical observer and macroscopic measurements violated the rules of quantum mechanics. The ensuing discussion ignored Feyerabend's account of measurement. Consequently, he felt obliged to repeat it at the end of the session: "my point is that...the strict account of [measurement] does not consider (a) that the measuring apparatus is a macroscopic system, and (b) that the observer who looks at it or the device (photographic camera) which is used for ascertaining its state cannot determine all its finer properties."[26] The company must have continued to ignore him as he soon gave up trying to teach quantum physics to quantum physicists.[27]

### 7.1.3 *Defense of Bohr*

During the 1960s Feyerabend came to see Popper's criticism of Bohr as a compound of ignorance and arrogance. One need not be a philosopher to regard Popper's charge that complementarity rests on a "muddle" or a "blunder" as irritating superciliousness. Feyerabend condemned it as a "piece of fiction" arising largely from Popper's mistaken insistence that quantum mechanics (QM) applied only to

[25] Ibid., 125–9.

[26] Ibid., 139–41, 146 (quote), 147.

[27] PF to Lakatos, 10 March 1968, in Motterlini, *Method* (1999), 134.

statistical ensembles and that the micro-entities making up these ensembles possessed well defined properties before measurement.[28] These errors derived from a deplorable ignorance of the developments that had prompted the invention of complementarity, such as the conundrum of the Stern-Guerlac experiment and the failure of the Bohr-Kramers-Slater approach. Nor did Popper grasp that Bohr had recognized from his first account of stationary states and quantum jumps that theories of the microworld were "formal" or "symbolic," that is, in Feyerabend's gloss, "instrumental."[29]

Popper and other philosophers found it easy to misunderstand Bohr because they confused or mingled his version of the Copenhagen Interpretation with Heisenberg's. In his paper introducing the uncertainty relations and later popular lectures, Heisenberg argued that because of the unavoidable disturbance of a micro-entity by a measuring instrument, even the cleverest physicist could not determine its momentum and position exactly and simultaneously. This formulation implicitly ascribed exact (though indeterminable) properties to the micro-entity before measurement. Bohr sometimes slipped into saying something similar, but eventually abandoned the notion that the puzzles of QM arise from disturbances during measurement. Had Popper been wiser or better informed, he might have seen that his propensity account of probability fit well with Bohr's basic postulate that the outcome of an experiment can only be calculated correctly when all the circumstances and apparatus employed in it are fully and precisely specified.[30]

These specifications must be stated in terms of the classical physics by which we describe the layout and functions of the measuring instruments. There is no way to derive this classical behavior from QM, for QM requires the concepts of classical physics for its formulation. And so, following Bohr, Feyerabend arrived at a pretty paradox. "[We] need to regard the quantum theory as an instrument for the prediction of classical results rather than as a universally valid description of the world." In this sophisticated rendition, the quantum mechanic designed statistical machinery to limit the operation of classical concepts in order to obtain results expressible only classically concerning micro-entities to which the concepts did not unconditionally apply.[31]

Since Popper perpetuated "depressing historical myths," that is, failed to value complementarity correctly because he did not know its history, Feyerabend thought it necessary to review Bohr's intellectual biography in detail. That required study of Bohr's struggle to come to a synthetic, comprehensive, and comprehensible understanding of the nature of QM and made the quasi-historical basis of Bohr's ongoing analysis clear to Feyerabend. For Bohr returned again and again to the inspiring challenge of Rutherford's atom and the subsequent development of atomic physics

---

[28] *PP1*, 248, 286–8, 290–1.

[29] *PPI*, 264–9, 279.

[30] *PP1*, 284–6, 250–2, 259n34. The logical positivist Martin Strauss, writing under Bohr's influence from Copenhagen, made similar criticisms of Popper soon after publication of Popper's *Logik der Forschung* (1935); Strauss, *Erkenntnis,* 6 (1926), 92, 109–13.

[31] *PP1*, 251–5, 259 (quote).

under the guidance of the Correspondence Principle; which, in Bohr's view and the fulness of time, realized its complete and final expression in QM. The mix of physics, epistemology, and reconstruction appealed to Feyerabend, but not, he thought, to most physicists and philosophers; as he observed with uncharacteristic mildness, "such an approach, which tempers physics with history and diffuses its success by a philosophical criticism, has occasionally annoyed the experts."[32]

For the benefit of Thomas Kuhn, whose colleague he became when both arrived in the Philosophy Department at Berkeley in the late 1950s, Feyerabend expressed his understanding of Bohr's frozen quasi-history in terms of the paradigms Kuhn was then developing for his historical analysis of scientific revolutions. "Quantum theory presents us with the empirical content of classical physics, rather than with an entirely new paradigm." Complementarity is not a new paradigm because—here Feyerabend invoked Popper—it is not refutable. "It is what remains of a former paradigm," "it is classical physics, disembodied from metaphysics." There can be no more revolutions in elementary quantum mechanics: like Aristotle's quintessence, it is exempt from change.[33]

From Copenhagen's point of view, Bohm's hidden variables were particularly odious because they represented a return to the vexatious old rhythms of paradigms, crises, and revolutions. Feyerabend pointed to Rosenfeld's evaluation of Bohm as a distinguished physicist and progressive thinker who regrettably wanted to do physics in the style of Descartes. Quantum mechanics is statistical, Rosenfeld declared; to plump for determinism is just irrational dogmatism, an atavistic adhesion to the old paradigm annihilated by Bohr and Heisenberg in the days of the founders. In fact, however, Feyerabend told Kuhn, dogmatism was rife in Copenhagen. Kuhn was then preparing to lead a group to Bohr's stables in the Carlsberg brewery to work on the history of quantum physics. Beware of the Copenhagen Geist, Feyerabend warned, "You will face a metaphysics." They think that history has stopped.[34]

## 7.2 The Historian

### *7.2.1 Against Kuhn*

Feyerabend had studied history at the university and retained a strong interest in it particularly when made amenable to philosophizing. He took an immediate, serious interest in Kuhn's work, which he criticized in detail and in writing; no doubt a safer way than talking to tell his new colleague, who claimed to be correcting philosophers by confronting their teachings about science with historical facts, that "what

[32] *PP1*, 271–2, 273–4 (quote), 279–81, 293–4.

[33] Hoyningen-Huene, *SHPS, 26* (1995), 378–9, 382.

[34] Ibid., 379–80.

you are writing is not history." What then? "It is ideology covered up as history," "a kind of doubletalk." It is neither true nor useful to claim (as Kuhn did in describing "normal science") that scientists cannot consider or hold more than one paradigm at a time; obviously they must be able to entertain at least two if they are to decide between them; Popper was right at least in this, that a scientific theory must be testable and when two or more contend, comparative testing may determine the better or best. "You are a mystic [Tom], an irrationalist." Presenting ideology as fact is a swindle, a surrender to powerful interests, undemocratic, dogmatic. "You really are like a witch doctor."[35]

Feyerabend directed his fire mainly against the dogmatism of Kuhn's paradigms and the pedestrian normal science developed under its autocratic control. Even if it were possible to proceed in this unvaried way, it would be as bad an idea as trying to maximize the production of children with only one woman. Have a harem! Alternative paradigms must be available and simultaneously comprehensible for the continuing health of science. Fortunately, alternatives are always available; it is unhistorical to think they come to be only when enough counter instances to the prevailing paradigm accumulate to convince its slavish followers that they must seek another.[36]

In any case, it is not easy to know whether a given doctrine is a new paradigm or a version of the old. Perhaps, Feyerabend went on, quantum physics in Bohr's interpretation is not new, but just classical physics without the metaphysics, that is, without the assumption of real entities independent of observation. But have your own way, let them be opposing paradigms; your notion that Bohr and Bohm therefore could not have a creative discussion, but (in your formulation) would have to "talk through one another," is nonsense. Let me give you an example. Bohr's proxy Pauli rejected hidden-variable theories; nonetheless he could convince himself that Bohm's theory was logically defensible. Pauli's objection then became redundancy: why complicate the Copenhagen way with a theory that gave results identical with its? But if Bohm could make and confirm a prediction different from what quantum mechanics required, history would revive in Copenhagen![37]

Feyerabend's attack on Kuhn main assertions did not change the structure of *The structure of scientific revolutions* but may well have accelerated Feyerabend's drift away from the program represented by his teacher Popper and his colleague Kuhn. In 1948, long before he came to espouse his reckless form of methodological anarchy, he had listened to a debate between Popper and Rosenfeld over the merits of the physics of Felix Ehrenhaft, who had stuck to his experimental finding that electrons could bear non-integral charges and other heterodoxies against the opinion of an entire profession. Feyerabend discovered what he thought was another case of mental closure in the supine acceptance of von Neumann's proof by quantum physicists

[35] Texts in ibid., 355–6, 360–1, 367–9, 375.

[36] Ibid., 369, 373, 375–8.

[37] Ibid., 381–2, 387.

and in the "dishonest" claim that complementarity provided the only possible interpretation of their science.[38]

Although as we know that Feyerabend withdrew the harsh estimate he made when influenced by Popper's negative attitude toward complementarity, he took away from his study of quantum mechanics an antipathy to the "chauvinism of science," to "ideologies that use the name of science for cultural murder," and to "miserable, unfriendly, self-righteous mechanisms," that is, scientists. Now Bohr appeared on the right side of the balance, for he had had to defend his initial atomic model against "precise and unshakeable contrary evidence." Without such courageous deviations from received wisdom, science would lack the "preconditions of progress."[39] It is easy to lose one's bearings in this morass, misguided by chauvinistic physicists and, "in the more backward regions of knowledge," by the Popperian school. Let us then turn again for help to history. "In most cases modern science is more opaque, and more deceptive, than its 16th- and 17th- century ancestors have ever been."[40] We are rewarded by finding clear cases of deception in the successes of the first of the moderns, the putative father of physics, Galileo.

### 7.2.2 *Against Galileo*

To help construct his version of Galileo as the exemplar and vindicator of anti-methodic practice, Feyerabend read Kuhn's *Copernican revolution*, first published in 1957. "A marvelous book," he informed Imre Lakatos, to which he would like to write a sequel that would emphasize change in standards of arguments and representations of the heavens. Altered standards would explain the acceptance of Galileo's freewheeling methods by people capable of reasoning, while pictures of the heavens that broke down the traditional distinction between celestial and terrestrial regions would have educated those for whom seeing is believing. Feyerabend admired Kuhn's book not only for its clear and unpretentious style (in contrast to his later pomposity) but also for its recognition that many of Galileo's claims about the Copernican system did not stand up.[41] Of Galileo's telescopic discoveries only the phases of Venus bore on the question whether the planets circled the sun and, although they did rule out a strict Ptolemaic universe, the phases left the Tychonic system unscathed. The information collected by the telescope, however suggestive, decided nothing. "It was no proof, but it was propaganda."[42]

---

[38] Gattei and Agassi, in *PP4*, xxv–xxvi, citing Feyerabend, *Science in a free society* (1978), 111, and *KT*, 65–7.

[39] *AM*, 4, 40, 154, 158, 163.

[40] *AM*, 150–1, 49, resp.

[41] PF to Lakatos, Mar 1968 and 12 Jan 1974, in Motterlini, *Method* (1999), 142, 349.

[42] Kuhn, *Copernican revolution* (1957), 224.

Kuhn's conclusion fit Feyerabend's views perfectly. "Galileo's theoretical achievements...involved a deceptive restructuring of fundamental ideas and relations." Why deceptive? Because he did not advertise a forbidding consequence of his concept of relative motion, to which his defense of Copernicus led him: repudiation of the conviction that the earth rests subverted commonsense judgments in all ordinary experience. Instead of raising an alarm, Galileo pretended that his doctrine was evident and natural, known even to philosophers, but obscured by habit or doctrine; which he demonstrated by forcing the good-natured Simplicio of the *Dialogue on the two chief world systems* to peel off the blinders he had grown through his study of Aristotle and recognize the truths about motion that he had always known. In a hard word, Galileo practiced anamnesis on defenseless Simplicio, the same technique by which Socrates had led an ignorant slave to realize that he knew without being taught that the square root of two is irrational. Feyerabend did not like the bullying imposed by this fake argument although he acknowledged that to gain a hearing Galileo had to use arguments neither logical nor rational, to resort to "propaganda."[43] The introduction of new concepts in flagrant contradiction to long accepted ones requires persuasion. Galileo was the Goebbels of science.

The episode that drew this analysis from Feyerabend was Galileo's treatment of a hackneyed problem in physics, the fall of a weight from a tower. According to everyday observation and received physics, the fall ends at the tower's base. Were the earth to spin as Copernicus required, from west to east, the weight would land west of the tower. Thus Simplicio, following an argument popularized by Tycho Brahe. Another of Galileo's creatures, his spokesman Salviati, answers in Galilean style: not only is that wrong, Simplicio, but just the opposite happens, as you already know, and we can be certain of without doing the experiment. Salviati brings Simplicio to admit that objects on a smoothly sailing ship behave exactly as they do on the ship at rest; a sailor running down the deck moves at the same rate relative to the keel irrespective of the ship's speed; only relative motion is discernible. The falling weight, released with the velocity of the tower's top, lands east of the tower because the initial easterly velocity of the weight is greater than that of the tower's foot. The argument assumes that the released weight retains the velocity impressed on it by the tower's rotation, a sort of circular inertia, which Salviati explains and forces Simplicio to "recall." The argument also assumes that the motion of objects on a moving ship is strictly analogous to that of objects on a spinning earth. Galileo did not bring this gratuitous assumption into the open, perhaps because he had no persuasive propaganda for it or perhaps because he did not recognize it. Feyerabend chose the first alternative and again accused Galileo of trickery.[44]

People willing to swallow Galileo's propaganda about falling bodies would think that it saved Copernicus from kinematic and dynamic objections based on theoretical physics and common sense. They would need more swindling to counter the optical evidence that Galileo conceded had kept him from a full acceptance of the

[43] *AM*, 65, 71–3; *KT*, 151.

[44] *AM*, 76n, 67–8, 74.

new system before the telescope arrived. To the naked eye Venus displays a constant size and brightness, whereas on the Copernican system it should show phases; and Mars, though varying in brightness, does not come close to displaying the robust range the system implied. The telescope revealed Venus's phases and resolved Mars' problem; but why believe the telescope rather than the naked eye? Galileo responded that he had checked telescopic observations of distant terrestrial objects with closeup unaided inspection a hundred thousand times and always found the instrument reliable: it did not falsify. But it was sleight of hand to transfer its terrestrial reliability to celestial observations. For this, according to Feyerabend, Galileo had no good warrant either in experiment or theory, for he could not visit Venus to confirm her phases and lacked a theory of image formation that justified accepting telescopic evidence at face value.[45]

And worse: without an adequate optical theory, Galileo could not explain why the telescope did not show all viewers the same thing, why it diminished the stars (by stripping off their haloes) and enlarged the planets, why the moon did not look the way Galileo drew it (his published image was a pastiche), and so on. More propaganda was required to compel belief that in observing the heavens Galileo's telescope gave a more reliable report than the straightforward evidence of the unimpeded eye. He practiced deception because he could not supply the new optics, dynamics, and meteorology his world view required.[46]

Despite the general unreliability of telescopic images, a few consistent sightings removed the objection to Copernican theory drawn from the brightness of the planets. Galileo turned this confirmation into a mutual support: ignore most of the optical and physical demerits of the heliocentric system and fix your attention on the astonishing fact that telescopic observation of Venus's phases and Mars's intensity removed a serious difficulty from Copernicus's cosmology. Is that not a strong indication that the alleged flaws in both are mutually resolvable? Not by standard methods of reasoning! But wily Galileo "contrived," "by ad hoc hypotheses and clever techniques of persuasion," to strengthen one inadequate theory by theories more inadequate still. "Galileo cheated and had to cheat." In his extremity he acted reasonably; fortunately, he did not know the philosophy of Popper et al. "Any attempt to enforce some of the more familiar methodologies of the 20th century would have had disastrous consequences."[47]

A persistent flaw in these methodologies, Feyerabend said, was impatience. Often, as in Galileo's case, the investigator must wait for the development of auxiliary sciences that will explain discrepancies between a new world view and the observations and deductions of the old. Meanwhile, the gaps must be filled by "irrational means such as propaganda, emotion, ad hoc hypotheses, and appeal to prejudice of all kinds." Without the "tricks, jokes, and *non-sequiturs*" that Galileo so masterfully deployed, reason would have destroyed Copernicus's cause. "Hence it

[45] *AM*, 75–85.

[46] *AM*, 87–102, 110–12.

[47] *AM*, 105; Feyerabend to Lakatos, 15 Jul 1971, in Motterlini, *Method* (1999), 257 ("had to cheat").

is advisable to let one's inclinations go against reason *in any circumstances*." Anticipating Feyerabend's dictum that "propaganda is of the essence" in theory change, Galileo manufactured experience, solidified it by fake connections with everyday experience, and brought falsified theories to support one another. "Proceeding in this way he exhibited a style, a sense of humour, an elasticity, an elegance, and an awareness of the valuable weaknesses of human thinking, which has never been equaled in the history of science. Here is an almost inexhaustible source of material for methodological speculation."[48]

This might have been a good place to stop. But Feyerabend thought to recast Galileo the clever propagandist of science into Galileo the dangerous subverter of the Church. He was a clear and present menace, according to Feyerabend, and met with a kinder reception than he deserved. Although an accomplished liar and cheat, Galileo was treated indulgently by a benevolent institution dedicated to protecting ordinary people and the wider culture against the hegemonic aspirations of crazed scientists. The Church banned Copernican theory not only because it contradicted Scripture, but also because, despite Galileo's clever propaganda, it destroyed the norms of accepted science. To fall for his defense of Copernicus amounted to admitting that "standards, facts and laws never decide a case and that an unfounded, opaque and incoherent doctrine can be presented as a fundamental truth." Nor was the argument from Scripture narrow or gratuitous, but rather expansive and moral. For the plain and convincing reason that an attack on the veracity of the Bible undermined its authority and thus the ethics it was supposed to inculcate, the church rightly moved to silence Galileo. "The judgment of the church experts was scientifically correct and had the right social intention, viz. to protect people from the machinations of specialists...from being corrupted by a narrow ideology that might work in restricted domains but was incapable of sustaining a harmonious life."[49]

We have yet to be told the full tale of Galileo's iniquities. He was not only a liar and a cheat; despite his brave challenge to the church, he must also be rated a coward. Not because he did not allow himself to be martyred, but because he needed the authority of someone else's mathematics before he allowed himself to accept the heliocentric view. That some ancient philosophers had entertained the possibility and saw the advantages of a sun-centered system was known to the moderns; if Galileo had been a man of bold conviction, he would have been able to develop this Pythagorean insight on his own. But no! He required certification by a methodical mathematician. "Galileo was a coward for whom the motion of the earth was acceptable only when it was thrown at him by...Copernicus." The faith confirmed, made irrefragable, by oppressive, unexamined mathematics. Where have we heard that before? Yes! "Copernicus was the von Neumann of the hypothesis of Philolaus." And we know that von Neumann's influence on quantum mechanics was stultifying. From which it followed, by a wild quantum jump, that methodology is inimical to physics. The ambitious investigator is better served by opportunism, irrationality,

[48] *AM*, quotes on 114,115, 116, 118, 121–2, resp.

[49] *AM*, 127 (1st quote), 129 (2nd), 131, 133 (3rd).

leaps of faith, dishonesty; in the making of physics, though not of its history, "anything goes."[50]

Feyerabend ends with the paradoxical notion that the better the propagandist and the weaker his case, the stronger or more useful his science might be. Had Galileo not resorted to tricks, jokes, non-sequiturs, false experience, and fake arguments, he could not have succeeded in obtaining a hearing for his physics. That is the price of progress. "*Rational views exist today only because reason was overruled at some time in the past.*" Feyerabend rose to an exaggerated non-sequitur that even Galileo would not have dared. "Hence it is advisable to let one's inclinations go against reason *in any circumstances*, for it makes life less constrained and science may profit from it."[51]

Feyerabend the philosopher, the righteous historian, the disciple of Popper, could not be content with this position. Galileo's propaganda may have been necessary to inculcate his physics in his time, but at a very high price. His tricks "postpone[d] for centuries the possibility of a reasonable philosophy." They hid the fact that he invented the "experiences" he called on and pretended that the unprejudiced mind needed only to attend carefully to notorious facts to perceive the truth. Let us face up to the true horror of his performance: "*Galileo invent*[*ed*] *an experience that has metaphysical ingredients*." And he did not say so! That unfortunately is the scientific way, or the way of Western science, a "power play" with only irrationality as a common structure. Yet it works. Galileo's assumptions and ad hoc hypotheses, doubtful and deceitful as they may have been, guided further fruitful research.[52] If you will the end you must will the means.

When read in the context of his criticisms of Kuhn, Feyerabend's antipathy to method amounted to antagonism to the despotism of a single point of view, to the abhorrent concept of a normal science that "could only give comfort to the most narrow minded and conceited of specialisms."[53] In fact, Feyerabend had a method: cast your mind widely, examine alternatives, and beware of the swindles and deceptions necessarily employed by enthusiasts. Feyerabend's account of Galileo as a maverick forced to employ unsavory tactics to place a needed alternative on the agenda answered Kuhn's account of science pursued under a paradigm; enthusiasts like Bohr and Galileo advance science by perpetually challenging the status quo. Debate between innovators and conservatives must therefore be possible. Kuhn's conclusion that fruitful discussion between inmates of opposing paradigms cannot occur may be logically appealing (how can persons who differ over whether the earth is a planet have a satisfactory discussion about planets?), it is violated every day in practice. The essential point is that there should always be alternatives. Kuhn's concept of a normal science pursued under well-worked recipes until enough

---

[50] Feyerabend to Lakatos, 15 Jul 1971 (1st quote), 27 Dec 1972 (2nd), and n.d. 1973, in Motterlini (1999), *Method*, 257, 307, 333; *AM*, 49n25. Cf. Popper, introduction to *Logic* (1959).

[51] *AM*, 68–73, 115, 116 (quotes, Feyerabend's italics).

[52] AM, 1–3, 65 (1st quote), 76 (2nd), 77.

[53] PF, in Lakatos and Musgrave, *Criticism* (1970), 197 (quote), 205.

instances of their failures have accumulated to cause despair and revolution does not fit modern science. He based it on his study of the Copernican revolution. Feyerabend had quantum mechanics in mind. Physicists developed Bohr's hodge-podge atom into quantum mechanics in 15 years; astronomers required 1500 to interchange the positions of the earth and the sun.

Feyerabend's method in *Against method* kept him from admitting that his message was pluralism and that Galileo practiced it. Galileo's *Dialogue* does abound in exaggerations, jokes, and tricks, and aims to persuade by whatever rhetorical means, including appeal to reason, that its author thought might succeed. But Galileo also wrote within the idiom and expectations of the science of the day; his *Discourses on two new sciences* offer propositions in Latin deducible rigorously from axioms and free discussions of them in Italian. The dialogue form allowed digressions and appeal to persuasive techniques other than geometry. Feyerabend's analysis of Galileo's method is propaganda; Galileo used many methods and deceit was not the main one.

The neglect of methods by the author of *Against method* appears again in his caricature of Galileo's relations with the Church. The committee of theologians who found the Copernican theory contrary to Scripture and absurd in philosophy in 1616 were puffed up with the spirit of the Council of Trent. Against their method stood another, discretion, which prudent people (including the future pope Urban VIII) preferred in view of the notorious difficulty of interpreting holy writ, the possibility that Galileo might be proved right, and the suspicion that the human mind could not penetrate to the truth of world systems. Galileo was certainly imprudent in his challenge to the Church; but so were the Holy Office and Pope Urban in the severity of their response. That they were unwise was recognized in their time and the Church backed away from its precarious salient to allow the teaching of heliocentric astronomy when properly purified. All that was required in the end was whitewashing the "theory" by calling it a "hypothesis."

Another dose of whitewash was applied to the case in 1992, when Pope John Paul II, on the advice of the surviving remnant of a committee of septuagenarians he had set up in 1979, declared that the contretemps of 1616, when the Inquisition condemned heliocentrism, arose from a profound difference of method. The difference was not what common sense would have supposed. According to the pope's advisors, Galileo had misunderstood the true method of science in thinking that he had proved Copernicus right, but he argued correctly that the bible was not a text on natural science, whereas the theologians understood what science required but had botched their hermeneutics. It occurred to some clear minds that Feyerabend's appraisal of Galileo and the theologians who evaluated him could be helpful in placing the disputants of 1616 on the same moral level.

Among those who applied Fayerabend's teachings in this way was Cardinal Joseph Ratzinger. He observed in a lecture at the Sapienza in Rome in 1990 that the new interpretation of Galileo helped to establish the church's "methodological profile" as legitimate, within certain bounds, in its criticism of science and its practitioners. He quoted Feyerabend's conclusion that the condemnation of Galileo had been "rational and just;" "the church at the time of Galileo was much more faithful to reason than Galileo himself, and also took into consideration the ethical and social consequences of Galileo's doctrine." Ratzinger glossed this teaching as "più drastico," that is, stronger and more effective, than the no-fault reading endorsed by John Paul. That proved a hostage to fortune. Years later, when Ratzinger was scheduled to address the students at the Sapienza again, this time as Pope, they declined to hear him. Central to their opposition were his views about Galileo. He cancelled the engagement.[54]

It was a humiliation for the Pope and, less painfully because posthumously, for Feyerabend. The students who would not listen to his views acted in a manner opposite to the practice he had intended to teach with his Galilean example. He had hoped to demonstrate that progress comes from breaking rules and the clashes arising from efforts to persuade others to accept new ones. For Galileo's tricks to work, his audiences had to be willing to consider views that made them uncomfortable. Some did. But the unwise students and professors of the Sapienza, unable to tolerate Feyerabend's account of Galileo as rendered by Ratzinger, closed their minds and their school to an important scholar who happened to be their Pope.

In so far as *Against Method* has a hero, it is the very methodical Cardinal Saint Robert Bellarmine. Feyerabend seems to have made a game of it. "Today in my lecture I defended the Church's stand for law and order against the anarchist Galileo and convinced everyone that Galileo was a bastard while Bellarmino was wise, scientific, humanitarian, etc., etc. Some people looked rather upset."[55] Feyerabend drew his favorable caricature of Bellarmine from a letter the saintly cardinal inquisitor sent to a theologian who inclined toward the Copernican system. You may entertain it, Bellarmine wrote, if you call it a hypothesis with no claim to truth; the Holy Office has found it "absurd" in philosophy for disregarding established physics and "formally heretical" for contradicting straightforward readings of scripture. If, however, apodictic proof is ever found, which Bellarmine reckoned infinitely unlikely, then, he conceded, the Church would have to reconsider its position.[56]

It should appear paradoxical that Feyerabend sided with the straightlaced administrator Bellarmine rather than with his fellow anarchist Galileo and accept that the tetragrammaton HYPO could convert a dangerous thesis into an innocent hypothesis.[57] Disputants in scholastic exercises used this talisman to enable them to wrangle over false doctrine with little risk; but the elite magic of HYPO could not protect the unsophisticated multitude whose salvation was supposedly the subject of Bellarmine's solicitous concern. Further to the paradox, Feyerabend had a personal reason to resent administrators: the powers that ran Berkeley forced him to comply with the teaching and grading norms they policed.[58]

---

[54] Allen, *National Catholic reporter*, 14 Jan 2008. See also Stefano Gattei's contribution to this volume.

[55] PF to Lakatos, n.d. 1973, in Motterlini, *Method* (1999), 333.

[56] PF, *Farewell* (1987), 250, 255.

[57] PF, *Farewell* (1987), 249, 251, 257–8.

[58] Motterlini, *Method* (1999), passim; *KT*, 121–2.

The resolution may be that Feyerabend disliked arrogant experts, "constipated" (a favorite pejorative) with their specialty and determined to foist their views on others, more than he disliked blinkered administrators honestly trying to preserve the institutions entrusted to them. It suited his argument to make Galileo into an "elitist authoritarian" (to use Lakatos's phrase), a combination of Feyerabend's caricatures of an arrogant constipated Popper and a paradigmatic Kuhnian slave.[59] Feyerabend's Galileo was a product of his efforts to come to terms with the philosophies of Popper and Kuhn. They raised serious issues about the functioning of science in a free society. He fought against trusting tyrannical scientists to find adequate answers.

## Works Cited

### *The following abbreviations of Feyerabend's books are used: AM, Against method; KT, Killing time; PP, Philosophical Papers*

Agassi, Joseph. 1991a. As you like it. In Munévar, *Beyond reason*, 379–387.

———. 1991b. Foreword. In Sachs, *Einstein versus Bohr. The continuing controversies in physics.* [1988], xvii–xxi. La Salle, IL: Open Court.

Allen, John L., Jr. 2008. Ratzinger's 1990 remarks on Galileo. *National Catholic reporter*, 14 Jan 2008.

Bohm, David. 1952. A suggested interpretation of the quantum theory in terms of 'hidden variables. *Physical Review* 85 (2): 166–179. 180–93.

Diecks, Dennis. 2017. Von Neumann's impossibility proof. Mathematics in the service of rhetorics. *Studies in the History and Philosophy of Modern Physics* 60: 136–148.

Feyerabend, Paul K. 1957. On the quantum theory of measurement. In Koerner, *Observation*, 121–30, = *PP1*, 207–18.

———. 1958. [Review of John Von Neumann, *Mathematical foundations of quantum mechanics*]. *British Journal for the Philosophy of Science* 8 (32): 343–7.

———. 1968. On a recent critique of complementarity. *Philosophy of Science* 35 (4): 309–31, and 36:1 (1969), 82–105, = "Niels's Bohr's world view," *PP1*, 246–97.

———. 1977. Consolations for the Specialist. In *Criticism and the Growth of Knowledge* [1970], ed. Imre Lakatos and Alan Musgrave. Cambridge: Cambridge University Press, 1977².

———. 1978. *Science in a free society*. London: New Left.

———. 1981–2016. *Philosophical papers*, 4 vols. New York: Cambridge University Press. Abbreviated *PP1-4*.

———. 1987. *Farewell to reason* [1985]. London: Verso, 1987².

———. 1991. Concluding unphilosophical conversation. In Munévar, *Beyond reason*, 487–527.

———. 1995. *Killing time*. Chicago: University of Chicago Press, Abbreviated *KT*.

———. *Against method* [1975]. London: Verso, 1997³. Abbreviated *AM*.

Gattei, Stefano, and Joseph Agassi. Introduction. In Feyerabend, *PP4*, x–xxix.

Hoyningen-Huene, Paul. 1995. Two letters of Paul Feyerabend to Thomas S. Kuhn on a draft of *The structure of scientific revolutions*. *Studies in the History and Philosophy of Science* 26: 353–387.

[59] Ibid., 27, 96–7.

Jammer, Max. 1974. *The Philosophy of Quantum Mechanics: The Interpretations of Quantum Mechanics in Historical Perspective*. New York: Wiley & Sons.

Koerner, Stephen, ed. 1957. *Observation and Interpretation in the Philosophy of Physics with Special Reference to Quantum Mechanics*. Proceedings of the Ninth Symposium of the Colston Research Society. London: Butterworth; New York: Dover, 1962.

Kuhn, Thomas S. 1957. *The Copernican Revolution*. Cambridge: Harvard University Press.

———. 1962. *The Structure of Scientific Revolutions*. Chicago: University of Chicago Press.

Motterlini, Matteo, ed. 1999. *For and Against Method. Imre Lakatos, Paul Feyerabend*. Chicago: University of Chicago Press.

Munévar, Gonzalo, ed. 1991. *Beyond Reason: Essays on the Philosophy of Paul Feyerabend*. Dordrecht: Kluwer.

Peat, F. David. 1996. *Infinite potential. The life and times of David Bohm*. New York: Addison Wesley.

Pinch, Trevor. 1977. What does a proof do if it does not prove? A study of the social conditions and metaphysical divisions leading to David Bohm and John von Neumann failing to communicate in quantum physics. In *The Social Production of Scientific Knowledge*, ed. Everett Mendelsohn et al., 171–215. Dordrecht: Reidel.

Popper, Karl. 1935. *Logik der Forschung*. Vienna: Springer.

———. 1957. The propensity interpretation of the calculus of probability, and the quantum theory. In Koerner. *Observation* 65–70: 88–89.

———. 1961. *The Logic of Scientific Discovery*. New York: Basic Books.

———. 1982. Quantum theory and the schism in physics. In *Postscript to the Logic of Scientific Discovery*, vol 3, ed. W.W. Bartley III. London: Unwin Hyman. Cited as *Postscript, 3*.

———. 2002. Unended quest. In *An intellectual autobiography [1974]*. Abingdon: Routledge.

Santo, Flavio del. 2019. Karl Popper's forgotten role in the quantum debates at the edge between philosophy and physics in the 1950s and 1960s. *Studies in the History and Philosophy of Modern Physics* 67: 78–83.

Seevinck, Michiel. 2017. Challenging the gospel: Grete Hermann on von Neumann's no-hidden-variable proof. In *Grete Hermann Between Physics and Philosophy*, ed. Elise Crull and Guido Bacciagaluppi, 107–117. Dordrecht: Springer.

Strauss, Martin. 1926. Ungenauigkeit, Wahrscheinlichkeit und Unbestimmtheit: zu K. Popper's 'Bemerkungen zur Quantenmechanik'. *Erkenntnis* 6: 90–113.

# Part III
# Feyerabend and His Contemporaries

# Chapter 8
# Feyerabend and Brecht

**Val Dusek**

**Abstract** Paul Feyerabend met Bertolt Brecht and frequently refers to Brecht in writing. Feyerabend's characterizing to his own writings as "collage" parallels Brecht's description of his own work as "montage," both forms of art and film respectively. Feyerabend's called himself a Dadaist referring to a post-WWI movement to which Brecht was originally sympathetic. Both Feyerabend and Brecht believed there was an ultimate unity of science and art. The two men also each had sympathy for Chinese culture, both ancient and Maoist. Feyerabend's criticisms of the scientific methodology advocated by Imre Lakatos parallel Brecht's criticisms of Georg Lukács in the "expressionism debate" in aesthetics.

**Keywords** Paul Feyerabend · Bertolt Brecht · Georg Lukács · Dadaism · Art

## 8.1 Feyerabend's Encounter with Brecht

Although the story of Feyerabend's encounter with Brecht is often mentioned, the Brechtian impact on Feyerabend's philosophy is rarely examined in detail. For instance, Eric Oberheim's very detailed and thorough account of influences on Feyerabend by, among others, Felix Ehrenhaft, Ludwig Wittgenstein, and Pierre Duhem, neglects the major influence of Brecht, mentioning him on only one page about the meeting with Feyerabend.[1]

Feyerabend was an excellent singer and took various voice lessons as a youth and voice training as a young adult, although one of his own auditions was a disaster because the melody in the music sheet was not at all what he expected.

[1] Cf. Oberheim (2006), p. 19.

V. Dusek (✉)
Department of Philosophy, University of New Hampshire, Durham, NH, USA
e-mail: val.dusek@usnh.edu

© The Author(s), under exclusive license to Springer Nature Switzerland AG 2024

S. Gattei, R. Corvi (eds.), *Feyerabend in Dialogue*, Boston Studies in the Philosophy and History of Science 346,
https://doi.org/10.1007/978-3-031-71938-7_8

In his autobiography he often comments on opera performances and plays that he attended.[2] One can hear his voice (mostly speaking but singing a few lines) narrating Verdi's opera *Falstaff* and other musical material in the recordings he sent to his wife Grazia Borrini-Feyerabend when they were apart in *Stories from Paolino's Tapes*.[3] He occasionally uses singing as a metaphor, as when he describes as Hegelians "singing arias",[4] and in the preliminary article version of *Against Method* uses singing to illustrate the bodily and emotional roles in learning.[5] He sometimes uses arias in a negative sense, as in *Science and a Free Society* concerning novel theories he speaks of rejection of novel theories with "arias called 'reasons'", and in *Farewell to Reason* he says, "We can of course sing moral arias. We may even write a moral opera where these arias hang together beautifully [...]".[6]

Through Feyerabend's Marxist friend and mentor Walter Hollitscher, Feyerabend got to know Berthold Viertel, a major theater director, who arranged Feyerabend's contact with Brecht. "We met Brecht at rehearsal [...]. It was a strange affair. [...] Brecht was complaining about the color of a barely visible pot [...]".[7]

Feyerabend refused Brecht's offer as he later refused Popper's offer to be an assistant. "I once thought (and said so in print) that this [refusing Brecht] was the biggest mistake of my life. Today I am not so sure [...]. But I would have detested the collective pressure of the partly fearful, partly dedicated, and certainly pushy and closely knit group that surrounded Brecht".[8] In a letter to Lakatos in the 1970s, Feyerabend wrote: "Why did I not join Brecht and become a producer? (Any fatherly comments on that?)".[9] Feyerabend at Berkeley frequently covered Brecht in his lecture: "[... G]iving my last lecture today on Brecht"; "PS. In my introductory lecture into philosophy I talk about drama, science, witchcraft, Aristotle, Brecht"; "All the while I am talking about Brecht, Homer, Festinger, Evans-Pritchard, and one of the main textbooks is the *Malleus Maleficarum* [a famous treatise on witches]".[10]

---

[2] Cf. Feyerabend (1995), pp. 82–85, 94, 102, 120–121, 123 and 132–133.

[3] Feyerabend (2001).

[4] Feyerabend (1991), p. 145.

[5] Feyerabend (1970a), p. 19.

[6] Feyerabend (1987a), p. 311.

[7] Feyerabend (1995), p. 73.

[8] Ibid., p. 7.

[9] Lakatos and Feyerabend (1999), p. 233.

[10] Ibid., pp. 246, 302 and 312, respectively; cf. pp. 212, 233, 245, 311, 384 and 407 as well.

## 8.2 Feyerabend's Collage, Dada, and Brecht's Montage

Feyerabend calls his later work a collage. Brecht calls his plays montages. The two artistic concepts usually apply to different media but have similar content and were developed around the same period. Both montage in film and collage in paintings involve juxtaposing disparate elements: temporally in films and spatially in visual art.

Feyerabend calls his *Against Method* a collage: "*AM* is not a book, it is a collage"; "[…] I started composing my collage"; "Science itself […] is a collage, not a system"; "I finished the collage in about a year"; "Having forgotten the details of my collage […]".[11]

Richard Rorty thinks that because Feyerabend calls his later work a collage that he can't be doing real philosophy.[12] It is ironic that Rorty, who flirts with French postmodern literature and relates philosophy to literary criticism and the arts, seems ignorant of the role of collage in art history. Similarly, many who find Feyerabend's description of himself as an unserious Dadaist miss or ignore the role of Dadaism in art history and its relation to Brecht via expressionism.

On Dada Feyerabend writes, "'Dada', says Hans Richter in *Dada: Art and Anti-Art,* 'not only had no programme, it was against all programs'. This does not exclude the skillful defense of programs to show the chimerical character of any defense, however 'rational'. (In the same way an actor or a playwright could produce all the outer manifestations of 'deep love' in order to debunk the idea of 'deep love' itself. Example: Pirandello)".[13] And again: "[…] I now prefer to use the term *Dadaism.* […] A Dadaist is unimpressed by any serious enterprise and he smells a rat whenever people stop smiling and assume that attitude and those facial expressions which indicate that something important is about to be said […]. A Dadaist is prepared to initiate joyful experiments even in those domains where change and experimentation seem to be out of the question (example: the basic functions of language). I hope that having read the pamphlet the reader will remember me as a flippant Dadaist and not as a serious anarchist".[14]

Dada arose as a reaction to the horrors of WWI. It was a general rejection of all traditional rules of art and morality. The movement seems to have originated in Zürich, Switzerland, and spread many countries including France and the United States. It included poems made of random words cut from a newspaper to make the reader newly aware of language, various performance art projects, and "found works" including Duchamp's famous urinal. Dada rejected expressionism, which emphasized subjective emotion.[15]

[11] Feyerabend (1995), pp. 139, 142 143, 144 and 145, respectively.

[12] Cf. Rorty (1995) p. 34.

[13] Feyerabend (1975a), p. 33, and (1993), p. 16 n3.

[14] Lakatos and Feyerabend (1999), p. 295.

[15] Richter (1965).

Collage, which Feyerabend describes his late work is originally an art form that attaches various cut-out objects to the surface of the painting or artwork. Braque and Picasso, the originators of cubism, produced collages under the term *papier collé*. Because three-dimensional objects can be attached to a painting, collage can fuse aspects of sculpture with painting. Later, Louise Nevelson would concentrate on making three dimensional wooden collages, large like statues. Though Brecht used the word montage for his plays, Jaroslav Hasek's *The Good Soldier Švejk*, telling the adventures of Švejk, which had a major impact on Brecht, was considered a model of the collage novel.[16]

A connection between collage and Dada as well as between montage and collage is the photomontage work of Hannah Höch. She did works such as "Cut with the kitchen knife DADA through the last Weimar beer belly culture epoch in Germany". She was highly critical of the sexism of the male Dadaists.[17]

Brecht frequently calls his dramas montages. Brecht says *Fear and Misery in the Third Reich* is a montage of 27 scenes, but Lukács, with awareness of only a fragment of it, treated the play as less modernist than Brecht's earlier works, and gave faint praise. Brecht wrote Lukács treated him "[…] as if I am a sinner who has returned to the bosom of the Salvation Army",[18] and totally missed the montage structure. Brecht said that for Georg Lukács "montage is viewed as a 'characteristic feature of decadence'".[19]

Brecht also treated his law case *The Threepenny Lawsuit*, where he sued for breach of a movie contract, as a montage. Brecht found the place to experiment with law precisely through a technique that one might call 'cinematic', namely, the practice of 'montage'.[20] In his lawsuit appeal, Brecht included a kind of tree diagram he made. This was meant to undermine what he claimed was the 'montage' of copyright law.[21]

Brecht also used the montage notion to contemporary social units. "But these new institutions which undoubtedly shape the individuals today are precisely-compared to the family—the products of montage, quite literally 'assembled'".[22] Brecht—like Feyerabend, associating collage with anarchism—associates anarchism with montage. He calls one work "a certain kind of *anarchistic montage* […]" (Brecht's italics),[23] and calls his own drama *Schweyk* a "straight Montage from the

[16] Cf. Willett (1996), p. 109.

[17] Makela (1996) and Sante (1997).

[18] Adorno et al. (1980), p. 58 n18.

[19] Brecht (1993), p. 21.

[20] Cf. Jürgens (1982).

[21] Brecht (2000), p. 178.

[22] Brecht (1993), p. 13.

[23] Ibid.

novel".[24] "What is needed most is documents, simple montage".[25] Brecht said that for Georg Lukács, montage is viewed as a "characteristic feature of decadence".[26]

A major difference between Brecht and the major early proponent and theorist of montage, Sergei Eisenstein, is that Eisenstein used various forms of montage to heighten the viewer's emotions, very much the opposite of Brecht's use of montage to make the audience critical, and of Brecht's criticisms of Aristotle and Stanislavsky and method acting to emotionally absorb the audience and accentuate emotions.[27]

Brecht's partner in dialogue Walter Benjamin and Brecht himself, as well as Feyerabend, emphasized the revival of earlier styles, plays, and works of science as serving progressive politics. In writings inspired by interaction with Brecht, Benjamin related his own defense of modern techniques such as montage to Benjamin's own earlier work on the then largely neglected German tragic drama,[28] centrally involving allegory, such as emphasized by Benjamin in German tragic drama,[29] was repurposed for contemporary drama. The literary critic Eric Auerbach's discussion of the Biblical technique of prefiguration (such the suffering servant in the book of Isaiah prefiguring Jesus) was revived in Brecht.[30] Feyerabend, in his response to Kuhn and in particular Lakatos, in "Consolations for the Specialist" and elsewhere, defended the view that degenerating research programs could "make comeback".[31] Also, Feyerabend wrote such works as "In Defense of Classical Physics" (1970). Feyerabend writes "There is no idea, however ancient and absurd, that is not capable of improving our knowledge."[32] The most extreme examples of this phenomenon were Greek atomism, which fell into unpopularity with Aristotle and Medieval philosophy, but revived in the early modern period, but fell into disfavor again in the late nineteenth century. Energeticism was revived and flourished again in the twentieth century.[33]

---

[24] Ibid.

[25] Ibid., p. 289.

[26] Ibid., p. 21.

[27] Eisenstein (1959).

[28] Benjamin (1977).

[29] Benjamin (2011).

[30] Cf. Auerbach (1984) and Suvin (1976).

[31] Cf. Lakatos and Musgrave, eds. (1970), pp. 197–230.

[32] Feyerabend (1993), p. 33.

[33] Feyerabend (1987a), pp. 33 and 263.

## 8.3 Feyerabend and Brecht on Galileo as Person

Feyerabend's controversial account of Galileo in *Against Method* obviously does not have to base its account of Galileo's science on Brecht's play, but Feyerabend's portrait of Galileo's personality, attitude to his research, and initial interest in making a science for the people does agree with Brecht's portrayal.

Many mainstream philosophers of science have been scandalized by Feyerabend's account of Galileo as using rhetorical tricks to justify his position. Larry Laudan, for instance, thinks he is discrediting Feyerabend by saying that Feyerabend bases his account of Galileo more on Brecht's play than on Galileo's texts. (Laudan, 1989, 303; 1996, 102). Laudan was probably counterattacking in reaction to Feyerabend's spot-on and devastating review "More Clothes from the Emperor's Bargain Basement", showing Laudan's lack of originality combined with claims to originality in producing among other things a form of warmed over Lakatos.[34] One might similarly and as misleadingly claim Kuhn's Galileo is more based on Alexandre Koyré than on Galileo. Brecht could have had a Marxist field day with Koyré's beginning his writing on the mysticism, and spiritualism of his *Mystiques, Spirituels, Alchimistes du XVI$^e$ Siècle Allemand*[35] as well as theosophy of his *La philosophie de Jacob Boehme*.[36] Despite making his intended insult both in a stronger and then in a toned-down version, Laudan does not bother to describe incidents, correct errors, or show passages where Feyerabend was allegedly misled by Brechts play. In fact, Feyerabend in his account of Galileo in *Against Method* has over forty citations of Galileo from ten of his works and over forty more from over two dozen contemporary historians of science, compared to only one citation of Brecht.

Brecht did read various translations of Galileo's *Dialogue*[37] as well as his *Dialogues Concerning Two New Sciences*[38] and his *Treatise on Floating Bodies*,[39] which appears in the demonstration in Scene 8 the play, the preparation of which was aided by his Los Angeles neighbor the logical empiricist philosopher of science Hans Reichenbach).[40] Brecht also read the text of Galileo's recantation and the exchange of letters with Kepler, both used in the play.

Brecht's secondary sources were hardly state of the contemporary art but they were serious works from the late nineteenth and first third of the twentieth century. Brecht read not only the *Dialogues* but also the text of Galileo's recantation and the exchange of letters with Kepler, both of which were used in the play. Brecht read several historical works and biographies of Galileo: Karl von Gebler's *Galileo*

---

[34] Feyerabend, 1981.

[35] Koyré, 1955.

[36] Koyré, 1979.

[37] Galileo, 1953.

[38] Galileo (1954).

[39] Galileo (2012).

[40] Brecht 1993, p. 339.

*Galilei und die Roman Curie*[41] and Wohlwill's *Galileo und sein Kampf für die Copernikanische Lehre*.[42] Brecht also owned a copy of Leonardo Olschki's *Galileo und Seine Zeit* (Schumacher, 40–42, and 390–394). Olschki wrote in detail inter alia about Galileo's pathbreaking literary style and its influence on later Italian literature, his lectures on Dante, and his appreciation and criticism of Ariosto's poetic form, topics that may have been of interest to Brecht as a poet.[43]

In contrast to Laudan, Peter Godfrey-Smith in his introductory philosophy of science text section on Feyerabend takes quite seriously his use of Galileo as a challenge to empiricism, despite Godfrey-Smith's own objection to Feyerabend's lack of a method for sorting the many alternative theories proliferated, and Godfrey-Smith titling one section on Feyerabend "The Ravings of a Madman." Nevertheless, Godfrey-Smith titles the previous section on Feyerabend's account of Galileo "An Argument from History that Haunts Philosophy"[44] and concludes "This is the Feyerabendian argument that haunts philosophy of science."[45]

Einstein replied favorably to Brecht's sending him the script of *Galileo*.[46] Scenes in *Life of Galileo* show science as pleasure as well as Galileo as sensuous eater and drinker which fits perfectly with Feyerabend's tying science (and criticism of science) with pleasure. Feyerabend, like Brecht, sees Galileo as something other than an academic. Brecht in the opening of the play introduces a naked Galileo taking a bath rather than as a dignified lecturer. Brecht emphasizes Galileo's sensuality and portrays his search for knowledge as a kind of erotic activity. Galileo's love of good food and wine is presented throughout the play. Feyerabend similarly even includes making better beer as part of making the good society.[47]

Feyerabend also emphasizes how Galileo differs from the contemporary "squares" and the constipated "searcher for 'truth.'... Various writers, whose lack of imagination and temperament is properly matched by their high moral standards, have been put off by the numerous signs of worldliness on the part of Galileo, and they have tried their best to explain his actions as the result of high (and dry) motives... [T]here is no reason why a man, and an extremely intelligent man at that, should have to conform to the standards of the academic squares of today... A Puritanical view like this surely is too naïve a background for understanding... Moreover, Galileo the mountebank is a much more interesting character than the constipated 'searcher for the truth' we are usually invited to revere."[48]

Similarly, in the notes to the play, Brecht wrote "...you shouldn't idolize Galileo...the stargazer, the pallid intellectualized idealist...My Galileo is a

---

41 Gebler (1879).

42 Wohlwill (1909).

43 Olschki (1967) in McMullin, pp. 140–159.

44 Godfrey-Smith, p. 113.

45 Godfrey-Smith 2003, p. 114.

46 Wizisla 2005, pp. 350–353.

47 Feyerabend, 1968, 130.

48 Feyerabend 1975a, b, pp. 105–6 n22.

powerful physicist with a tummy on him, a face like Socrates, a vociferous full-blooded man with a sense of humor, the new type of physicist, earthy, a great teacher."[49] "Galileo of course is not a Falstaff. He insists on sensual pleasures….He *works* in a sensual way.…A great part of his sensuality is of an intellectual kind…a little theatrical performance….he test good words and tests them like a spice."[50] Except for the tummy, this sounds like Feyerabend himself.

In Brecht's comments on Charles Laughton's acting Galileo, Brecht also several times notes the sensuous side of Laughton's portrayal of Galileo. "The first thing L [Laughton] did was to rid the figure…of the pallid, spiritual, Stargazing aura of the textbooks. Above all the scholar must be made into a man."[51] "…[I]t was just this mixture of the physical and the intellectual that attracted L… his taste for wine… His sensual walking, the play of his hands in his pockets while he is planning new research came close to being offensive."[52] Galileo "…blamed his indominable urge to research like a sex offender blaming his glands."[53] This sensual side of personality and its tie to science fits well with Feyerabend's ideal of the scientist and of science.

Jameson[54] correctly notes Brecht's connection of science with practical activity and with pleasure. Scenes in *Life of Galileo*, show science as pleasure as well as Galileo as sensuous eater and drinker, which fits perfectly with Feyerabend's tying science (and criticism of science) with pleasure. The literary critic Jameson ignorantly criticizes Brecht for making science "popular mechanics and chemistry sets" apparently thinking that an experimental practice conception of science is only fit for youth and amateurs.[55] Jameson also claims Brecht's experimental practice notion of science is appropriate only to a "primitive" stage of capitalism, not our supposedly postindustrial, postmodern capitalism.[56] Jameson was unaware of the turn to practice in science studies or the notion of technoscience.

Despite his Marxism, Brecht did not ridicule the Roman Catholic Church and warned directors and actors not to portray the churchmen in a highly negative manner. Here there is some similarity to Feyerabend's defense of Cardinal Bellarmine who rejected literal understanding of Copernicus' theory. The later, anti-Popper, Feyerabend calls Popper's presentation of Bellarmine's instrumentalism in "Three Views of Human Knowledge" "bombastic."[57] Galileo, by publishing his *Dialogue* in Italian rather than Latin, wished to make his work popularly known. Brecht shared with Feyerabend the wish that there be a science for the people. This is a phrase

---

[49] Brecht 1994, p. 119.

[50] Brecht 1994, p. 120.

[51] Brecht 1994, pp. 138–139.

[52] Brecht 1994, p. 141.

[53] Brecht 1994, p. 147.

[54] Jameson 1998, pp. 204–205.

[55] Adorno et al, 156., Jameson, 2.

[56] Adorno et al, p. 165.

[57] Feyerabend 1999, pp. 155n55.

Brecht uses in his *Galileo* and thinks was originally fostered by Galileo but then betrayed and stifled by his recantation of belief in Copernicanism).

Feyerabend from the beginning was involved in science for the people. At age eleven, he presented at public outdoor astronomy lectures organized by an educator.[58] Brecht wanted his theater to be one in which the audience participated. He also wished to have working class audience and attempted to adapt some plays for this purpose. Despite these attempts, in East Germany it was mostly upper middle-class people who attended. Brecht described as one of those goods produced mostly for export and available only in small amounts domestically in the GDR.[59]

## 8.4 Drama and Science: Brecht and Feyerabend

There is an apparent contradiction or ambiguity of Feyerabend's account of the relation of art and science. There is "the possible identity" between the two. In two articles of the 1960s on the sciences and the arts, Feyerabend makes two apparently contradictory assertions. First, that the gap between science and arts should not close—should even grow greater, and second, that the desirable principle of proliferation and holding multiple views or theories unifies the sciences and the arts. I think he supports the gap in the present society of extreme specialization as a corrective to the religion of scientism and overestimation of science, but sees the major desirable procedures of the two to be identical.

On the one hand, "The Sciences and the arts are different manifestations, or different subdisciplines of one comprehensive enterprise. Strictly speaking, there is no differences between the two of them. Admittedly, there are some obvious differences between the two as they exist here and now. It is, however, possible and beneficial to eliminate these differences".[60] On the other hand, "[...] I for one would prefer if the gulf between the sciences and arts would never close and why I would hope that it would even increase in the future",[61] and in a later article, "I do not think separation of the two is a bad thing".[62] He despises contemporary attempts to unify art and science by "creativity", which he thinks is a myth,[63] or "aesthetic aspects", metaphor, etc. "Where is the [scientific] team that gets an award for its aesthetic achievements?".[64] But perhaps Feyerabend thinks the opposition in our present situation of professional, specialized science is a good thing, but that there might be unity in a future good society. He does remark occasionally how science should be

58 Feyerabend 1995, pp. 28–29.

59 Bradley 2011, p. 4.

60 Feyerabend (1968), p. 127.

61 Feyerabend (1967), p. 412.

62 Feyerabend, 2011, p. 7.

63 Cf. Feyerabend (1987b).

64 Feyerabend (2011), p. 7.

taught using drama and movies as well as texts and lectures. However, he advocates disunity and pluralism, include plurality of ways of life and favorite activities. He also suggests that scientists can learn from poets' experimentation.

In the later *Farewell to Reason*, Feyerabend writes that "The scientific enterprise seems to be much closer to the arts than most logicians and philosophers of science (myself among them) once thought [...]",[65] and emphasizes in italics "*good sciences are arts, or humanities, not sciences in the textbook sense*".[66] "In his later *Three Dialogues on Knowledge*, he emphasizes the unity of science and art without the statement of support for separation. "What I do deny is that there is *one big line* with all the sciences on one side and all the arts on the other".[67] Doing elementary particle physics "[...] is very much like writing a poem. Poems are not without constraints [...]. Twistor theory or superstring theory uses mathematical formulae. That is the only difference".[68]

He sees Brecht as an advocate of proliferation in art as he is in science. "Brecht has been the first writer to give a concise *theoretical* account of this art of multiple representation (which he has extended even to physics: cf. his remarks on Galileo and the chandelier in section 44 of the *Organon*)".[69] "In the arts, the idea of alternatives has been introduced with excellent arguments by Bertolt Brecht".[70] "For Brecht's idea of a critical theater making use of alternative interpretations of the action on stage, see *Schriften zum Theater* [...]".[71] "The principles that Brecht develops for the theater [...] coincide with the principles we have already outlined for different subjects. We have pointed out that partisan ideas can be judged only with the help of contrary hypotheses which provide a new viewpoint [...]".[72]

In one of his earliest and longest statements about the relation of drama and science, Feyerabend claims that Brecht's alienation effect can be used to advantage in science education. "[...The alienation effect] is exactly the same thing that is needed in the domain of *epistemology* [...]. I am afraid that only very few scientists have ever been aware of the need for such a method, and that most of them are to be compared to the transfixed audience of one of the familiar pieces of the 'classical' repertoire which has been described so vividly by Brecht—textbooks and scientific journals replacing the images projected by the actors".[73]

Feyerabend writes, "Closed societies disappeared long ago, but their methods of support are still among us, in the guise of aesthetic theories, theories of acting, philosophies of science, rules of faith [...]. Nobody has described these phenomena

---

[65] Feyerabend (1987a), p. 293.

[66] Ibid., p. 295.

[67] Feyerabend (1995), p. 144.

[68] Feyerabend (1991), p. 145.

[69] Feyerabend (1967), p. 411.

[70] Ibid., p. 260 n180.

[71] Feyerabend (1965), p. 218.

[72] Feyerabend (1967), p. 410.

[73] Feyerabend (1962), pp. 71–72.

more clearly than Bertolt Brecht: 'The theater with which we are confronted shows the structure of society (represented on the stage) as being removed from the influence of society (which is situated in the audience). [...] Human sacrifice is wherever one looks. Barbaric entertainment! We know that barbarians produce some kind of art. We must make a different one!".[74]

Brecht on poetry emphasizes experiment and revision. Feyerabend states, "Every poet who is not completely irrational compares, improves, argues until he finds the correct formulation of what he wants to say. Would it not be marvelous if this process played a role in the sciences also?". He further suggests that poets could aid the language of philosophy and science: "[...] Shall we continue using outmoded terms to describe novel insight or would it not be better to start using a new language? And wouldn't poets and journalists be of great help in finding such a language?".[75]

"In the arts, the idea of alternatives has been introduced with excellent arguments by Bertolt Brecht. For Brecht's idea of a critical theater making use of alternative interpretations of the action on stage see *Schriften zum Theater* [...]".[76] Much later, Feyerabend moves to apply his proliferation not only theories but to reality. "[...T]he manifest worlds themselves demonstrate their fragmentary character; they harbour events which should not be there, and which are classified away with some embarrassment (example: the separation of the arts and the sciences)".[77]

Like Feyerabend, Brecht often emphasizes the process of a plurality of approaches. "With the people struggling and changing reality before our eyes, we must not cling to tried rules of narrative, venerable literary models, eternal aesthetic laws".[78] "They [workers] did not believe in such things as 'the' method. They knew that many methods were necessary [...]. [It] is the tireless energy of science that is needed to investigate in each individual case [...]".[79] Brecht's more explicit defense of proliferation is, "A man with only one theory is lost. He must have several, four, many." [80]

"Occasionally Brecht, like Feyerabend, criticizes some attempts to unify art and science. He writes in criticism of modernist authors, "They fall into line with the 'progress' of physics. They abandon strict causality and switch to statistical causality, [...] adopt Schrödinger's [*sic*] uncertainty principle".[81] In *Refugee Conversations*, Brecht correctly identifies the author of Heisenberg's uncertainty principle. Here the physicist refugees in the dialogue uses the Heisenberg's principle as an analogy to the lack of understanding by contemporary economists of the origin of the crash

[74] Brecht (1964), p. 189.

[75] Feyerabend (1993), p. xii.

[76] Feyerabend (1965), p. 180 n26.

[77] Feyerabend (1999), p. 204.

[78] Brecht (1980), p. 81.

[79] Ibid., p. 85.

[80] Brecht (2003). p. 13.

[81] Ibid., pp. 68–69.

of the Great Depression.[82] Nevertheless, Brecht is almost always positive about a science-like approach to drama. Note that Brecht emphasizes the active approach to science, and also (*pace* Feyerabend) its method as approach to science and not in terms of the results of science.

The place he says the most about Brecht's *Galileo* is in "Let's Make More Movies", in *The Owl of Minerva*, a work containing short articles by various philosophers of the 1970s discussing their conception of philosophy. Feyerabend, along with a few others, does not present an overall view of his philosophy, but discusses a single topic. The title might mislead the prospective reader as it is not meant as a frivolous shocker like some of his remarks elsewhere. He writes,

> We must try to revive mythical ways of presentation and we must try to adopt them to contemporary needs [...]. a form of life in which the constituents of older myths—theories, books, images, emotions, sounds, institutions—an attempt to create such a form of life. He did not entirely succeed. I suggest we try movies instead [...]. One of the advantages of film is that the number of elements which are at the disposal of the director and their degrees of freedom are vastly greater than in any other medium. [...] Of course, it will need a new generation of *thinking* directors to exploit all the possibilities of this medium. But their rise will be the beginning of mythologies that will continue the work of the older philosophers and put an end to the strange business that has lived off their results in the last few centuries.[83]

Feyerabend's hopes for the future use of movies for the communication of philosophical ideas differ from those expressed by Adorno, that scourge of popular culture and the mass media, about advantages and disadvantages of film. The camera has a single point of view unlike the multiple points of view of the stage by the audience. There is also the problem of quality of the sound recording's lack of stereo. The big issue in film for Brecht is lack of feedback from the audience. He rightly thinks the purely technological problems will be solved. More problematic for his conception of drama, Brecht complains that film appears as completed product that cannot be modified by audience reaction.[84]

There is a literary device present in the Bible and elsewhere used by Brecht. According to Darko Suvin, it is the notion of *figura*.[85] Medieval treatments of Biblical stories and characters prefiguring later figures is like Brecht's use of prefigurations in his plays. Feyerabend also uses something like prefiguration in his use of Mach and of defense of classical mechanics which serve as a model for his own hopes for the future philosophy of science. The present philosophy of science he describes disparagingly as "a subject with a great past" in contrast to the best nineteenth century scientist-philosophy. Feyerabend against Kuhn and Lakatos in "Consolations for the Specialist" and elsewhere, argues that formerly "degenerating research programmes" can "make a comeback" and that much older theories (such as atomism) can be revived and take new life in much changed circumstances.[86]

---

[82] Brecht (2020), p. 36.

[83] Feyerabend (1975b), p. 210.

[84] Brecht (1993), p. 214.

[85] Cf. Suvin (1976), as well as Auerbach (1984), pp. 11–76.

[86] Cf. Feyerabend (1970b), p. 118.

## 8.5 Brecht and the Vienna Circle vs. Feyerabend's Early Popperianism and Later Anarchism

Brecht's ideas about drama are congenial to those of Feyerabend on science. Both emphasize practice and plurality of mutually contradictory approaches, whether in Brecht's dramas or in Feyerabend's scientific theories. Feyerabend after his youthful positivism influenced by Victor Kraft in Vienna Feyerabend summarized the work of the later Wittgenstein, worked with Popper, and moved to anarchism. Victor Kraft, unlike Carnap and several other logical empiricists who mainly had been physicists, had studied with leading theorists of interpretive understanding in sociology, Max Weber and Georg Simmel. Kraft also studied and taught geography, another subject quite different from the physics that was the background of most Vienna Circle members. Kraft also rejected the verification criterion held by the leading members of the Vienna Circle. Also, Kraft was virtually the only member of the circle who thought that claims about ethics and values were cognitively meaningful. Kraft's unusual background and positions may have helped Feyerabend to go beyond his own earlier positivism, where he along with fellow students enjoyed debunking all sorts of statements as meaningless.[87] Later, as he began his teaching career, Feyerabend returned to the Alpbach seminars in Austria and found "Philipp Frank was a delight".[88] Frank's lectures planted the seed of Feyerabend's decades later *Against Method* by arguing that Galileo's Aristotelian opponents were empiricists while Galileo was not. Frank was unusual among the Vienna Circle members by incorporating social context in his account of verification of scientific theories,[89] perhaps something that also later influenced Feyerabend. So, both Kraft and Frank, though Vienna Circle members in good standing, gave Feyerabend a picture of science quite different from that of, say, Carnap and Reichenbach.

Brecht's own account of science is almost entirely based of that of the logical positivists,[90] though, like Kraft, he doesn't seem to have accepted the verification criterion. However, his Marxist theory was largely guided by his friend Karl Korsch, who presented a version of Marxism like that of philosophical pragmatism.

Brecht had a strong but purely amateurish interest in physics. He often immodestly claimed, "I am the Einstein of the new stage form".[91] One biographer suggests Brecht's confident self-importance, even if justified, was off-putting to producers and directors in Hollywood to whom he was unsuccessfully attempting to market his plays.[92] Greatly condensing Einstein's biographical time sequence by decades, and segueing into Marx's "Eleventh Thesis on Feuerbach", Brecht wrote, "Einstein says ever since his childhood, thought only about the man hurrying after a ray of

---

[87] Cf. Feyerabend (1995), p. 68.

[88] Ibid., p. 103.

[89] Frank (1961).

[90] Cf. Danneberg and Müller (1995), Sauter (1995), Giles (1995).

[91] Gorelik (1957), p. 73.

[92] Cf. Lyon (1980).

light and the man hurrying after a ray of light and the man falling in an elevator. And just look at how complicated this grew! I wanted to apply to the theater that one should not just interpret but change the world".[93] In Walter Benjamin's conversations with Brecht there is twice the comparison of non-Euclidean geometry with non-Aristotelean theater.[94] In *Refugee Conversations* one of the partners in discussion is a physicist who had been educated and worked in Germany compares the situation of economists failing to explain the German economic depression with the Heisenberg Uncertainty Principle.[95] Throughout March 1942, Brecht fills his journal with reflections on physical indeterminism.[96] One scene in the play *Fear and Misery in the Third Reich* portrays two German Nazi physicists nervously discussing a letter from Einstein, worrying that their use of a Jewish physicist may be overheard by an informer.[97] Several tensor equations are included in the script. Brecht later says that they don't represent a real physical theory, but he wrote to collaborator that he was worried they didn't look authentic and asked for recommendations for a physicist to correct them. Brecht received a letter from Einstein himself in appreciation of the script of *Life of Galileo* that he received.[98] Brecht planned to write a play about Einstein, but he composed only a few pages.

In 1938–1939, in Denmark, Brecht had contacts with people at Bohr's Copenhagen institute where some of young post-doctoral visiting scholars helped Brecht with the construction of an astrolabe. While there, he heard from Christian Møller the relativity theorist and chemist about new discoveries about uranium.[99] Møller's 1938 suggestion that excess energy could be produced in nuclear fission led others to the theory of the chain reaction and eventually to the development of the atomic bomb. Thus, Brecht was present where the preliminary notion of source of the atom bomb was first broached, and it is warned that the atom bomb "may make an appearance" as a prologue to the American production of *The Life of Galileo*.[100]

Brecht's contacts with philosophers, both in Europe and later in America were mainly with members of the Vienna Circle and logical empiricists. In Europe, Brecht's contact with the Circle was mainly with Karl Menger. In the United States he had a good deal of personal interaction with Hans Reichenbach, a member of the Berlin group of empiricists. Brecht wrote to Karl Korsch, "I can talk to Reichenbach next week [...] though I'll see him before that".[101] Reichenbach complimented

---

[93] Brecht (1964), p. 248.

[94] Cf. Benjamin, (1977) pp. 18 and 38.

[95] Cf. Brecht (2020), p. 33.

[96] Cf. Brecht (1993), pp. 208–216.

[97] Cf. Brecht (1958) and Schucking (1999).

[98] Cf. Wizisla (2005), pp. 350–353.

[99] Cf. Kragh (2022).

[100] Cf. Brecht (1994), p. 127.

[101] Brecht (1990), p. 358, and (1993), p. 208.

Brecht on the physics and the history in *The Life of Galileo*.[102] Reichenbach later in a lecture on indeterminism wiggled his finger to represent random path. This gesture appears in an exchange between Galileo and Cardinal Barberini (soon to be Pope) in the play. "If the stars moved like that there would be no astronomy (although they would no doubt have good reasons for doing so)".[103] He wrote: "I like the world of the physicists. We can appear as the gamblers we are. With our approximations, with the to-the-best-of-our-abilities, with our dependence on others [...] more than just one path is open".[104] Brecht read some non-positivist works on philosophy of science such by Eddington,[105] of which Brecht had a copy with the first three chapters heavily annotated with inscribed "Brecht 1931",[106] as well as another somewhat spiritualistic British popularization of astrophysics, James Jeans' *The Mysterious Universe*.[107] Brecht also owned a copy of Max Planck's *Positivism and the Real External World* that criticized the sense datum version of positivism and was countered by Moritz Schlick.[108] Despite reading these works, most of Brecht's knowledge about physics came from the logical empiricists.

In contrast to Popper and Feyerabend, Bertolt Brecht supported inductivism and the inductive method. Brecht liked Bacon for his rejection of Aristotle's *Organon* or logical writings, replacing Aristotelian deduction with induction. Brecht called his major theoretical work on drama *Little Organon* mimicking the title of Bacon's anti-Aristotelian *New Organon*. Brecht several times praises "the splendid, inductive method".[109] Brecht's praise of inductivism would have pleased Hans Reichenbach with whom Brecht talked in Los Angeles. Brecht also talked of use of induction in directing plays. He and the director Erwin Piscator called their method of directing "inductive rehearsal". This involved the actors learning their social relations with the other actors/characters in the play. It also left the actors more leeway, though not total freedom, to find their situation in the scene.

By contrast, Feyerabend retained his Popperian attitude to Bacon and induction into his last writings (unlike his later rejection of Popper's account of Cardinal Bellarmine on Copernicus). Feyerabend rejected Bacon's ideal of observation undistorted by preconceptions and compared it to Protestant rules of faith and he continues to accept this comparison.[110]

One logical positivist work to which Brecht paid particular attention was "Sociology in the Framework of Physicalism" by Otto Neurath,[111] whose attack on

[102] Cf. Brecht (1993), p. 167.

[103] Brecht (1966), p. 78.

[104] Brecht (1993) p. 209.

[105] Eddington (1928).

[106] Cf. Brecht (1990), p. 609 n333.

[107] Jeans (1948).

[108] Planck and Moritz (2020).

[109] Brecht (2000), p. 162.

[110] Feyerabend (1985 [1970]), 34–41.

[111] Neurath (1983 [1932]).

empathy in understanding human behavior was particularly congenial to Brecht. This criticism of interpretive understanding is like that found in the earlier German controversy concerning "verstehen" of understanding versus explanation, though Brecht seems not to have discussed that dispute, despite his lifelong opposition to empathetic understanding. He corresponded with Neurath about the latter's book *Empirical Sociology*,[112] a copy of which Brecht owned. Despite his interest in and study of logical positivist writings, Brecht did not follow the verification theory of meaning but a pragmatist conception of truth reinforced the version of Marxism he received from his mentor Karl Korsch.[113] Brecht was not an uncritical follower but probed Korsch, asking what Korsch meant by science and from what stance he was approaching questions.[114]

Hans Reichenbach moved nearby to Brecht in Los Angeles. Brecht invited Reichenbach to join a committee of intellectuals against fascism, but "Reichenbach wants to consult the officers of his university" before joining.[115] The logical empiricists in the USA as non-citizens during WWII and the Cold War had to be extremely cautious with good reason. Former Vienna Circle members dropped the political goals of their movement and focused purely on the logic of science. In his book subtitled *To the Icy Slopes of Logic*, George Reisch reports that the FBI engaged in an extensive investigation of Carnap's memberships and interviewed several of his contacts.[116] Brecht wished to meet with Rudolf Carnap who is referred to as "the definer".[117] Writing to Paul Tillich about a proposed committee of German refugees opposing fascism, Brecht mentioned that Reichenbach had to check with administration before associating with the group.[118] The particular focus by the California Unamerican Activities Committee on UCLA (where Reichenbach taught) and the California Plan to root out subversives, which is described by John McCumber[119] as well Reichenbach's socialist activity in Germany and earlier confinement to his house in Los Angeles as an enemy alien would have made Reichenbach especially cautious.

Brecht read and underlined copies of *Erkenntnis* 3 (1932/3) and he had programs for the 1932/3 winter and summer sessions of the "Gesellschaft für Empirische Philosophie", marking lectures he wished to attend. Brecht made notes on logical positivism in his American exile but had also studied logical positivism during the 1930s. Karl Korsch, Brecht's mentor in Marxism had studied with Richard von Mises and was himself an acquaintance of Reichenbach and Walter Dubislav who

---

[112] Neurath (1973 [1931]).

[113] Giles (1995), pp. 85–86.

[114] Brecht (1990), p. 238.

[115] Ibid., p. 339.

[116] Cf. Reisch (2005), pp. 272–276.

[117] Cf. Brecht (1990), p. 206 and n703.

[118] Cf. ibid., pp. 378–379.

[119] McCumber (2016).

with Reichenbach helped found *Berliner Gesellschaft für Empirische Philosophie*.[120] Korsch gave two lectures at the "Gesellschaft für Empirische Philosophie". Korsch also "got to know" Neurath, Carnap (from 1909), and Philipp Frank (from before 1929), according to Korsch's mentor and widow, Hedda.

Kurt Lewin was an associate and sometime collaborator of Hans Reichenbach's, and participant at the Erlangen Conference of 1923, where discussions began that led in 1930 to the founding of the logical positivist journal *Erkenntnis*. Lewin is recognized as the source of Reichenbach's concept of genidentity or identity through time.[121] The *Gestalt* psychologist Lewin developed what he called "topological psychology". Lewin's conception of topological psychology likely influenced Brecht's theory of the stage setting as a "field".[122] Despite Lewin's earlier theoretical as well as organizational collaboration with Reichenbach his topological psychology was totally rejected by Gustav Bergmann, a younger member of the Vienna Circle.[123] Ironically, Bergmann himself was later excommunicated by the logical positivists for writing *The Metaphysics of Logical Positivism*, since metaphysics was anathema to the positivists.[124] When Lewin emigrated to the University of Iowa, he planned to work on a mathematization of his topology with Gustav Bergmann. However, Bergmann thought Lewin's topological psychology was nonsense.[125] and turned from Lewin to work with the behaviorist Kenneth Spence, much to Lewin's disappointment.[126] Bergmann later denounced *Gestalt* theory in general along with holism in general in his *Philosophy of Science*.[127] Actually, decades later mathematicians such as Ralph Abraham and Luciano Boi have worked on genuinely mathematical topology and mathematical field approaches to psychology.[128] Lewin instead eventually collaborated with Brecht's tutor in Marxism Karl Korsch in social psychology.[129]

Brecht had a view of human action like that of psychological behaviorism. Various logical positivists favored behaviorism, though there were sometimes mutual misunderstandings between the psychologists and the positivist philosophers.[130] Brecht writes, "Our times has seen amazing developments in all the sciences. We have a acquired and entirely new psychology: viz. the American

[120] Cf. Danneberg and Müller (1995).
[121] Padovani (2013).
[122] Langenmeyer (2017).
[123] Bergmann (1957).
[124] Bergmann (1954).
[125] Van Elternen (1992).
[126] Gentry, personal communication (1994).
[127] Bergmann (1957).
[128] Abraham (1990), Boi (2004).
[129] Van Elternen (1992).
[130] Smith (1986).

Dr. Watson's behaviorism".[131] Likewise, Brecht liked Menger's behaviorist approach as we have seen.

In the "Threepenny Lawsuit" brief by Brecht (suing a film company that backed out of agreement to produce a film version of *Threepenny Opera*), we find "Behaviourism is a psychology that, based on the needs of commodity production, seeks to develop methods to influence the customer, an active psychology, therefore, quintessentially progressive and revolutionary".[132] Although he sees behaviorism used by capitalism to persuade customers, he also sees its revolutionary potential. In the Brecht's brief for the "Threepenny Lawsuit" a behavioral, pragmatic approach to law is used rather like that of Oliver Wendel Holmes in the US. Brecht describing his characters writes, "allows no work to express a person. It [film] provides (or could provide) useful conclusions about the details of human actions. The splendid, inductive method [...] at least facilitates this approach".[133] For film "Applied to the sciences, for example, in medicine, biology, statistics, etc., in order to document visible behaviour [...] they would have learned more quickly to document human interactions".[134] Supreme Court justice and pragmatist Oliver Wendel Holmes' "predictive theory of law" is somewhat like Brecht's treatment of law.

In contrast to his interest in and sympathy for the logical positivists, Brecht had contempt for the Frankfurt School critical theorists except as material for a proposed satirical (never finished) novel. "I also see, with much less profit [than Reichenbach] – except for my 'Tui Novel' – Marcuse, Horkheimer, and Pollack".[135] In contrast to Theodor Adorno, Brecht found Marcuse "quite approachable".[136] Though Feyerabend appreciated Marcuse's earlier *Reason and Revolution* about Hegel and Marxism, he reacted very negatively at Joseph Agassi's very negative review of *Against Method*.[137] Agassi claimed that when Feyerabend writes Lenin "he means Marcuse". Feyerabend replies that he used Hegel's earlier *Reason and Revolution,* for his later anti-democratic essay "repressive tolerance", that Feyerabend rejected along with Robert Paul Wolff's *The Poverty of Liberalism*.[138]

Although Brecht praised Watsonian behaviorism and used it in his account of acting, and his theoretical work on society was influenced by Karl Korsch's Marxism, he held a pragmatic account of Marxism which has relations to behaviorism but puts more emphasis on human agency.

Feyerabend, too, has a behavioristic approach with his "pragmatic theory of observation". "An observation sentence is distinguished from other sentences of a theory, not, as was the case in earlier positivism, by its content; but by the cause of

[131] Brecht (1964), p. 67.
[132] Brecht (2000), p. 166.
[133] Ibid., p. 162.
[134] Brecht (2000), p. 162.
[135] Brecht (1990), p. 339.
[136] Ibid., p. 337.
[137] Agassi (1974).
[138] Feyerabend (1970a), p. 109, fn. 117.

its production, or by the fact that its production conforms to certain behavioral pattern".[139] Even more explicitly behavioristic, "An observation sentence is distinguished from other sentences of a theory not by its content but by the cause of its production, by the fact that its production conforms to certain behavioral patterns. It is claimed he "acquired during his formative years: a radical behaviorist strategy applied to philosophical problems",[140] and "[...H]e uses a behaviorist account of philosophical beliefs, which makes these beliefs amenable to pragmatic warrant".[141] For Feyerabend observations involve a disposition to verbal behavior. Kuby argues that Feyerabend cites and uses ideas of Carnap concerning humans as detectors treated physicalistically.[142]

In Brecht's alliance mainly to traditional logical positivist and logical empiricist theory, his dramatic theory shares with Feyerabend a behaviorist orientation to human action. In Brecht's theory of drama, despite his positivist and Baconian influenced general philosophy, Brecht has an account of experimentation and proliferation in his theory of drama that approaches that of Feyerabend, and which Feyerabend recommended as such.

## 8.6 Feyerabend and Brecht's Interest in China: Chinoiserie or Sinology?

Both Feyerabend (in his later life) and Brecht had interest and support for China, though the interest in China is almost certainly totally independent, although Feyerabend was surely aware of Brecht's plays that are set in China and essays about Chinese acting. Brecht's lifelong appreciation and study of China was far more extensive than that of Feyerabend. While a famous photo of Feyerabend in his later anarchist incarnation shows a poster of King Kong on the wall, in contrast Brecht on the wall above his desk had a scroll of a Chinese poem, another poem by Mao Zedong, a picture of Confucius, a Chinese picture Brecht called "The Thinker", and some Japanese No masks. Brecht carried these throughout his moves to Denmark. East Germany, and America. Brecht studied the philosophy of Mozi, Zhuangzi, Confucius, and other Chinese philosophers as well as Mao. In his exhaustive study of Brecht's dramas, poems, and prose concerning China, Tatlow claims, "Brecht's response to Chinese was not a marginal factor at the periphery of his thought. It leads to central questions of the nature of society and the nature of man".[143]

[139] Feyerabend (1962), p. 36.

[140] Kuby (2000), p. 3275.

[141] Ibid., p. 3276.

[142] Kuby (2018).

[143] Tatlow (1977), p. 348.

Although Brecht's study and use of Chinese works goes far back, Feyerabend's interest goes back mainly to an acupuncturist curing his urinary disease and other physical difficulties while Western doctors had all failed. In the 1960s, at Berkeley, he encountered the radical student movement. Because of the student movement at Berkeley, he came to appreciate Maoism and praise some of Mao's political policies, especially in the (relatively short-lived) Hundred Flowers period of encouragement of popular criticism of government policies. Feyerabend also praises the union of traditional Chinese medicine with Western scientific medicine in Communist China. "The Chinese Communists of the fifties who forced traditional medicine back into the universities and thereby started most interesting lines of research the world over had only little knowledge of the intricacies of scientific medicine".[144]

Because of the resounding success of his acupuncture cure Feyerabend supported non-invasive medicine in general, such as the use of pulse and excretion in diagnosis. "No painful interference with the Organism";[145] "when reading the *Nei Ching*, which develops the philosophy behind acupuncture, I found that in China this was intentional: the human body must be treated with respect".[146] Feyerabend often refers to the *Nei Ching*,[147] the major classic of traditional Chinese medicine as well as to twentieth century Chinese and Western and Chinese discussions of and research on Chinese medicine. A good part of the Second of his *Three Dialogues* is devoted to defense of non-invasive medicine against Western scientific medicine.[148]

Later in life, Feyerabend read the works of Daoism. He mentions Laozi of the *Dao Te Ching* and Daoism,[149] and how Daoism among other beliefs supplied a successful way of life. In the posthumously edited *The Conquest of Abundance*, Chuang Tzu is listed along with Brecht, Plato, and Tolstoy and later with Homer, Hesiod, Aeschylus, Lessing, and Beckett (the latter "not at all my favorite").[150] In *The Conquest of Abundance* Feyerabend emphasizes how Chinese thinkers accepted the diversity of nature.[151] He discusses the contrast of Chinese and Western science in general.[152] Unlike Western thinkers, the Chinese collected anomalies and did not try to fit nature into the procrustean bed of universal laws of nature. He cites the great historian of Chinese science Joseph Needham,[153] who claimed, following Edgar Zilsel, that the concepts of laws of nature arose in the West from the extension to non-human nature of concepts from the Western legal system while China lacked the conception of purely formalistic law, whether human or natural law.

[144] Feyerabend (1982), pp. 78–79 and 88–89.

[145] Feyerabend (1991), p. 57.

[146] Feyerabend (1982), p. 137.

[147] Cf. Veith 1972.

[148] Cf. Feyerabend (1991), pp. 57–59, 72–75 and 117–120.

[149] Feyerabend (1987a), p. 61.

[150] Cf. Feyerabend (1991), pp. 133–135.

[151] Lakatos and Feyerabend (1999), pp. 7, 165, 192 and 237.

[152] Feyerabend (2011), pp. 76–77.

[153] Cf. Feyerabend (1999), pp. 131–132n1, as well as Feyerabend (1987a), p. 24 n38, and p. 88n.

Feyerabend quotes Needham's words "Chinese civilization had been much more effective than the European in finding out about nature and using natural knowledge for the benefit of mankind for fourteen centuries or so before the scientific revolution".[154] and notes that in the Sung dynasty in 1054-1056 a supernova was observed and recorded in China but in the West as it did not fit Western medieval conceptions of the eternity of the celestial bodies.[155]

Feyerabend's interest in China also included twentieth century Maoism. In *Against Method* he cites the essay "On Contradiction" by Mao, which he says "splendidly explains" the background of Marxism-Leninism, and Brecht refers to the same Mao essay in his later writing.[156] Feyerabend also makes several references to Lenin in the earlier article version of *Against Method*,[157] as well as in four citations in the book *Against Method*.[158] In his reply to what he dubs "Marxist Fairytales from Australia" Feyerabend writes, "I take seriously Chairman Mao's "Third Inditement Against Stereotyped Party Writing".[159]

Chinese themes appear in many of Brecht's plays, most notably the manuscript of *Mi Ti*,[160] the children's play *The Ginger Pot*,[161] the 1938 poem "Legend of the origin of the book *Tao Te Ching*",[162] and, of course, the major and famous play *The Good Woman of Szechuan*, which, along with *The Life of Galileo* and *The Caucasian Chalk Circle* are generally thought to be his greatest plays. Brecht uses his extensive but unfinished *Mi Ti* dialogues to introduce (under invented ancient Chinese names) opinions of Lenin, Trotsky, Stalin, Karl Korsch and himself.

Brecht's very positive reaction to the actor Mai Lan-Fang in Moscow led him to see Chinese acting as involving his alienation effect though some writers have criticized Brecht's understanding of Chinese acting as manifesting his alienation effect, though some Chinese writers have claimed he misinterpreted the Chinese style.[163] In the USSR, Brecht saw for the first time a Chinese actor. "Mai Lan-Fang is here, the greatest Chinese actor.[164] "He plays girl's parts and is really splendid".[165] Brecht later went to see a Chinese actor Tsiang in two plays and found the acting, blending Chinese and Western acting styles and found it extremely interesting. Brecht later accompanied the China expert Karl August Wittfogel to several Chinese plays in New York, despite the latter's anti-Stalinism and growing anti-Marxism, which

[154] Feyerabend (1987a), p. 88n.

[155] Feyerabend (1991), p. 94.

[156] Brecht (1964), pp. 261 and 278 n1.

[157] Feyerabend (1970a), pp. 101 n27, 104 n35, 110 n48, 113 n53 and 114 n68.

[158] Feyerabend (1975a), pp. 17 n1, 18 n4 and 147 n2.

[159] Feyerabend (1982), p. 162 and n22.

[160] Brecht (2016).

[161] Brecht (1958).

[162] Benjamin (1977), pp. 70–72.

[163] Tian (1997).

[164] Brecht (1990), no. 251, p. 201.

[165] Ibid.

Brecht must have found uncongenial, but he must have appreciated Wittfogel's vast knowledge of China.[166]

Critics Theodor Adorno and Frederic Jameson dismiss Brecht's use of ancient Chinese settings. Adorno sneered at Brecht's use of archaic Chinese setting and characters and called it primitive. Adorno's similar denigration of African American jazz as castration is well known. The Marxist historian and jazz aficionado Eric Hobsbawm wrote that Adorno's writing about jazz was "Some of the stupidest pages ever written about jazz [...]".[167]

Brecht, in contrast, liked American jazz, both artistically and politically: "There was a scene in an American movie in which Fred Astaire did a tap dance to the noises of an engine room; the tap rhythm showed the startling similarity between the new noises and jazz. Jazz meant a broad influx of popular musical elements into recent music, no matter what was later made of it in our commercial world. The relationship of jazz to the emancipation of the Negro is well known".[168]

Brecht believed treating contemporary issues in the form of an historically and culturally distant setting allowed people to gain the critical distance he desired. In contrast two leading Marxist literary critics mentioned above also totally reject Brecht's use of this cultural distancing method. Adorno sneers at the use of Chinese settings as "[T]he fiction of the old peasant sated with epic experience [...], an echo of archaic social relations lost beyond recall [...]", and says "What his classical predecessors [Marx] denounced as the idiocy of rural life, Brecht, like some ancient ontologist [Heidegger] treats as ancient truth".[169] Likewise, Frederic Jameson with all his praise for Brecht's theater as the high point of Marxist literature combining popularity with radical message, like Adorno, dismisses Brecht's use of Chinese philosophers Mozi and Confucius with a dismissive footnote about Brecht's "uncritical cult of Chinese philosophers [...] as repositories of Oriental wisdom". And claims his interest in Chinese philosophy is "superficial" and "mythopoeic".[170] Both critics were apparently unaware of Brecht's serious assimilation of Chinese philosophy and drama.

Brecht's behaviorism has affinity to Chinese thought similar to the more recent interpretation of ethics and action to be found in Fingarette's *Confucius: The Secular as Sacred,* which presents a behavioristic account of Confucian moral thought, denying that Confucius' ethics and ritual avoids subjectivity.[171] There is likewise a behavioristic aspect of Feyerabend's epistemology, the pragmatic theory of observation.[172]

---

166 Brecht (1993), p. 277.

167 Witkin (2000), p. 145.

168 Brecht (1957), p. 37.

169 Adorno et al. (1980), p. 188; bracketed identifications are mine.

170 Cf. Jameson (1980), as well as Brecht (1980), p. 75 and n6.

171 Fingarette (1998).

172 Cf. Kuby (2000).

Brecht had been progressively disillusioned with the East German Communist government, especially after the suppression of a genuine workers' rebellion there, about which Brecht a later wrote a cynical poem:

> ...Would it not in that caseBe simpler for the governmentTo dissolve the peopleAnd elect another?[173]

Because of this disillusionment with Communist East Germany, toward the end of his life Brecht life hoped that Communist China would actualize the popular participation in decisions of a Communist regime that he had been sxpecting in the USSR and the German Democratic Republic.

## 8.7 Feyerabend's Criticism of Lakatos and Brecht's Criticism of Lukács

In the 1930s, Brecht engaged in the so-called Expressionism Debate. His main opponent was the Hungarian Marxist Georg Lukács whose most famous and influential work was his *History and Class Consciousness*.[174] Many later European cultural Marxists have thought him the greatest Marxist since Marx. However, the book was condemned by orthodox Communists as not sufficiently materialistic, too idealistic. Initially he wrote a defense of his position against the objectivist position of orthodox Marxism, but out of caution and political pressure did not publish it.[175] After the condemnations, Lukács changed his position to more orthodox Communism. Almost immediately he wrote a brief work praising Lenin in attempt to pacify his critics.[176] He rejected his own book and even attempting to prevent its later reprinting in France. When Lukács emigrated to the USSR, he had moved from his earliest highly subjective literary criticism *Soul and Form*[177] based on the religious existentialist Søren Kierkegaard and the extremely subjective idealist Gottlieb Fichte to the materialist position of Engels and Lenin and to an aesthetic position close to so-called Socialist Realism, the official literary line in Stalinist Russia. Lukács had always been an enemy of literary modernism, whether as precocious literary aesthete or later Stalinist critic. For him writers such as Brecht along with Proust, Kafka, and James Joyce exemplified decadence and degeneration. Balzac and Tolstoy, various Soviet writers, as well as Thomas Mann (despite Mann having fictionalized Lukács as a Jesuit fanatic in *The Magic Mountain*) exemplified a healthy realism in contrast to the stream of consciousness and expressionist writers.

---

173 Brecht (2006), p. 118.

174 Lukács (1970a [1923]).

175 Lukács (2002).

176 Lukács (1970b [1924]).

177 Lukács (2010 [1911]).

Lukács entitled a chapter "Franz Kafka or Thomas Mann".[178] Again, this chimed with Stalinist artistic doctrines, but it also fit with Lukács' sincere anti-modernist beliefs by that time. He wrote articles with titles such as "Healthy or Sick Art"[179] and "Marx and the Problem of Ideological Decay",[180] as well as the later huge book, *The Destruction of Reason*.[181] It traces a claimed line from the German Romantic idealists such as Friedrich Schelling and the early existentialists such as Søren Kierkegaard as well as possibly more plausible precursors such as Nietzsche and Spengler to Hitler. In an afterword, Lukács portrayed Bertrand Russell cheek to jowl with the Pope as an irrationalist Cold Warrior.[182] Progress versus degeneration remained the major theme in Lukács' writings of his later period.

On the other side of the Expressionism Debate, defending modern literature, was Brecht, the religious and utopian thinker Marxist Ernst Bloch, who defended expressionism in *Heritage of Our Times*.[183] Also on the pro-expresionist side was the combiner of Jewish messianism and Marxism Walter Benjamin (who supported modernist literature and was a friend of Brecht), as well as several expressionist writers. Lukács made his initial contribution to the debate with "Expressionism: Significance and Decline",[184] in which he made several attacks on Brecht in the course of his denunciation of the expressionist movement. Brecht by then was not an expressionist but still did appreciate the contribution of past expressionism. By then Brecht had developed his "alienation effect", totally rejecting expressionism's emotionalism for his own theory of estrangement by which the actor should not immerse himself in emotions but should have and affect a kind of detachment from her own delivery to influence the audience to maintain a critical attitude. Nevertheless, Brecht felt that expressionism had contributed to the development of theater and had earlier been closer to the expressionist movement but not uncritical of it. He ironically noted that in Lukács' essay, "I discover that decadence includes me".[185] Brecht argues that Lukács denunciation of "formalism" in literature was itself extremely formalist. Brecht's most trenchant critiques of Lukács were not published. It is not clear if they were rejected by the Communist journals or if Brecht did not publish them out of caution.

Feyerabend's criticism of Lakatos parallels Brecht's criticism of Lukács. Lakatos had studied with Lukács in Moscow when he was a dedicated Communist. Lakatos had risked his life in attempts to convert peasants and was part of a group attempting to resist the Nazi during their occupation of Hungary. He was victim of a later purge and tortured. Afterwards, as a librarian, Lakatos began to read the works of the

178 Lukács (1964).
179 Lukács (1971 [1934]).
180 Lukács (1981b).
181 Lukács (1980).
182 Cf. Lukács (1980), pp. 806–809.
183 Bloch (1991 [1935]).
184 Lukács (1981a).
185 Brecht (1993), p. 16.

free-market capitalist Friedrich Hayek and Karl Popper, which were not open to ordinary visitors but kept in a "poison cabinet". During the Hungarian revolution Lakatos escaped to Britain and there, became, a student of Karl Popper. During the student revolts at the London School of Economics of the 1960s, Lakatos became more and more conservative (or reactionary) and issued a "Black Paper" against the student rebels as a counter to conciliatory "White Paper" of the administration.

Despite his new free market conservatism Lakatos retained his Marxist terminology. Like Lukács in literature, Lakatos evaluated scientific research as "progressive" or "degenerating". In his "Methodology of Scientific Research Programmes", Lakatos analyzed scientific theories in terms of a "hard core" of the main hypotheses and a "protective belt" of auxiliary hypotheses, recapitulating the Leninist talk about the hard core of the part and transmission belts. Lakatos' castigations of irrationalism and accusations of followers of historian Thomas Kuhn (whose picture of "normal science" in some ways Lakatos logically reconstructed as his research programs) inspired "religious maniacs ('student revolutionaries')", and Kuhn's process of paradigm change is a "mystical conversion" and is "*irrational, a matter of mob psychology*".[186] This attack on Kuhn and others as irrationalists is reminiscent of Lukács' depiction of numerous thinkers as irrationalists, including his own former teachers, the German sociologists Max Weber and Georg Simmel, whose ideas Lukács had used earlier in life, as paving the road to Nazism.

Lakatos was a critical evaluator of science rather than a participant or producer to science, just as Lukács as a literary critic was an evaluator and not a producer of literature. At times, Lukács probably hoped to be a cultural commissar. Lakatos similarly thought his evaluation of research programmes as progressing or degenerating could be used by journal editors to refuse publication and administrators of science to refuse funding to clearly degenerating programmes.[187] Feyerabend wished philosophy not merely to evaluate but to positively contribute to the nature of scientific research. Lakatos opposed the non-scientist public evaluating and directing scientific research, while Feyerabend in his later, more political writings, wished the non-specialist public to participate in and chose what science should be pursued. In childhood, Feyerabend had participated in the popular astronomy movement and on his thirteenth birthday he even gave a lecture to a large outdoor sky watchers' gathering.[188]

Lakatos' methodology has formal rules while Feyerabend claims no such rules apply to all of science. Similarly, Lukács evaluated novels in terms of an ideal aesthetic, while Brecht in his early 1927 essay wrote that aesthetics should be abolished.[189] This was rather like Feyerabend's notorious "anything goes" in scientific research. However, Brecht had already advocated "epic theater" and went on to elaborate on his non-Aristotelian theory of drama moving from his earlier abolitionism.

---

[186] Lakatos (1970a), pp. 93 and 178.

[187] Cf. Lakatos (1970b), p. 105.

[188] Feyerabend (1995), p. 28.

[189] Brecht (1964), pp. 20-21.

Feyerabend liked Brecht's earlier writings on theater and not Brecht's later more systematic theories.

Brecht's theorizing was meant for the production of dramas rather than than for after-the-fact aesthetic evaluation of them. This contrasted with Lukács' role purely as literary critic. With respect to science, the contrast parallels that between Feyerabend and Lakatos. For Feyerabend, philosophy of science should be like that of the early modern scientists, such as Galileo, Descartes, or Leibniz and the renaissance artists such as Alberti, or, in the more recent past, the work of Ernst Mach criticizing science and contributing directly to research and theorizing. For Lakatos, the philosopher of science is an external arbiter of research programs.

Brecht's plays themselves and his theory of estrangement for the actors were meant for the audience to evaluate the action critically. Feyerabend early suggested that this method should be used in the teaching of science and would help science students to become more critical and creative.[190]

Unlike Lukács, Brecht believed popular culture, with such techniques as those of the circus and the cabaret, should be incorporated into the theater. He used photographs and recordings along with minimal scenery in his productions. Brecht, unlike Lukács, believed that both popular culture and modern art should be taken seriously as they are expressions of modern society. As with Brecht's appreciation of popular culture, Feyerabend appreciated popular mystery novels and movies as well as the high culture of opera. In one of the tapes he sent to his wife, he enthuses about the power of the 1936 Spencer Tracy movie *Fury* he had seen the previous night.[191] He once enticed a party including the famous logician Tarski to leave a high culture performance at the Old Vic to attend the comedy movie *Cat Ballou* and Tarski enjoyed it.[192]

Shared positions of Lukács and Lakatos opposed those of Brecht and Feyerabend:

1. Traditional concept of objectivity of science opposed to the subjectivity of art.
2. Holding of formal rules of method in Lakatos and aesthetics in Lukács.
3. Role as critic and evaluator rather than producer of science (Lakatos) or art (Lukács) as possible advisor to authorities.
4. Use of essences or generalizations overriding the particularity and variety of individual works.
5. Seeing the public as recipients rather than participants in science or literature.
6. Extreme rationalism coupled with accusation of all opponents as irrationalist.

Shared positions of Brecht and Feyerabend opposed to those of Lukács and Lakatos:

1. Holding that science and art have similar features, involving experimentation and proliferation.
2. Claiming that formal rules are insufficient to characterize art or science.

---

[190] Feyerabend (1962), pp. 71–72.

[191] Feyerabend (2001).

[192] Feyerabend (1995), p. 130.

3. Advocating that critic or philosopher act as contributor to the production of art or science.
4. Claiming that works of art and of science have individuality and variety that cannot be grasped by universals.
5. The public should have role of judge and contributor in art and science.
6. Rationality is not a self-evident concept and irrationality has a role.

## 8.8 Conclusion

From his initial contact with Brecht as a young man, Feyerabend often referred to Brecht's ideas. Feyerabend called his own work a collage and wrote sympathetically of the Dada movement. Brecht similarly referred to his own work as montage and had sympathy for Dadaism though he felt his own approach went beyond it. Feyerabend recognized that Brecht's use of multiple contradictory positions in his plays, of his characters having opposite sides to their personalities. Brecht's extreme version of this is the protagonist of one of his major works, *The Good Woman of Szechuan*, who alternates between two opposite personalities, one kind and one ruthless because of social pressures. Similarly, Brecht himself was a very complex character who had opposite, contradictory personality traits, idealistic, and cynical.

Brecht's emphasis activism, experimentation, and correction of errors had great similarity to Feyerabend's ideal in the philosophy of science. Both emphasized discontinuity with their montages and collages. Both focused their perhaps most notable composition on the character and activities of Galileo. Both liked and admired Chinese thought, in both its ancient and its contemporary Communist versions, though their sources of introduction to Chinese thought differed. Finally, Feyerabend's criticisms of Lakatos' methodology parallel Brecht's criticisms of Lukács' literary criticism.

## References

Abraham, Frederick D., et al. 1990. *A Visual Introduction to Dynamical Systems Psychology*. Santa Cruz: Aerial Press.

Adorno, Theodor, et al. 1980. In *Aesthetics and Politics*, ed. Fredric Jameson. London: Verso.

Agassi, Joseph. 1974. Review of *Against Method*. *Philosophia* 6 (1): 166–177.

Auerbach, Erich. 1984. *Scenes from the Drama of European Literature*. Minneapolis: University of Minnesota Press.

Benjamin, Walter. 1977. *Understanding Brecht*. London: New Left Books.

———. 2011. *The Origin of German Tragic Drama*, translated by John Osborne and Introduction by George Steiner. London: Verso.

Bergmann, Gustav. 1954. *The Metaphysics of Logical Positivism*. London: Longmans Green & Co.

———. 1957. *Philosophy of Science*. Madison: University of Wisconsin Press.

Bloch, Ernst. 1991 [1935]. *Heritage of Our Times* London: Polity Press.

Boi, Luciano. 2004. Questions Regarding Husserlian Geometry and Phenomenology. A Study of the Concept of Manifold and Spatial Perception. *Husserl Studies* 20 (3): 207–267.
Bradley, Laura. 2011. Introduction. In *Edinburgh German Yearbook 5: Brecht in the GDR: Politics, Culture, Posterity*, ed. Laura Bradley and Karen Leeder. Rochester: Camden House.
Brecht, Bertolt. 1957. On Unrhymed Lyrics in Irregular Rhythms. *Tulane Drama Review* 2 (1): 33–38.
———. 1958. The Life of Confucius. *Kenyon Review* 20 (3): 393–398.
———. 1964. *Brecht on Theater: The Development of an Aesthetic*, ed. and trans. John Willet. New York: Hill and Wang.
———. 1966. In *Galileo*, ed. Charles Laughton. New York: Grove Press.
——— (1980). Against Georg Lukács, translated by Stuart Hood, in Theodor Adorno et al., *Aesthetics and Politics*, ed. Fredric Jameson, London: Verso, 68-85.
———. 1990. In *Letters 1913-1956*, ed. John Willet. London: Routledge.
———. 1993. *Journals* [1934-1955], ed. John Willett and Ralph Manheim and trans. Hugh Rorrison. London: Routledge.
———. 1994. *Life of Galileo,* translated by John Willet. In , ed. John Willet and Ralph Manheim. New York: Arcade Publishing.
———. 2000. The Threepenny Lawsuit. In *Bertolt Brecht on Film and Radio*, 147–201. London: Methuen.
———. 2003. In *Brecht on Art and Politics*, ed. Tom Kuhn and Steve Giles. London: Bloomsbury Publishing.
———. 2006. *Bertolt Brecht: Poetry and Prose*, edited and translated by Reinhold Grimm. New York: Continuum.
———. 2016. In *Me-Ti: The Book of Interventions in the Flow of Things*, ed. Antony Tatlow. London: Methuen.
———. 2020. *Refugee Conversations*. Yorkshire: Methuen Drama.
Danneberg, Lutz, and Hans-Harald Müller. 1995. Brecht and Logical Positivism. *Brecht Yearbook* 15: 151–161.
Eddington, Arthur S. 1928. *The Nature of the Physical World*. New York: The Macmillan Company.
Eisenstein, Sergei. 1959. *Film Form and Film Sense*. New York: Meridian Books.
Feyerabend, Paul K. 1962. Explanation, Reduction, and Empiricism. In *Scientific Explanation, Space, and Time*, ed. Herbert Feigl and Grover Maxwell, 28–97. Minneapolis: University of Minnesota Press.
———. 1965. Problems of Empiricism. In *Beyond the Edge of Certainty*, ed. Robert G. Colodny, 145–260. Englewood: Prentice Hall.
———. 1967. The Improvement of the Sciences and the Arts, and the Possible Identify of the Two. In *Proceedings of the Boston Colloquium for the Philosophy of Science, 1964-66: In Memory of Norwood Russell Hanson*, ed. Robert S. Cohen and Marx W. Wartofsky, 387–415. Dordrecht: D. Reidel.
———. 1968. Science, Freedom, and the Good Life. *The Philosophical Forum* 1 (3): 127–135.
———. 1970a. Against Method: Outline of an Anarchistic Theory of Knowledge. In *Analysis of Theories and Methods in Physics and Psychology*, ed. Michael Radner and Stephen Winokur, 17–130. Minneapolis: University of Minnesota Press.
———. 1970b. Consolations for the Specialist. In *Criticism and the Growth of Knowledge*, ed. Imrc Lakatos and Alan Musgrave, 197–240. Cambridge: Cambridge University Press.
———. 1975a. *Against Method: Outline of an Anarchistic Theory of Knowledge*. London: New Left Books.
———. 1975b. Let's Make More Movies. In *The Owl of Minerva: Philosophers on Philosophy*, ed. Charles J. Bontempo and S. Jack Odell, 201–210. New York: McGraw-Hill.
———. 1981. More Clothes from the Emperor's Bargain Basement. *British Journal for the Philosophy of Science* 32 (1): 57–71.
———. 1982. *Science in a Free Society*. London: Verso.

———. 1985 [1970]. Classical Empiricism. In *Problems of Empiricism: Philosophical Papers Volume 2*, 34–54. Cambridge: Cambridge University Press
———. 1987a. *Farewell to Reason*. London: Verso.
———. 1987b. Creativity: A Dangerous Myth. *Critical Inquiry* 13 (4): 700–711.
———. 1991. *Three Dialogues on Knowledge*. Oxford: Blackwell.
———. 1993. *Against Method: Outline of an Anarchistic Theory of Knowledge*. 3rd ed. London: Verso.
———. 1995. *Killing Time*. Chicago: University of Chicago Press.
———. 1999. In *The Conquest of Abundance: A Tale of Abstraction Versus the Richness of Being*, ed. Bert Terpstra. Chicago: The University of Chicago Press.
———. 2001. *Stories from Paolino's Tapes (1984–1993)*. Berlin: Supposé.
———. 2011. In *The Tyranny of Science*, ed. Eric Oberheim. London: Polity Press.
Fingarette, Herbert. 1998. *Confucius: The Secular as Sacred*. Long Grove: Waveland Press.
Frank, Philipp G. 1961. The Variety of Reasons for the Acceptance of Scientific Theories. In *The Validation of Scientific Theories*, ed. Philipp G. Frank, 3–18. Boston: The Beacon Press.
Galileo, Galilei. 1953. *Dialogue Concerning the Two Chief World Systems*. Berkeley: University of California Press.
———. 1954. *Dialogues Concerning Two New Sciences, translated by Henry Crew and Alfonso de Salvio*. New York: Dover.
———. 2012. *Treatise on Floating Bodies*. Oxford: Benediction Classics.
Giles, Steve. 1995. Bertolt Brecht, Logical Empiricism and Social Behaviorism. *Modern Language Review* 90 (1): 83–93.
Godfrey-Smith. 2003. *Theory and Reality: An Introduction to the Philosophy of Science*. Chicago: The University of Chicago Press.
Gorelik, Mordecai. 1957. Brecht: I am the Einstein of the New Stage Form. *Theater Arts* 41 (3): 71–75.
Jameson, Fredric (1980). "Reflections in Conclusion", in Theodor Adorno et al., *Aesthetics and Politics*, ed. Fredric Jameson, London: Verso, pp. 196-213.
———. 1998. *Brecht and Method*. London: Verso.
Jeans, James. 1948. *The Mysterious Universe*. Cambridge: Cambridge University Press.
Jürgens, Martin. 1982. Zum Prinzip der Montage in Bertolt Brechts 'soziologischen Experimenten'. *Zeitschrift für Literaturwissenschaft und Linguistik* 46 (12): 88–103.
Koyré, Alexandre. 1979. *La Philosophie de Jacob Boehme*. Paris: Librairie Philosophique J. Vrin.
———. 1955 [1929]. *Mystiques, Spirituels, Alchemistes: Schwenkfeld, Seb. Frank, Weigel, Paracels*. Paris: Librarie Armand Colin.
Kragh, Helge. 2022. Brecht, Galileo, and Møller: A View from Copenhagen, 1938–1939. *Physics in Perspective* 24 (2–3): 99–124.
Kuby, Daniel. 2000. Decision-Based Epistemology: Sketching a Systematic Framework of Feyerabend's Metaphilosophy. *Synthese* 199 (1–2): 3271–3299.
Kuby, D. 2018. Feyerabend, Carnap, and the Pragmatic Theory of Observation. *HOPOS: The Journal of the International Society for the Philosophy os the History of Science* 8 (2): 432–470.
Lakatos, Imre. 1970a. Falsification and the Methodology of Scientific Research Programmes. In *Criticism and the Growth of Knowledge*, ed. Imre Lakatos and Alan Musgrave, 91–196. Cambridge: Cambridge University Press.
———. 1970b. History of Science and Its Rational Reconstructions. In *PSA 1970: Proceedings of the Biennial Meeting of the Philosophy of Science Association*, ed. Roger C. Buck and Robert S. Cohen, 91–136. Dordrecht: D. Reidel.
Lakatos, Imre, and Paul K. Feyerabend. 1999. In *For and Against Method*, ed. Matteo Motterlini. Chicago: The University of Chicago Press.
Langemeyer, Ines. 2017. The Field Concept in Psychology, Gestalt Theory, Physics, and Epic Theatre – Brecht's Adaptations of Kurt Lewin. *Journal of New Frontiers in Spatial Concepts* 9: 1–16.

Laudan, Larry. 1989. For Method: Or Against Feyerabend. In *An Intimate Relation: Studies in History and Philosophy of Science Presented to Robert Butts on his 60th Birthday*, ed. James R. Brown and Jürgen Mittelstrass, 299–317. Dortrecht: Kluwer. esp. p. 303.
———. 1996. *Beyond Positivism and Relativism: Theory, Method, and Evidence*. Boulder: Westview Press.
Lukács, Georg. 1964. *Realism in Our Time: Literature and the Class Struggle*. New York: Harper and Row.
———. 1970a [1923]. *History and Class Consciousness*. Cambridge, MA: MIT Press.
———. 1970b [1924]. *Lenin: A Study in the Unity of His Thought*. Cambridge: MIT Press.
———. 1971 [1934]. Healthy or Sick Art?. In *Writer and Critic & Other Essays*, ed. György Lukács and trans. Arthur D. Kahn, 103–110. New York: The Universal Library.
———. 1980 [1953]. *The Destruction of Reason*. London: Merlin Press.
———. 1981a [1934]. Expressionism: Its Significance and Decline. In György Lukács, *Essays on Realism*, ed. Rodney Livingstone and trans. David Fernbach, 76–113. Cambridge, MA: MIT Press
———. 1981b [1938]. Marx and the Problem of Ideological Decay. In György Lukács, *Essays on Realism*, ed. Rodney Livingstone and trans. David Fernbach, 114–166. Cambridge, MA: MIT Press.
———. 2002. *In Defense of History and Class Consciousnes: Tailism and the Dialectic*, Afterword by Slavoj Žižek. London: Verso.
———. 2010 [1911]. *Soul and Form*. Cambridge, MA: MIT Press.
Lyon, James K. 1980. *Bertolt Brecht in America*. Princeton: Princeton University Press.
Makela, Maria. 1996. *The Photomontages of Hannah Höch*. Minneapolis: Walker Gallery.
McCumber, John. 2016. *The Philosophy Scare: The Politics of Reason in the Early Cold War*. Chicago: The University of Chicago Press.
Neurath, Otto. 1973 [1931]. Empirical Sociology. In *Empiricism and Sociology*, ed. Marie Neurath and Robert S. Cohen, 319–422. New York: Springer.
———. 1983 [1932]. Sociology in the Framework of Physicalism. In Otto Neurath, *Philosophical Papers 1913-1946*, ed. Robert S. Cohen and Marie Neurath, 58–90. New York: Springer.
Oberheim, Eric. 2006. *Feyerabend's Philosophy*. Berlin: De Gruyter.
Olschki, Leonardo. 1927. *Galilei und seine Zeit*. Halle: Max Niemeyer.
———. 1967. Galileo's Literary Formation. In *Galileo: Man of Science*, ed. Ernan McMullin, 140–159. New York: Basic Books.
Padovani, Flavia. 2013. Genidentity and Topology of Time: Kurt Lewin and Hans Reichenbach. In *The Berlin Group and the Philosophy of Logical Empiricism*, ed. Nikolay Milkov and Volker Peckhaus, 97–122. Boston: Kluwer.
Planck, Max, and Schlick Moritz. 2020. *Positivism and the Real External World & Positivism and Realism*. Montreal: Minkowski Institute Press.
Reisch, George A. 2005. *How Cold War Transformed Philosophy of Science: To the Icy Slopes of Logic*. Cambridge: Cambridge University Press.
Richter, Hans. 1965. *Dada: Art and Anti-Art*. London: Thames & Hudson.
Rorty, Richard. 1995. *The New Republic* 213 (5): 32–36.
Sante, Lucy. 1997. Dada's Girl: Hannah Höch Thumbs Her Nose at Modern Art. *Slate*. 10. https://slate.com/culture/1997/04/dada-s-girl.html. Accessed 6 Dec 2022.
Sauter, Ulrich. 1995. Ich selber nehme kaum noch an einer Diskussion teil. *Deutsche Zeitschrift für Philosophie* 43 (4): 687–709.
Schucking, Engelbert L. 1999. Jordan, Pauli, Politics, Brecht ...and a Variable Gravitational Constant. *Physics Today* 52 (10): 26–32.
Schumacher, Ernst. 1968. *Brecht's Leben des Galileo: Drama und Geschichte*. Berlin: Henschelverlag.
Smith, Laurence D. 1986. *Behaviorism and Logical Positivism: A Reassessment of the Alliance*. Stanford: Stanford University Press.

Suvin, Darko. 1976. Brecht's *Caucasian Chalk Circle* as Marxist Figuralism: Open Drama. In *The Weapons of Criticism*, ed. Norman Rudich, 341–357. Palo Alto: Ramparts Press.

Tatlow, Antony. 1977. *The Mask of Evil Brecht's Response to the Poetry, Theatre and Thought of China and Japan*. Berne: Peter Lang.

Tian, Min. 1997. 'Alienation-Effect' for Whom? Brecht's (Mis)interpretation of the Classical Chinese Theatre. *Asian Theatre Journal* 14 (2): 200–222.

Van Elternen, Mel. 1992. Karl Korsch and Lewinian Social Psychology: Failure of a Project. *Journal of the History of the Human Sciences* 5 (2): 33–61.

Veith, Ilse, editor and translator. 1972. *Nei Ching Su Weng, The Yellow Emperor's Classic of Internal Medicine*, Chs. 1–34. Berkeley: University of California Press.

von Gebler, Karl. 1879. *Galileo Galilei and the Roman Curia*, trans. George Sturge. London: C. Kegan Paul.

Willet, John. 1996. *Art and Politics in the Weimar Period, 1917–1933: The New Sobriety*. New York: Da Capo Press.

Witkin, Robert W. 2000. Why Did Adorno Hate Jazz. *Sociological Theory* 18 (1): 145–170.

Wizisla, Erdmut. 2005. Vortreffliches für die verbildeten Zeitgenossen: Einstein schreibt Brecht über Galilei. In *Albert Einstein–Ingenieur des Universums*, ed. Jürgen Renn, vol. 3, 350–353. Weinheim: Wiley.

Wohlwill, Emil. 1909. *Galileo und Sein Kampf für die copernicanische Lehre*. Vol. 2 Vols. Hamburg: Leopold Voss.

# Chapter 9
# Feyerabend and the Vienna Circle

**Friedrich Stadler**

**Abstract** Vienna-born Paul Feyerabend (1924–1994) is one of the best-known philosophers of the twentieth century, who has been the subject of controversial debates among the general public beyond the world of academic scholars right up to the present day. He gained this image after the publication of his book *Against Method* (1975)/*Wider den Methodenzwang. Skizze einer anarchistischen Erkenntnistheorie* (1976), which radically criticized the normative analytical philosophy of science. Before this career as a provocateur and *enfant terrible* of the philosophy of science, Feyerabend experienced a little-noticed intellectual socialization in Vienna, which shaped him until the end of his life, as can be grasped from his autobiography *Killing Time* (1994)/*Zeitverschwendung* (1995). After WW II Feyerabend studied at the University of Vienna art history and natural sciences before devoting himself to philosophy. Here he joined the discussion circle around the philosophy professor Viktor Kraft (the so called "Third Vienna Circle"), who, as the last member of the Vienna Circle, tried for a short time to continue this tradition. Ludwig Wittgenstein also participated in this circle shortly before his death, as did his later his literary heir Elizabeth Anscombe from England. All these conversations flowed into his dissertation *Zur Theorie der Basissätze* (1951). Feyerabend then worked as an assistant to visiting professor Arthur Pap from the USA in 1953/54, for whom he prepared the book *Analytische Erkenntnistheorie* (1955), dedicated to the Vienna Circle. From 1948 Feyerabend was a member of the Austrian College / Forum Alpbach, where he also met Karl Popper, and translated bestseller *The Open Society and its Enemies* (2 volumes) into German. In London, he became friends with Imre Lakatos, who had been exiled from Hungary and with whom he planned the joint book project "For and Against Method". This remained unfinished due to Lakatos' early death, so that his original essay "Against Method" (1970) in "Minnesota Studies" (ed. by Herbert Feigl's "Minnesota Center for Philosophy of Science") became an international bestseller as a later monograph in 1975. In it, he argued against a

F. Stadler (✉)
Institute Vienna Circle, University of Vienna, Vienna, Austria
e-mail: friedrich.stadler@univie.ac.at

© The Author(s), under exclusive license to Springer Nature Switzerland AG 2024

S. Gattei, R. Corvi (eds.), *Feyerabend in Dialogue*, Boston Studies in the Philosophy and History of Science 346,
https://doi.org/10.1007/978-3-031-71938-7_9

privileged scientific method in favor of a historically oriented pluralistic philosophy of science. These forgotten episodes of early Feyerabend and his intellectual return to Vienna with the appreciation of Ernst Mach are described and contextualized.

**Keywords** Paul Feyerabend · Ernst Mach · Logical Empiricism · Vienna Circle · "Third Vienna Circle" · Viktor Kraft-Circle · Forum Alpbach · Herbert Feigl · Minnesota Center for Philosophy of Science · Karl Popper · Imre Lakatos

## 9.1 Paul Feyerabend (1924–1994): "The Worst Enemy of Science"?

Even though many years have passed since his death, Paul Feyerabend continues to be discussed—by philosophers and the scientific community—but he is also the subject of a broader public debate. Especially in the German-speaking world, his image as an *enfant terrible* of philosophy continues to be nourished by fragments that have appeared posthumously. (Feyerabend 2005) His image remains a complex and contradictory one. As an icon of 'anti-science' or 'worst enemy of science' (cf. *Nature* 1987/*Scientific American* 1993) his oeuvre has assumed a life of its own, and his name has become a popular instrument for polarizing different camps in intellectual debates.

While his work continues to be studied since the publication of his successful book *Against Method* (Feyerabend 1975), it is surprising that the period preceding it has hardly been examined with the exception of a few studies (e.g., Haller 1997; Hochkeppel 2006; and only after completion of my German version: Radler 2006; Oberheim 2006, and esp. most recently Collodel and Oberheim 2024). This is even more extraordinary in view of the fact that this phase was decisive for his intellectual development. One can even say that the later Feyerabend returned to the early subjects of his Vienna period. In short, there are more continuities than ruptures in his biography and more consistencies than contradictions in the theoretical aspects of Feyerabend's philosophical work.

My main argument is directed against the usual research trend of an exclusive poststructuralist interpretation (as evident, for instance, in a workshop in Paris on *Against Method*: Grebovicz 2006). I claim that throughout his whole life Feyerabend was actually a philosopher from Vienna or a Viennese philosopher in spite of some significant intellectual developments—even if he himself may not have subscribed to this statement. Moreover, I would also argue that even after he left Vienna he remained closely bound to Austrian philosophy and the Central European tradition of science, so that it is difficult to understand his intellectual development without this context. This was already confirmed in his selective autobiography *Killing Time* (1994), which given his focus and his diction by means of which he sought to consolidate his identity—like any other self-portrayal—is problematic if it is the *sole* source of information and often results in misleading clichés. However, there can be no doubt that the main themes and issues of his writings are already present here, even if the impressionistic arrangement of fragmentary memories culminates in an autobiography with a teleological slant. In the following this will be briefly illustrated from a biographical and philosophical perspective.

## 9.2 Feyerabend in Vienna: The Forgotten "Third Vienna Circle" 1947–1954

After a brief sojourn in Germany (Weimar) during which he pursued artistic inter ests, Feyerabend returned to Vienna and began studying at the University of Vienna in 1947. He first enrolled in history and sociology, and then later began studying physics with Hans Thirring, Karl Przibram and Felix Ehrenhaft and philosophy with Alois Dempf and in particular with Viktor Kraft. (See Feyerabend 1995, 87–110).

The young student's intellectual development was significantly influenced by an important institutional international platform, i.e., the "Austrian College/Forum Alpbach" founded in 1945 by the two brothers Fritz and Otto Molden in the context of the US cultural policy in the Cold War period (with OSS/CIA and the "Congress of Cultural Freedon"). This forum, which still exists today, initiated a number of activities and events (Molden 1981; Auer 1994). Here Feyerabend for the first time came into contact, as of 1948, with philosophers of science, most of whom lived in exile, and with the rest of the scientific community, including Karl Popper, Friedrich A. von Hayek, Hans Albert and with members of the former Vienna Circle, e.g., Rudolf Carnap, Herbert Feigl, and Philipp Frank. Even for Wolfgang Stegmüller, who did not succeed in getting a position in Austria, Alpbach provided an entry to the international network of philosophers and scientists.

Within the context of this unique forum that contributed to making scientific and cultural life less provincial at the beginning of the Second Republic, a separate "working group for natural philosophy" was founded. The so-called "Kraft Circle" organized regular discussions and publications from 1948 to 1954. Viktor Kraft was the academic director and Feyerabend served as the students' speaker. If one looks at the list of participants from Austria and abroad and the reception of this group it is not an exaggeration to refer to it as the "Third Vienna Circle". To date, this circle has hardly received any recognition (Stadler 2006). The members of this circle included Béla Juhos, Walter Hollitscher, Ernst Topitsch as well as students such as Johnny Sagan, Heinrich Eichhorn, Goldberger de Buda, Peter Schiske, Erich Jantsch as well as visitors from abroad such as Elisabeth Anscombe, Emil J. Walter, Georg Henrik von Wright, Edgar Tranekjaer-Rasmussen and the prominent visitor (at least one time) Ludwig Wittgenstein (in 1950). In his memoirs Feyerabend comments on this as follows:

> Kraft was a leading member of the Vienna Circle. Like Thirring he was forced to retire after the "Anschluss" (annexation) of Austria. He was a mediocre speaker but a brilliant and thoughtful thinker. He anticipated some ideas that were later ascribed to Popper …
>
> He knew most of us from the seminar and expressed a desire to meet regularly. This is how the Kraft Circle evolved, a sort of student counterpart to the old Vienna Circle. We were given a room on Kolingasse, the office of the Austrian College, and met twice a month. We discussed concrete scientific theories. For instance, we devoted five sessions to non-Einsteinian interpretations of the Lorentz transformations. Our main topic was the question as to the existence of an external world. (Feyerabend 1995, 104)

In retrospect, Feyerabend criticized the view of science as a system of statements based on his reading of the journal *Erkenntnis*. Broadly speaking, this position was related to the attempts to legitimize a critical and constructive realism with a hypothetic-deductive methodology, which Kraft had already addressed in his monograph *Die Grundformen der wissenschaftlichen Methoden* (1925). (On Kraft, in comparison to Feyerabend, see recently Radler 2006).

All of these discussions are very clearly reflected in Feyerabend's (still unpublished) dissertation titled *Zur Theorie der Basissätze* (On the Theory of Basic Statements, 1951), for which Kraft had served as advisor. In his CV, which Feyerabend had submitted with the papers for his dissertation defense, he describes his personal motivation and the background of the subject. In the following the translation of this important document reads:

> I, Paul Feyerabend, was born in Vienna on January 13, 1924, where I attended primary and secondary school. My interest in philosophy was based on my intense interest in the natural sciences. The books I read included those by Duhem, Mach and Dingler. 1942 to 1945: military service, followed by a year in the military hospital; one year of study at the state Music University in Weimar. I then moved back to Vienna. One semester of history and art history, six semesters of astronomy, physics and mathematics, before finally switching to philosophy. In this connection the discussions conducted in a small circle directed by Prof. Kraft in the style of the Vienna Circle which dealt primarily with issues related to theory of science were very useful. Here, with the inspiration I received from Prof. Popper (London School of Economics), I became interested in the subject I was to deal with in my dissertation. Since 1948 I had had the opportunity to take part in discussions from which I was able to benefit a lot for the final version of this dissertation. I benefited the most from my discussions with Prof. Walter Hollitscher (Berlin). If not from his arguments then through his repeated encouragement to verify more precisely my philosophical views and to substantially correct them (from the Machian positivism to the position I took here). I also benefited from my discussions with Mrs. Anscombe (Cambridge) on the problems of UG [= Untersuchungsgegenstand/subject of investigation, F.S.]. At that time she presented to me a number of formulations that struck me as being completely incomprehensible and which accompanied me for a long time in an unresolved fashion (as well as several formulations that I had picked up from a discussion with L. Wittgenstein who participated in an evening discussion in Kraft's circle.)
>
> Over the course of time, I found ways to understand these ideas—less through reflection than through an unconscious process of development. They are presented in my dissertation. Today they seem to me to be the right interpretation of those formulations (which does not have to meant that this interpretation is historically correct). I was able to discuss the basic ideas of an earlier version in a lecture I gave to the philosophical society in Uppsala (Sweden) and in small circle in Copenhagen with Prof. Marc-Wogau and Prof. Joergensen

> (with the latter I also had some private discussions. I am much indebted to both discussions.
>
> Prof. Kraft drew my attention to some dreadful confusion in an earlier version as well as to several ambiguities.
>
> I am very indebted to Prof. Tranekjaer-Rasmussen (Copenhagen) for some significant aspects of the basic position. He allowed me to read two still unpublished manuscripts, which elaborate on what he only alluded to in a lecture given at Alpbach in 1948 (to which there have also been numerous references.).
>
> I hope that I will soon be able to discover a theory of physical knowledge on the basis of this (still incomplete) preliminary study." (signed: Paul Feyerabend)

According to the dissertation defense file no. 18.107 of the University of Vienna, this decisive study by Feyerabend, which to date has hardly received recognition was fully approved by the first reader Viktor Kraft who wrote the following:

> The dissertation shows extraordinary talent. This is already reflected in the curriculum vitae which does not fit the normal mold at all. The subject of the dissertation is the role that perception statements play in the empirical sciences which has been discussed in modern empiricism. The studies of the dissertation, however, go far beyond this fundamental task in that from there address the definition of the subjects of physics and of the non-physical sciences, in particular, psychology, and thus resolve the opposition of phenomenalism and physicalism. It is the fundamental vantage point of the study to clearly distinguish between two different types of perception statements. On the one hand, the description of perception statements as characterized by an immediately given central function, and on the other hand, its verifying function. With regard to the former, the role of perception is defined as prompting a certain statement; it is exhausted in merely exerting a function of motivation. This way, the content-statement also becomes understandable from experience. However, a perception assertion is not logically distinguished from another one, it does not obtain this way any absolutely certain validity as the theory of "Konstatierung" assumes. Like any other scientific statement it has to be verified. This procedure of verification is analyzed in detail and a separate theory of reliable observation developed. Thus the author opposes, with extensive criticism of the dominant view that the perception statement constitutes the logical foundation of empirical knowledge. He points to the indispensable condition that it is always one theory that establishes the foundation for making use of perception statements. Only within a theory does it assume a specific logical function. Basically, the results of this study deserve recognition for the following reasons: they are new and are really of lasting value. The reflections are on an exceptionally high level, and especially in the final part of the dissertation they show logical perspicacity, and are often developed logistically. The author is extremely well versed with the relevant Anglo-Saxon and Scandinavian literature. However, he also refers to the original passages of classical philosophers. Given the wealth of questions addressed, his presentation is very condensed and often based on so many prerequisites that his ideas are not always easily accessible. The study thus stands out from the average dissertations and must thus be given the highest mark.
>
> Vienna, June 12, 1951
> (signed: V. Kraft, F. Kainz)

The final exams in the subjects of philosophy and psychology with the examiners Viktor Kraft, Friedrich Kainz and Hubert Rohracher received "outstanding mention" and his one-hour exam in physics was also deemed as being "outstanding" by Hans Thirring and E. Schmid.

What stands out here, in epistemological terms, is Feyerabend's unique reconstruction of the basic statement problem, including the protocol statement debate, in the Vienna Circle, taking into account the contemporary experimental-psychological research with a precise reformulation of the Duhem-Neurath-Quine thesis regarding the fundamental theoreticity of all empirical statements, that is to say a modern discussion of the Quinean "Two Dogmas of Empiricism" (1951). Indeed, the archives of the Austrian College contain a manuscript by Feyerabend titled "Die Dogmen des Logischen Empirismus" (The Dogmas of Logical Empiricism) on which a lecture given the same year was based. (cf. Stegmüller 1983).

In the years 1949–1952, Feyerabend traveled abroad—first to Scandinavia where he met with Louis Hjelmslev, Tranekjaer-Rasmussen, Joergen Joergensen, Konrad Marc-Wogau, Anders(?) Wedberg and others. In this connection it must also be noted that the intense scientific communication between Central Europe and the Northern countries since 1900—as opposed to the Anglo-Saxon world—shows a considerable, yet hardly noticed reciprocal continuity before, during and after World War 2. This communication thus represents an exception in connection with the *Cultural Exodus of Logical Empiricism.* The major role played by Eino Kaila and Arne Naess is illustrative of this "Nordic connection" which has been neglected by historians. It was also evident in the journal edited by Ake Petzäll (*Theoria,* 1935 ff) and is continued to the present day by Jaakko Hintikka and his students (Hintikka 2006). It is also no coincidence that Kaila visited the Schlick Circle on several occasions and in his writings criticized the so-called "logistic neo-positivism" (1930) in the spirit of an epistemological realism, but also in agreement with Kraft. By the same token, Arne Naess' early study of Neurath and Carnap in his dissertation and in *Erkenntnis und wissenschaftliches Verhalten* (1936) seems to be an anticipation of Feyerabend's skepticism vis-à-vis science as a system of statements. Nevertheless, Feyerabend was not exactly accepted as a "hero of the greens" by Naess, who was later to become the philosopher of the ecological movement. (Naess 1980). Naess also happened to be the one who made a futile attempt to organize the sixth—and last—"International Congress for the Unity of Science" in Oslo after the Second World War erupted. (The second congress had taken place in Copenhagen in 1936 under the aegis of Niels Bohr and Joergensen). The themes Nordic philosophers were interested in—such as realism/materialism vs. phenomenalism and the relationship of psychology and philosophy ("psychologism")—were those that also played an eminent role in the classical Vienna Circle of Schlick's group and the Third Vienna Circle. Feyerabend was also to a lesser or greater extent interested in these themes.

At the initiative of Kraft, who already had emeritus status at the time, Arthur Pap (1921–1959) was invited to the Philosophy Department of the University of Vienna as a Fulbright visiting professor in 1953/54. This permitted a short-lived revival of analytic philosophy in the tradition of the exiled logical empiricism. (Pap 1955). The promising philosopher hired the highly talented Paul Feyerabend as his assistant. The latter edited the Pap's lecture manuscripts for the book *Analytische Erkenntnistheorie. Kritische Übersicht über die neueste Entwicklung in den USA und England,* which was published in 1955 by Springer Verlag. This book—"dedicated to the Vienna Circle as a commemoration and a revival"—was a reconstruction and critical further development of scientific and analytical philosophy in the wake of the exodus of Logical Empiricism—by the way, also written by an emigrant whose family had been forced to move from Switzerland (Zurich) to the US because of their Jewish background and for political reasons.

There, Pap studied at Columbia University and at Yale, also with Ernst Cassirer, Ernest Nagel and Charles Stevenson. Later he came into contact with Rudolf Carnap and Herbert Feigl, with whom he maintained a life-long friendship. Before this, he has also translated Kraft's *Der Wiener Kreis* (1950) into English (1953) and after his return to the States in 1955, he had a research stint at Feigl's "Minnesota Center for Philosophy of Science" before he succeeded Carl G. Hempel at Yale. His last book only appeared after his premature death under the title *An Introduction to the Philosophy of Science* (1962).

Arthur Pap who at that time, even at his young age, was already a renowned representative of analytic philosophy, had hoped—in vain—to receive a permanent academic position at the University of Vienna. His unexpected death at the age of 28 was due to an insidious kidney illness. (Keupink and Shie 2006)—by the way, a similar tragic life as that of F. P. Ramsey.

If he had been able to stay in Vienna his presence may have ensured a longer influence of the "Third Vienna Circle" and possibly a position for Feyerabend. The latter (who had originally wanted to study with Wittgenstein) had studied with Karl Popper at LSE in 1952 who wanted to hire him as an assistant. Feyerabend did not accept his offer but he did translate Popper's voluminous oeuvre *Open Society* into German. One of the first reviews of Wittgenstein's *Philosophical Investigations* in the context of Austrian philosophy (Ernst Mach, Heinrich Gomperz) is another result of these activities. His first academic position in Bristol (UK)—together with Stephan Körner—in 1955 marked the beginning of Feyerabend's international career and also de facto the end of the short-lived revival of the Vienna Circle during the Second Republic, even though his contacts with Vienna and Alpbach were never to be severed. Before he left he wrote an unpublished, informative and balanced manuscript on the "Geisteswissenschaften in Österreich" (Humanities in Austria, 1954), in which, in a chapter on philosophy, he presents a very detailed description of the specificities and deficits of the post-war era.

That same year, Feyerabend was able—through Pap's intervention—to meet Herbert Feigl (1902–1988) in Vienna, which was to be decisive for his further career and the "Vienna Circle in America" (Feigl 1968/1969) at the Minnesota Center. In the *Festschrift* for Feigl that he edited together with Grover Maxwell, he writes:

> I first met Herbert Feigl in 1954, in the pleasant and stimulating atmosphere of a Vienna coffeehouse. I was then an assistant to Arthur Pap, who had come to Vienna to lecture on analytic philosophy and who hoped, perhaps somewhat unrealistically, that he would be able to revive what was left from the great years of the Vienna Circle and the analytic tradition there… After a lecture, which frequently turned into a heated debate with the attending metaphysicians, we would both retire to the professors' room in the philosophy department and discuss what had just happened. Pap was alternately depressed and incensed what he thought was the impertinence of those who approached philosophical problems without any

> knowledge of logic and of analytical techniques, and he contrasted their easily produced *Sprachträumereien* with the much more modest results which analytic philosophers had achieved by hard work. (1966, 3).

One final manifestation of the philosophical subculture of the Kraft Circle was his *Festschrift* which—edited by Ernst Topitsch (1960)—for the last time brought together students and disciples of Kraft in a remarkable volume on *Problems of Philosophy of Science*, including, Hans Albert, Franz Austeda, Rudolf Freundlich, Béla Juhos, Hubert Schleichert, Wolfgang Stegmüller, Emil J. Walter. Even Feyerabend figured among the contributors, dealing with a central theme of the Kraft Circle: "The Problem of the Existence of Theoretical Entities" (published in Feyerabend 1999). This discussion to be conducted over many years was triggered by Herbert Feigl's article "Existential Hypotheses: Realistic vs. Phenomenalistic Interpretations" (1950) which was largely a justification of "empirical realism". In Feigl's postscript (1981) a continuity of realism in the philosophy of science becomes visible as a variant of Logical Empiricism since the early Schlick and can be traced in contemporary science all the way up to Wilfrid Sellars, whose relevant contributions he favored the most. Feyerabend, who already received a call to Berkeley in 1958, once again addressed the relation of theoretical and observation concepts on the basis of a correspondence theory with a causal theory of experience: "This is the solution which we suggest for the problem of theoretical entities: each observation language contains theoretical elements (this is a result of the renunciation of sense data)." (Feyerabend 1960, 70)

## 9.3 Austrian College/Forum Alpbach and the "Minnesota Center for Philosophy of Science" 1945–1965

The parallel philosophical initiative accompanying the Kraft Circle was the aforementioned "Forum Alpbach" of the "Austrian College". There, in the first decade, an innovative renaissance of exiled scientists took place in a dialogue between Austrian philosophers and visiting scholars from abroad. Thus, for a short time, there was a sort of alternative intellectual forum which represented a contrast to the conservative-clerical "long fifties" of the Cold War. (Hanisch 1995)

The contemporaneous presence of both traditional philosophers and liberal (Hayek, Machlup, Haberler) and neo-Marxist (Adorno, Horkheimer, Marcuse, Bloch) thinkers resulted in a strong revival of the emigrant's science—including nine Nobel prize laureates. Highlights of the annual meetings regarding philosophy of science were the appearances of Rudolf Carnap (in 1964), Herbert Feigl (in 1961 and 1964), Philipp Frank (in 1955) and Karl Popper (regularly from 1948 on), with Erwin Schrödinger figuring as a regular guest in the midst of the "*Intellectual Province*" (Österreichische - Geistige Provinz? 1965) of the post-war years. While it is not easy to reconstruct the history of reception, there is at least one striking manifestation: the book series titled "Wissenschaftstheorie, Wissenschaft und Philosophie" (Philosophy of Science, Science and Philosophy) edited by Simon

Moser together with Otto Molden, the founder of the Forum Alpbach, published by Vieweg Verlag (Braunschweig). In this book series important selected writing by scholars such as Hans Reichenbach, Imre Lakatos, Heinz v. Foerster, as well as Feyerabend's *Philosophical Papers* appeared in two volumes: *Der wissenschafts-theoretische Realismus und die Autorität der Wissenschaften* (1978b) and *Probleme des Empirismus* (1981). In his autobiography (97ff.) Feyerabend described the role of Alpbach and of the Austrian College which he attended for the first time in 1948 for his intellectual development:

> The society had been founded in 1945 on the initiative of Otto Molden, Fritz Molden… and other members of the Austrian resistance. In his book *Der andere Zauberberg* (The Other Magic Mountain) Otto Molden describes the ideas and events which led up to the first meeting in Alpbach, a small village near Brixlegg in Tyrol. Alpbach soon became an international center of intellectual, artistic and economic exchange. A student who had lunch here could find a seat next to Lise Meitner, Bruno Kreisky or Dirac. He could bump into Arthur Köstler, Anneliese Maier or Ernst Krenek … I visited Alpbach about fifteen times, first as a student, then as an instructor and finally, on three occasions, as the head of a seminar.

On his first visit in August 1948, the young student already met Karl Popper who made a great impression on him compared to the German professors. This early appreciation was also reciprocal as was evidenced by Popper's later offer to Feyerabend to serve as his assistant at LSE.

In Alpbach the radical thinker also met two Austrian philosophers and communists—Hans Grümm (1992) and Walter Hollitscher (Klahr Gesellschaft 2001)—who were to win over Feyerabend for realism. It was mainly with Schlick's student Hollitscher, who at that time was still a psychoanalyst and philosopher, that he maintained a deep intellectual friendship until the end of his life in spite of political differences. Feyerabend recalls the influence (1995, 100 f.):

> Walter Hollitscher needed two years to convince me that circularity represents a gain and a practical enrichment and not a disadvantage. Walter showed me that scientific research can be conducted with realistic concepts. I answered that scientists unfortunately had not given up their metaphysical eggshells. Independent of metaphysics scientists had results to show which were accepted by everyone, Walter said: this is also true for the positivists, whereas an antiseptic language and a rigid logic would not lead us forward. The argument made me fall silent for a while but a bit of doubt still remained. Walter moved from physics to politics and that meant: Marx and Lenin. On this issue I resisted his deliberations like a stubborn donkey.

Thanks to Hollitscher Feyerabend also came into contact with Bertolt Brecht whose offer to become his assistant in Berlin he rejected—something he later described as the greatest mistake of his life.

In a letter to Hollitscher (Sept. 5, 1977) Feyerabend defended himself against the reproach of irrationalism in *Against Method:*

> All professional critics of science have jumped at my jokes and analyzed these in philosophical terms, but no one has discussed what prompted these jokes, i.e., the historical material, and some have even claimed that they do not have anything to do with the subject at hand (i.e., theory of science, history of science). … The main argument of the book, which I never formulate explicitly is: the appraisal of methodological (and logical) rules should be left to concrete research. Scholars invent and study material as well as intellectual

instruments of measurement. Of course, the abstract reflections of philosophers are taken into account but they are not considered to be useful because there are philosophical or logical arguments for their "rationality". This idea is not new at all, but apparently many philosophers of science are not familiar with it.

In Alpbach Feyerabend was very quickly involved in the conference activities. In 1955, the physicist and philosopher of science Philipp Frank, once a prominent member of the Vienna Circle and an Einstein biographer came to Alpbach from Harvard (Reisch 2004) to lead the "Erkenntnis and Handlung" (Knowledge and Action) study group. Feyerabend describes this as follows (1955, 140):

> Philipp Frank was a joy. He was very educated, intelligent, witty and a skilful narrator. When he had the choice between explaining a problem with a story or an analytic proof he always opted for the story. Some philosophers did not like that at all. But they always failed to recognize that science is also a story, not a logical problem. Frank elaborated that the Aristotelian objections against Copernicus converged with empiricism, whereas Galilei's law of inertia does not. As in other cases, this remark lay nascent in my mind for years. Then it began to grow. The Galilei chapters in *Against Method* are a late outgrowth of this idea.

Even though, in the case of Alpbach—as opposed to the Kraft Circle—we cannot speak of a uniform trend in philosophy of science, but the regular discussion of philosophy, science and weltanschauung, and its reciprocal influences and foundations, in the context of an international forum was certainly one of the prime factors determining the re-transfer or the new beginning of modern philosophy of science in Austria/Germany after World War 2:

> Particularly important was the fact that representatives of philosophy and of the individual disciplines who had emigrated from Austria and other European countries in the 1930s especially because of the political developments and had, in the meantime, found international recognition, came to Alpbach—for instance, the philosophers Karl Popper, Herbert Feigl, Rudolf Carnap, Philipp Frank, Walter Kaufmann, Karl Löwith, Theodor W. Adorno, Max Horkheimer, Herbert Marcuse, and Ernst Bloch, the legal expert and legal philosopher Hans Kelsen, the sociologist Theodor Geiger, the economists Friedrich August von Hayek, Fritz Machlup and Gottfried Haberler. They met there with younger scholars such as Wolfgang Stegmüller, Ernst Topitsch, Paul Feyerabend, Bernulf Kanitscheider, Paul Weingartner, Karl Acham, and Rudolf Wohlgenannt, who showed interest for their works. And they could also talk with representatives from the Eastern Bloc who were happy to take advantage of this occasion. (Albert 1994, 18).

In his description of a second generation philosopher of science who himself had been a participant it becomes plausible how the local culture of science had, thanks to a younger generation of Austrian philosophers, exerted a rather strong influence—which in the case of Feyerabend and Stegmüller can be illustrated particularly well. (Stegmüller 1979; see also: http://univie.ac.at/ivc/stegmueller)

After leaving Austria in 1956 Feyerabend led a study group "Stetige und unstetige Veränderungen in der Natur" (Permanent and non-permanent changes in nature) and, together with Herbert Feigl, a study group on "Grundlagenforschung und Einzelforschung" (Basic research and individual research) (1964). Still in 1965, he conducted a study group with the title "Philosophy of the exact natural sciences".

Especially Herbert Feigl, Schlick's student, who had emigrated to America already in 1931, was to become one of Feyerabend's most important philosophical influences. As mentioned above, Feyerabend had met Feigl through Pap in Vienna in 1954. After Feyerabend's appointment in Berkeley, Feigl was particularly important for Feyerabend as his host at the "Minnesota Center for the Philosophy of Science" (which Feigl had founded and run from 1953 on). Feyerabend's sojourns at the MCPS (in 1957 and 1959) had brought forth his essays "Explanation, Reduction and Empiricism" (1962) and—most important enough—the first version of "Against Method" (1970) which went unnoticed for a long time. During this time he also wrote his biographical sketch of Feigl in the *Festschrift* that he co-edited with Grover Maxwell: *Mind, Matter and Method. Essays in Philosophy of Science in Honor of Herbert Feigl* (1966). All of these texts are informative documents of this reception to be explored. Herein Feyerabend confirms the impact of Feigl's "Existential Hypotheses" in the Kraft Circle after even Wittgenstein's personal appearance could not convince him on the problem of the external world (Feyerabend and Maxwell 1966, 4):

> It was at this stage of confusion and uncertainty that our attention focused on Feigl's "Existential Hypotheses". Our debates now took a completely different turn. This paper, taken together with Kraft's own contributions and with the ideas which Popper had explained to us on the occasion of his visits to the Alpbach Summer University in the summers of 1948 and 1949, greatly diminished our doubts about realism. There were still some points, which were not entirely clear, and I hoped for an opportunity to discuss the matter with Feigl in person. Another problem that had come up with the realism-positivism issue concerned the application of the calculus of probabilities.

Feyerabend's first meeting with Feigl in Vienna in 1954, where he discussed the issue of applying probability impressed the former because of Feigl's convincing common sense style in the context of a philosophical realism (preferring Popper as opposed to Wittgenstein.) It was also not surprising that Feyerabend was to be a guest at the Minnesota Center twice, whose intellectual climate he describes as follows (Feyerabend and Maxwell 1966, 9):

> The atmosphere at the Center, and especially Feigl's own attitude, his humor, his eagerness to advance philosophy and to get at least a glimpse of the truth, and his quite incredible modesty, made impossible from the very beginning that subjective tension that occasionally accompanies debate and that is liable to turn individual contributions into proclamations of faith rather than into answers to questions chosen. The critical attitude was not absent, on the contrary, one now felt free to voice basic disagreements in clear, sharp, straightforward fashion. The discussions were, and still are, in many respects similar to the earlier discussion in the Vienna Circle. The differences are that things are seen now to be much more complex than was originally thought and that there is much less confidence that a single, comprehensive empirical philosophy might emerge.

This general description obscures how productive Feyerabend was at the MCPS under the direction of Feigl with whom he corresponded from 1957 to 1968. The following manuscripts can be found at the center (cf. "The Book", Walter Library, University of Minnesota):

| | |
|---|---|
| 1955: | Carnap's Theory of Interpretation of Theoretical Systems |
| | A Note on Carnap's New Criterion of Empirical Science |
| 1957: | Replies to Hempel and Carnap |
| 1957: | On the Quantum Theory of Measurement ... Appendix |
| 1958: | An Attempt at a Realistic Interpretation of Theoretical Knowledge Complementarity; Reply to A. Grünbaum; Some further Comments on Conventionality in Geometry (ad Reichenbach); Further Notes on Conventionalism; Comments on Rozeboom |
| 1959: | Reply to Hanson |
| 1960: | Explanation, Reduction, and Empiricism |

From the older friend's perspective the acquaintance with the younger Viennese philosopher was as follows (Feigl 1968, 668):

> I met Feyerabend on my first visit to Vienna after the war (my last previous visit was in 1935.) This was in the summer of 1954 when Arthur Pap was a visiting professor at the University of Vienna. Feyerabend had been working as an assistant to Pap. Immediately, during my first conversation with Feyerabend, I recognized his competence and brilliance. He is, perhaps, the most unorthodox philosopher of science I have ever known. We have often discussed our differences publicly. Although the audiences usually sided with my more conservative views, it may well be that Feyerabend is right, and I am wrong.

This, perhaps, also alludes to Feyerabend's first publication of "Against Method" (15 chapters and 1 appendix comprising 114 pages as opposed to the German edition of 1983: 19 chapters with 3 appendixes) in volume IV of the *Minnesota Studies for the Philosophy of Science,* a still existing series formely edited by Feigl and Maxwell. But the origins of *Against Method* are not mentioned in the biographical and autobiographical literature on Feyerabend. The editors of this special volume IV of MSPS on *Analyses of Theories and Methods of Physics and Psychology*, Michael Radner and Stephen Winokur (1970) describe the contribution together with the essays by Feigl, N. R. Hanson, Carl G. Hempel, Mary Hesse, Grover Maxwell, Joseph Margolis and William W. Rozeboom as a thoroughly "normal science" result of related conferences and discussions at the MCPS. Against this backdrop, it is now surprising to see how simplified and vaguely Feyerabend writes about the evolution of what was to become a bestseller—its history has meanwhile been excellently reconstructed in the volume edited by Matteo Motterlini, *For and Against Method* (1999). In the preface to the German edition of *Against Method* (*Wider den Methodenzwang*) Feyerabend writes:

> In 1970 Imre Lakatos, one of the best friends I ever had, pulled me over to the side and said: "Paul, you have such strange ideas. Why don't you write them down, and I'll write an answer, we'll publish the whole thing and have a great time." I liked the suggestion and I began sat down to work. The manuscript of my part of the planned book was finished in 1972 and I sent it off to London. There it disappeared mysteriously. Imre Lakatos who loved dramatic gestures notified the Interpol, and in the event Interpol actually found my manuscript and sent it back to me. I read it again and rewrote most of it. In February of 1974, only a few weeks after I had completed my revision, I published my part without his reply. (Feyerabend 1989, 11)

This de-historicizing description seems to be motivated by the fact, that for the successful *agent provocateur* of philosophy of science (or criticized as "Salvador Dali of Philosophy") the context of the third Vienna Circle associated with Kraft and the Vienna Circle in America around Feigl (1968) was not being opportune anymore. The success of his—largely justified—criticism of the normative philosophy of science ("received view") had apparently prompted him to this move even without the planned additional contribution from the critical-rational Lakatos. He now advocated a new image as an icon of post-modernity and of epistemological and cultural relativism. It was only at the end of his life that he returned to the Vienna roots of his intellectual development, i.e., the programmatic unity of philosophy and history of science, with Mach, Boltzmann, Alois Riegl and the historical tradition of Logical Empiricism, even if he later criticized the Vienna Circle in an undifferentiated way in connection with Karl Popper.

## 9.4 Feyerabend's Philosophical Return Home after 1980: Ernst Mach and the Historical Tradition in the Theory of Science

By the time Feyerabend received a call to the ETH in Zurich in the 1980s, he had returned both physically and mentally to his early intellectual socialization by rereading Mach and elaborating his relativism and pluralism, after he had already earlier (in 1962) advocated a concept of incommensurability (although critically of Kuhn) and his contextual theory of meaning. This process seemed to take place in parallel to the edition of his two volumes *Philosophical Papers* (1981) and in parallel *Ausgewählte Schriften* (Selected Writings) (1978 and 1981), which he reworked and added postscripts to. It continued with the appearance of his book *Farewell to Reason* (1987), which was published in the German translation as *Irrwege der Vernunft*. (1989).

In his essay "Mach's Theory of Research and its Relation to Einstein" (Haller/Stadler eds. 1988), which he dedicated to Adolf Grünbaum, he lauded—without any reference to Machians Philipp Frank or Richard von Mises—his heuristic and historical approach to research (cf. Stadler 1982, 123ff. and 1988, 40f.). At stake here was preference of a historico-critical theory of research as opposed to the rationalist, abstract-theoretical tradition in philosophy of science. He continued this reconstruction of Mach's oeuvre which he already had formulated in *Erkenntnis für freie Menschen* (1980, 273f.) the following way:

> Mach's critique was part of a general reform of science in the sense that it linked criticism with new results. However, the positivists and their relentless opponents, the "critical rationalists" proceeded from some petrified components of science, which are no longer accessible to research, and reinforced them with the help of philosophical arguments (Popper's "contributions" to realism.) Mach's critique was dialectic and productive, the philosophers' critique was dogmatic and without results.

Feyerabend saw Einstein, Bohr and Otto Neurath as also belonging to this tradition, without retracting his global criticism of the Vienna Circle. Following this rehabilitation, Feyerabend sought to pay homage to Mach's theory of research between abstraction and fantasy independent from his phenomenalistic epistemology, by means of which the dualisms of theory vs. experience, philosophy vs. science, materialism vs. idealism as well as history of science vs. theory of science were to be overcome:

> Together, Mach's arguments result in a philosophy of science that differs from positivism, converges with Einstein's approach to science and moreover present some sensible objections to the 19th century notions of the atom and the theory of relativity. … It is also proven that Mach's 'epistemology' is no epistemology. Rather, it is a general scientific theory (or a draft of a theory) which, in terms of form (but not in terms of content) is comparable to an atomism which, however, differs from all positivist ontology. (1988, 435f.)

To sum up, as this case study in the history of science nicely shows, Feyerabend finally recommends skepticism vis-à-vis established opinions. Reading original texts, for him, serves as a corrective, and he also recommends recognizing the simplification of established opinions as well as "pseudo-disputes" like the one between positivism and realism and the confusion resulting from purely philosophical systems. Finally: systems are a blessing for those who want to remain philosophers if they do not want to recognize the histories of science with relate "myths" as a salvation.

## 9.5 Concluding Remarks

Surveying the above findings on the early Feyerabend between 1945 and 1955, one obtains the following image which both complements and revises the results of previous studies.

The intense focus on the period following the publication of the book *Against Method* (1975) deflects attention from the strong impact, and the reciprocal influence of the "Third Vienna Circle", the Kraft Circle in Vienna of the post-war years and the Austrian College/Forum Alpbach up through the 1960s. Moreover, this intellectual constellation revealed a certain continuity in connection with the Minnesota Center for the Philosophy of Science. Here Herbert Feigl was most influential for the "Vienna Circle in America" both as an individual and in terms of the subjects he dealt with, e.g., materialism-realism-idealism, relativism, theory and experience, confirmation and validation as well as explanation and prediction, which continued to form the range of the pluralistic discussions within philosophy of science.

Even if this philosophical development which can be easily traced in Feyerabend's *Selected Writings* in English and German and his autobiography, as fragmentary and selective as it may be, reveals significant elements of this development, this process has hardly been taken into account in particular in the German-speaking countries. This was also backed by Feyerabend himself who in view of the international bestseller of *Against Method* (1976) and *Science in a Free Society* (1978a/German 1980) let himself be celebrated as the post-modern relativistic thinker of "anything goes" and of "anarchistic epistemology".

Only in the last decade of his life in Zurich, his philosophical roots that ranged from Mach to the Vienna Circle and the "Scandinavian connection" became more visible in a systematic way. The reasons for his neglect of autobiographic factors and historiographic lacunae could be found in his vehement and even polemical break with Popper's critical rationalism but also in the simplistic and holistic perception of logical empiricism (as an ahistorical analytic theory of science.) But even the awareness of the non-reductive naturalism and relativism advocated by Otto Neurath and Philip Frank whose positions were thus often criticized in the Cold War years, should have mitigated this opposition precisely in view of the rediscovery of Ernst Mach. This to be claimed even if a critical stance towards rationalism and post-enlightenment philosophy of science remained a difference. In no way does any of these results limit Paul Feyerabend's undisputed originality and intellectual autonomy, the qualities of this often so misunderstood creative thinker and convinced democrat.

It thus does not appear exaggerated to refer to Paul Feyerabend as a philosopher from Vienna and to align him with the Vienna Circle, as someone who was so strongly influenced by this group (from the first to the third Vienna Circle, including Wittgenstein and Popper, from the "other Magic Mountain" to the "Minnesota Center) and who towards the end of his life returned to these philosophical traditions in a both constructive and provocative way.

In his last interview Feyerabend (1995) himself gives an answer to the question why he did not return to Vienna and whether he associated unpleasant experiences with this city in his typical way:

> No, no. My experiences there were excellent, it was wonderful. There were good people there. You know, we physics students stormed the philosophy lectures, stood up in the middle of them and said: "That's all nonsense what's being spoken here." Then we were kicked out of the lectures.

What the person being interviewed here so elegantly remained silent about was the deplorable fact that both as a highly talented young scholar and as internationally renowned philosopher he was never asked officially to return to Vienna. He himself returned in spirit—but the potential and results of this move have yet to be appraised.

## 9.6 Personal Reminiscence

The author of this text unfortunately did not have a chance to meet Feyerabend in person. However, he did have the great pleasure of hearing him speak at the old Vienna city hall. Before that event he sent me a nice postcard, after I had sent him a copy of my book on the reception of Ernst Mach (Stadler 1982) with a reference to his own Mach reading. He then wrote me the following card (originally in German), in his typically capricious and provocative style anticipating some arguments I have just presented:

> Dear Mr. Stadler—many thanks for your nice collection of gossip—by which I mean your report on the positivists, their precursors and successors. I have only taken a quick glance at various parts of the book and I was delighted to find so many familiar names, above all, my teachers Thirring, Kraft and Hollitscher. (THEY were my teachers and not this windbag Popper.) A further teacher, unfortunately no longer alive when I was a student, was Mach and as a convinced follower of Mach I am presenting a further exegesis which more or less contradicts everything which erudite gentlemen such as Holton and others presently have to say about Mach (Gereon Wolters however agrees with me.) Yes, and perhaps we'll bump into each other some time. (signed Paul Feyerabend).

*Note*: The "further exegesis" alluded to here were the enclosed proofs of Feyerabend's above cited essay "Mach's Theory of Research and its Relation to Einstein" (first in: *Studies in the History and Philosophy of Science*, 1984, vol. 15, no.1.) The text also appeared in *Farewell to Reason* (Feyerabend 1987) and in subsequent German translations in Haller/Stadler (Hrsg.), *Ernst Mach—Werk und Wirkung* (1988), Feyerabend, *Irrwege der Vernunft* (1989, with an addendum 1988), and Gereon Wolters (1987). This article is a slightly revised and updated version of (Stadler 2010). This topic will be the focus of the proceedings of the international conference "Paul Feyerabend and Austrian Philosophy - His Formative Years in Postwar Vienna" in Vienna 2024, to be ed. by Matteo Collodel, Daniel Kuby, Friedrich Stadler (forthcoming with Springer Nature Switzerland 2025).

## References


Albert, Hans. 1994. Wissenschaft in Alpbach. In a.a.O. *Forum Alpbach*, Alexander Auer (Hrsg.), 17–22.

Auer, Alexander, ed. 1994. *Das Forum Alpbach 1945–1994*. Wien: Ibera Verlag-European University Press.

Collodel, Matteo, and Oberheim, Eric. eds. 2024. *Feyerabend's Formative Years. Volume 2. Feyerabend on Logical Empiricism, Bohm & Kuhn. Correspondence and Unpublished Papers*. Cham: Springer.

Feigl, Herbert. 1950. Existential Hypotheses. Realistic Versus Phenomenalistic Interpretations. In *Philosophy of Science,* Vol. 17, 35–62. Auch in: *Ebda.,* 192–223.

———. 1968. The *Wiener Kreis* in America, vol. 69. In *Ebda*., 57–94.

———. 1981. In *Inquiries and Provocations. Selected Writings*, ed. Robert S. Cohen. Dordrecht/Boston/London: Reidel.

Feyerabend, Paul. 1951. *Zur Theorie der Basissätze*. Unveröffentlichte Dissertation. Universität Wien.

———. 1954. Die Geisteswissenschaften in Österreich. Unveröffentlichtes Manuskript, Wien (163pp).

———. 1955. Wittgenstein's Philosophical Investigations. In *The Philosophical Review,* Vol. 64, 449–483. Auch in Feyerabend 1981, 293–325.

———. 1960. Das Problem der Existenz theoretischer Entitäten. In Topitsch (Hrsg.), 35–72.

———. 1970. Against Method: Outline of an Anarchistic Theory of Knowledge. In *Analyses of Theories and Methods of Physics and Psychology*, ed. Michael Radner and Stephen Winokur,

17–130. Minneapolis: University of Minnesota Press. (Minnesota Studies in the Philosophy of Science, Vol. IV, ed. Grover Maxwell and Herbert Feigl).
———. 1975. *Against Method. Outline of an Anarchistic Theory of Knowledge*. London: New Left Books.
———. 1976. *Wider den Methodenzwang. Skizze einer anarchistischen Erkenntnistheorie*. Frankfurt/M.: Suhrkamp. Revidierte und erweiterte Fassung 1983.
———. 1978a. *Science in a Free Society*. London: New Left Books. Paul Feyerabend. 1980. *Erkenntnis für freie Menschen*. Frankfurt/M.: Suhrkamp.
———. 1978b. *Der wissenschaftstheoretische Realismus und die Autorität der Wissenschaften. Ausgewählte Schriften*, Band 1. Braunschweig/Wiesbaden: Vieweg & Sohn.
———. 1981. *Probleme des Empirismus. Schriften zur Theorie der Erklärung, der Quantentheorie und der Wissenschaftsgeschichte. Ausgewählte Schriften*, Band 2. Braunschweig/Wiesbaden: Vieweg & Sohn.
———. 1989. *Irrwege der Vernunft*. Aus dem Amerikanischen von Jürgen Balsius. Frankfurt/M.
———. 1995. *Zeitverschwendung*. Übersetzt von Joachim Jung. Frankfurt/M.: Suhrkamp.
———. 1999. *Knowledge, Science and Relativism. Philosophical Papers*, Vol. 3, ed. John Preston. Cambridge University Press.
———. 2005. *Die Vernichtung der Vielfalt. Ein Bericht*. Aus dem Englischen von Volker Bönigk und Rainer Noske. Hrsg. Von Peter Engelmann. Wien: Passagen Verlag.
———. Das Problem der Existenz theoretischer Entitäten. In Topitsch (Hrsg.), a.a.O., 35–72.
———.. *Herbert Feigl: A Biographical Sketch*. In *Ebda.*, 3–16.
Feyerabend, Paul, and Grover Maxwell, eds. 1966. *Mind, Matter, and Method. Essays in Philosophy and Science in Honor of Herbert Feigl*. Minneapolis: University of Minnesota Press.
Grebowicz, Margaret. 2006. Organizer of the Symposium *Against Method*: Thirty Years Later. In Abstracts of the 6th International History of Philosophy of Science Congress, Paris, 14–18 Juni 2006. *Abstracts*, 64–67.
Grümm, Hans. 1992. *Drei Leben. Krieg, Partei, Atom*. Wien: Löcker.
Haller, Rudolf. 1997. In memoriam Paul Feyerabend. In *Bausteine Wissenschaftlicher Weltauffassung*, ed. Friedrich Stadler, 91–100. Wien/New York: Springer.
Hanisch, Ernst. 1995. *Der lange Schatten des Staates. Österreichische Gesellschaftsgeschichte im 20. Jahrhundert*. Wien: Ueberreuter.
Hintikka, Jaakko. 2006. *The Philosophy of Jaakko Hintikka*, ed. Randalle E. Auxier and Lewis Edwin Hahn. Chicago/LaSalle: Open Court 2006 (Library of Living Philosophers, Vol. XXX).
Willy Hochkeppel, Propheten im eigenen Lande ... Über den Umgang mit einem großen philosophisch-wissenschaftlichen Erbe in der 2. österreichischen Republik. Impressionen eines Außenseiters. In Gerbert Frodl, Paul Kruntorad, Manfried Rauchensteiner (Hrsg.), *Physiognomie der 2. Republik von Raab bis Bruno Kreisky*. Wien 2006, 115-151.
Keupink, Alfons, and Sanford Shie, eds. 2006. *The Limits of Logical Empiricism. Selected Papers of Arthur Pap*. Dordrecht: Springer.
Klahr Gesellschaft, Alfred (Hrsg.). 2001. *Zwischen Wiener Kreis und Marx. Walter Hollitscher (1911–1986)*. Wien.
Kraft, Viktor. 1925. *Die Grundformen der wissenschaftlichen Methoden* ( = Sitzungsberichte der Österreichischen Akademie der Wissenschaften, philos.-histor. Klasse 203). Wien, 1–104.
———. 1950. *Der Wiener Kreis. Der Ursprung des Neopositivismus*. Wien/New York: Springer. 3. Auflage 1997.
Molden, Otto. 1981. *Der andere Zauberberg. Das Phänomen Alpbach*. Wien/München/Zürich/ New York.
Motterlini, Matteo, ed. 1999. *For and Against Method. Imre Lakatos and Paul Feyerabend. Including Lakatos's Lectures on Scientific Method and the Lakatos-Feyerabend Correspondence*. Chicago/London: The University of Chicago Press.
———. 2005. Paul Karl Feyerabend. In *The Philosophy of Science. An Encyclopedia*, ed. Sahotra Sarkar and Jessica Pfeifer, 304–310. New York/London: Routledge.

Naess, Arne. 1936. *Erkenntnis und wissenschaftliches Verhalten*. Oslo.
———. 1980. Paul Feyerabend – Ein Held der Grünen? In Hans-Peter Dürr (Hrsg.), a.a.O., 184–199.
Oberheim, Eric. 2006. *Feyerabend's Philosophy*. Berlin/Wien: De Gruyter.
Österreich – Geistige Provinz?. 1965. Wien/Hannover/Bern: Forum Verlag.
Pap, Arthur. 1955. *Analytische Erkenntnistheorie. Kritische Übersicht über die Entwicklung in USA und England*. Wien: Springer.
Preston, John. 1997. *Feyerabend. Philosophy, Science and Society*. Cambridge: Polity Press.
Willard Van Orman Quine. 1951. Two Dogmas of Empiricism. *Philosophical Review* 60: 20–43.
Radler, Jan. 2006. *Victor Krafts konstruktiver Empirismus. Eine historische und philosophische Untersuchung*. Berlin: Logos Verlag.
Reisch, George. 2004. *From the Icy Slopes of Logic. How the Cold War Transformed Philosophy of Science*. Cambridge University Press.
Stadler, Friedrich. 2006. Organizer of the Symposium "On the Banishment and Return of the Philosophy of Science after World War II" im Rahmen von HOPOS 2006, 6th International History of Philosophy of Science Congress, in Paris, 14–18. Juni 2006. Vgl. *Abstracts*, 82–86.
———. Philosophie – Zwischen 'Anschluss' und Ausschluss, Restauration und Innovation. In *Zukunft mit Altlasten. Die Universität Wien 1945 bis 1955*, ed. Margarete Grandner, Gernot Heiss, and Oliver Rathkolb, vol. 2005, 121–136. Innsbruck/Wien/München/Bozen: Studien Verlag.
———. 2010. Paul Feyerabend and the Forgotten 'Third Vienna Circle'. In Friedrich Stadler (Hrsg.), *Vertreibung, Transformation und Rückkehr der Wissenschaftstheorie. Am Beispiel von Rudolf Carnap und Wolfgang Stegmüller*. Wien: LIT Verlag, 169–188.
Stegmüller, Wolfgang. 1979. *Rationale Rekonstruktion von Wissenschaft und ihrem Wandel*. Mit einer autobiografischen Einleitung. Stuttgart: Reclam.
———. 1983. Vom dritten bis zum sechsten (siebten?) Dogma des Empirismus. In *Information Philosophie* 1983/4, 4–24.
Topitsch, Ernst, ed. 1960. *Probleme der Wissenschaftstheorie. Festschrift für Victor Kraft*. Wien: Springer.
Wolters, Gereon. 1987. *Mach I, Mach II, Einstein und die Relativitätstheorie*. Berlin/New York: de Gruyter.

# Chapter 10
# Feyerabend and Wittgenstein

**Vasso Kindi**

**Abstract** In this chapter I consider the ambivalent relation between Feyerabend and Wittgenstein. I first recount the facts that pertain to this relation and then focus on Feyerabend's philosophical work and the influence it received from Wittgenstein's. I argue that Feyerabend remained much closer to Popper, and his Viennese positivist roots, than to Wittgenstein in the greater part of his career, but was idiosyncratically influenced by Wittgenstein's later philosophy in his own early and late writings. In his early work, Feyerabend used Wittgenstein's discussion of perception to challenge the theory-observation dichotomy while in his late work he appropriated Wittgenstein's focus on particulars to combat abstract theorizing and Wittgenstein's understanding of meaning to contest the view that meanings are determined by phenomena. Yet, despite the influence, I argue that the orientations of the two philosophers diverge. Feyerabend, a staunch reformist, criticized Wittgenstein as a conformist and contrasted his pluralism to Wittgenstein's considerations of alternative games. Finally, I question whether Feyerabend's anti-scientism and emphasis on the sense of wonder at the mysteriousness of life and the world match ostensibly similar remarks made by Wittgenstein.

**Keywords** Paul Feyerabend · Ludwig Wittgenstein · Karl Popper · Theory-observation dichotomy · Meaning

The Feyerabend-Wittgenstein relation is a contentious and, yet, marginal issue in the Feyerabend literature. It is non-existent in the Wittgenstein literature. Why is it marginal in the Feyerabend literature? Because Feyerabend's philosophy is most commonly discussed in relation to the philosophy of science figures and literature, and, more particularly, in relation to Popper's philosophy. Feyerabend used to be in

V. Kindi (✉)
Department of History and Philosophy of Science, National and Kapodistrian University of Athens, Athens, Greece
e-mail: vkindi@phs.uoa.gr

© The Author(s), under exclusive license to Springer Nature Switzerland AG 2024

S. Gattei, R. Corvi (eds.), *Feyerabend in Dialogue*, Boston Studies in the Philosophy and History of Science 346,
https://doi.org/10.1007/978-3-031-71938-7_10

Popper's circle for quite some time, either, initially, as his student, or later as his associate, for a long time as his defender, and finally as his vociferous and unsparing critic. It is also marginal because Wittgenstein was phased out from philosophy of science, especially in the United States where, and during the time, Feyerabend was active publishing and lecturing. The logical positivists who had emigrated to the United States and influenced the development of philosophy of science there, left behind most of their European roots and forgot their indebtedness to Wittgenstein, while Wittgenstein's influence on the historical turn in philosophy of science in the 1950s and 1960s, was played down and rarely appreciated (Glock 2008, 2013, 586; Kindi 2017, 592 – 593). Why is the Feyerabend-Wittgenstein relation contentious? Because both Feyerabend and the major Feyerabend scholars do not concur as to whether Feyerabend was indeed influenced by Wittgenstein and his philosophy. In his "Autobiographical Remarks" (2020, 28), Feyerabend admitted that Wittgenstein had influenced him "more than [he, Feyerabend] perhaps realise[d]," and thought that there were "sometimes objectionable, but often enlightening traces" of Wittgenstein's influence in his later work.[1] And yet, he said that it was a lucky coincidence that he did not have Wittgenstein as a supervisor because, in that case, he would have become a Wittgensteinian for the rest of his life (ibid, 28–29). Feyerabend both praised and belittled Wittgenstein's philosophy. On the one hand, he said that he admired Wittgenstein's philosophy and he regarded Wittgenstein "as one of the great philosophers of the twentieth century" (Feyerabend 1991a, 50). On the other, he said that Wittgenstein's philosophy "contains important (though by no means original) contributions" (Feyerabend 1960, 247), and wrote in a letter to Thomas Kuhn, that he refused to talk about Wittgenstein when asked. He said that when asked whether one should read the *Philosophical Investigations*, he used to reply: "If you want to waste your time, yes" (Hoyningen-Huene 1995, 384).

The Feyerabend scholars, on their part, disagree among themselves as to whether there was any significant influence and, if any, what the affinities between the two philosophers are. Eric Oberheim, devotes a chapter in his *Feyerabend's Philosophy* (2006, 43–76) to the Feyerabend-Wittgenstein relation and contends that Wittgenstein influenced Feyerabend's semantic views even though Feyerabend was critical of Wittgenstein's account of meaning as use. Oberheim also contends that "[t]he deepest and most lasting impact that Wittgenstein's ideas had on Feyerabend's philosophical development concerns nominalism … [i.e.,] that our concepts are not uniquely correct descriptions of the real structures of the world" (ibid., 76). Feyerabend says in his autobiography (1995, 94) that he became a Wittgensteinian nominalist, while Joseph Agassi (2008, 200n56) thought that Feyerabend, mistakenly, took Wittgenstein to be a nominalist, i.e., an anti-essentialist about meaning. Ian James Kidd (2017), on his part, claims that what the two philosophers most significantly share is their rejection of scientism which is blamed for eliminating a sense of mystery and wonder from our worldview.

---

[1] Cf.: "They [i.e., discussions with Elizabeth Anscombe about Wittgenstein's later writings] had a profound influence upon me though it is not at all easy to specify particulars." (Feyerabend 1982, 116).

We've seen why Wittgenstein is marginal in the Feyerabend literature and why the Feyerabend-Wittgenstein relation is contentious. But why is it virtually non-existent in the Wittgenstein literature, especially given the fact that Elizabeth Anscombe, Wittgenstein's student and literary executor, was very encouraging and very instrumental in getting Feyerabend engaged with Wittgenstein's philosophy? Feyerabend himself thought that Wittgensteinians were mere followers and he, a free spirit, was not to be welcomed in their midst.[2] That is why he said that it was a lucky accident that he did not become a Wittgensteinian (Feyerabend 2020, 28–29). But, the fact is that Feyerabend, apart from a couple of early papers and his review of the *Philosophical Investigations* in 1955, did not work on Wittgenstein's philosophy either approvingly or disapprovingly. He may have made use of some Wittgensteinian ideas that he adapted to his purposes, but he did not really dwell on it. Maybe, the fact that he met the philosopher himself and studied Wittgenstein's philosophy very early on, made him think that he knew it too well and assumed a cavalier attitude towards it. Yet, despite some perceptive comments regarding Wittgenstein's philosophy,[3] the truth is that Feyerabend interpreted it in an idiosyncratic way and had a different philosophical orientation than Wittgenstein. He said, for instance, that, "Popper's ideas were similar to Wittgenstein's," and that he [Feyerabend] was inclined towards the dogmatic and the abstract in contradistinction to Wittgenstein's emphasis on the concrete and the particular (Feyerabend 1982, 115–116).

In what follows, I will, first, give some facts regarding Feyerabend's relation to Wittgenstein and then proceed to discuss some themes that have been pointed out as showing affinities between Feyerabend's and Wittgenstein's philosophy.

## 10.1 The Facts

Matteo Collodel (2016) and Eric Oberheim (2006) have given us a detailed account of Feyerabend's relation to Wittgenstein. Feyerabend met Wittgenstein only once in early 1950 in Vienna, when Wittgenstein's student, Elizabeth Anscombe, who was in Vienna to learn German to translate the *Philosophical Investigations*, mediated, and Wittgenstein was invited and gave a lecture at the "Kraft Circle,", which was formed by Viktor Kraft's students. Viktor Kraft, a founding member of the Vienna Circle, was Feyerabend's dissertation advisor, and Feyerabend had a leading role in these meetings. Feyerabend remembers from Wittgenstein's lecture that

---

[2] Cf.: "… Wittgensteinians neither seek nor are in need of great numbers of followers and so they do not claim me as one of their own" (Feyerabend 1991a, 50). Cf. also what Feyerabend says in a letter of 1974 to Imre Lakatos: "… Popperianism is very much like Wittgensteinianism: in both cases philosophy seems to start with the master …" (Motterlini 1999, 365).

[3] For instance, Feyerabend writes that Wittgenstein had "emphasized that science contains not only formulae and rules for their application but entire traditions" (Feyerabend 1982, 66).

> Wittgenstein was very impressive in his way of presenting concrete cases, such as amoebas under a microscope (I cannot now recall the reason this example was used), but when he left we still did not know whether or not there was an external world, or, if there was one, what the arguments were in favor of it. 'You philosophers' said one of the participating engineers in despair, 'are all alike. There you tell us that Wittgenstein turned philosophy upside down. He talks just as much as everyone in this profession and can't give a straightforward answer to a straightforward question.' (Feyerabend 1966, 4)[4]

Subsequently, Feyerabend went to England, allegedly to study, as a post doc, with Wittgenstein at Cambridge. Since Wittgenstein had in the meantime passed away, Feyerabend had to find a new post and moved to London to work with Popper at LSE (see, for instance, Feyerabend 1995, 86; 2020, 28). Collodel, however, has shown that, from the beginning, as it is stated in Feyerabend's application to the British Council for a scholarship at the end of 1951, i.e., after Wittgenstein's death, Feyerabend's first choice was to work with Popper at LSE (Collodel 2016, 36).

In 1949 Elizabeth Anscombe gave Feyerabend manuscript copies of the *Philosophical Investigations* and of the *Remarks on the Foundations of Mathematics* (Feyerabend 2020, 28),[5] and in 1955 Feyerabend published a review of the *Philosophical Investigations* (Feyerabend 1955), which Norman Malcolm partly approved of (Feyerabend 2020, 161), Feyerabend found "*much* too talkative (ibid., 185, his emphasis), and Ryle said it was an "efficient condensation, not a review" (Feyerabend 1995, 93). Feyerabend (1995, 93) also thought that his rendering of Wittgenstein's ideas in the review "falsified Wittgenstein's intentions". Anscombe herself translated the review from German to English. Feyerabend also had the opportunity to present his views on incommensurability (without yet the term), at a meeting at Anscombe's house with Peter Geach, H.L.A. Hart, and Georg von Wright because he thought that Wittgenstein's ideas of meaning "were a very essential ingredient of the idea of incommensurability" that he had eventually developed (Feyerabend 2020, 28).

## 10.2 The Philosophy

Feyerabend (2020, 28) confessed in his "Autobiographical Remarks" of 1976, that Wittgenstein set him afire while Popper left him cold. Yet, Juliet Floyd (2005, 105) has claimed that "Feyerabend remained to the end a Popperian," while John Preston (2000, 99) maintained that Feyerabend "*does* deserve to be thought of as Wittgensteinian" (emphasis in the original). I will argue that Feyerabend remained much closer to Popper, and his Viennese positivist roots, than to Wittgenstein in the

[4] Feyerabend, in his biographical sketch of Herbert Feigl (1970, 4), says that Wittgenstein's visit to the Kraft Circle took place in 1952. Wittgenstein passed away in 1951 and Matteo Collodel (2016, 36) has persuasively shown that Wittgenstein gave a lecture at the Kraft Circle in early 1950.

[5] Eric Oberheim (2006, 65n72) writes that Anscombe gave Feyerabend German manuscripts of the *Investigations* in 1952 in London.

greater part of his career, but I will agree with Preston that, Wittgenstein's philosophy was, perhaps idiosyncratically, appropriated by Feyerabend, especially in his late writings.

### *10.2.1 Feyerabend's Statement View of Theories*

Feyerabend may have bragged about his "Wittgensteinian upbringing" (Motterlini 1999, 272), but his upbringing was actually logical positivist. His doctoral dissertation, entitled *Zur Theorie der Basissätze*, which was supervised by Viktor Kraft, intervened in the protocol sentence debate and challenged the sharp theory-observation dichotomy, so crucial for testing theory against experience. He defended the view that observation sentences are thoroughly theoretical and that there is no unadulterated *Given* (Feyerabend 2020, 25).[6] In two unpublished letters to Thomas Kuhn in the 1980s, when Kuhn was working again on the concept of incommensurability trying to address the criticism of those "who made a mess of it," Feyerabend explains how his concept of incommensurability was tied to the so-called "statement view" of theories. In the first letter, dated April 4, 1983,[7] after he refers Kuhn to chapter 17 of his *Against Method*, Feyerabend writes:

> …I came to incommensurability (one narrow aspect of it) via a critique of the double language model. In Minnesota where I spent two years (in installments) the situation was presented thus: there is a theoretical language, there is an observation language and meaning flows from the observation language to the theoretical language by "upward seepage"[8] (this is how Paul Meehl, the psychologist expressed the matter). In my thesis (1951) I tried to show that observations taken by themselves cannot determine meanings. I therefore suggested a "downward seepage" theory, i.e. I suggested that all there is are theoretical meanings, all terms are, semantically speaking, theoretical terms and the difference between observation and theory has nothing to do with meaning. (Kuhn Papers MC 240, Box 21, Folder 28)

Feyerabend repeats the same claim in his letter of February 21, 1988:

> You [Kuhn] say that quantum theory must have played a role in my way towards incommensurability. I am not so sure, for I don't remember all the details. The idea that there exists no semantic distinction between observation statements and theoretical statements (observation statements are not merely theory-laden but fully theoretical) came from the protocol-statement discussion of the Vienna Circle; just good old positivism. What I added was a discussion and refutation of what I called the 'stability thesis', viz. that some meanings are unaffected by theoretical change. In my 1958 paper [Feyerabend 1958], I discussed two ways of defending that thesis, one being connected with Bohr's version of

[6] Feyerabend defended a pragmatic theory of observation. For more on this see Feyerabend (1958, 1962), and Kuby (2018).

[7] Feyerabend's own letter bears the date "4.4.83" but, on the copy in the archive there is also a seal stamp with the date "April 14 1983". Probably the latter is the date the letter reached Kuhn.

[8] Cf. Feigl's diagram (1970, 6–7) for the "upward seepage" of meaning from the observational terms to the theoretical concepts.

> complementarity. But this was only an example and I don't remember how much it meant to me at the time. The objection that crucial experiments would now be incomparable because theories would be incomparable came from Feigl; it did not bother me at all for I interpreted crucial experiments as practical decisions not as theoretical comparisons of statements with stable meanings. (Kuhn Papers MC 240, Box 21, Folder 28)[9]

In both excerpts, Feyerabend traces the origins of his idea of incommensurability and shows how connected it is to the positivist debates. In a handwritten note in the second letter, Feyerabend clarifies this issue further:

> All this was carried out within the anti- (or a-) historical framework of Neopositivism up to Quine. The historical approach I adopted only later—and there *you* were my teacher. (Kuhn Papers MC 240, Box 21, Folder 28, emphasis in the original)

The statement view of theories was matter-of-factly adopted in Popper's philosophy and, in that respect at least, Popper was very close to the positivists. Feyerabend shared this view about theories for quite some time.[10] In his influential paper "Explanation, Reduction and Empiricism" (1992), Feyerabend discusses and criticizes the formal accounts of explanation and reduction, then dominant in philosophy of science, which considered relations between statements. In the same paper, Feyerabend explicitly draws upon Popper's philosophy and is critical of linguistic

---

[9] Feyerabend gives the same account in his "Concluding Unphilosophical Conversation" (1991b, 526–7). Retrospectively, after Kuhn's emphasis on the practice of science, Feyerabend criticized the approach that saw only statements in science:

> For the logical positivists, ... science was a system of statements and theories ... Can you imagine that? There are these philosophers, and they are intelligent people and all they see when looking at science is *statements*. They do not see laboratories; they do not see the fights scientists and politicians engage in to settle financial questions; they do not see the large telescopes, the observatories, the staff buildings, the staff conferences, the effects of an asshole in power on his underlings—they see only statements. [...] the thesis that science is system of statements, i.e., a grandiose text. It is an absurd caricature of what science is. (Feyerabend 2011, 64, 129)

[10] Feyerabend says that, contrary to Kuhn's approach to science, his own research "concerned theories only" (1982, 67). Yet, in his "Reply to criticism: comments on Smart, Sellars, and Putnam" (1985a, 105-6n5), Feyerabend compares his 'theories' to, among other things, Wittgenstein's 'language-games' and Kuhn's 'paradigms,' and says that, by 'theories,' he also means myths, political ideas, and religious systems. So, it may be thought that he does not share the logical positivist and Popperian idea of theories as sets of statements. I believe that in comparing his 'theories' to Wittgenstein's 'language-games' and Kuhn's 'paradigms,' Feyerabend underestimates or ignores the practical dimension of Wittgenstein's and Kuhn's concepts. Language-games and paradigms are not just larger, more comprehensive, or more spacious entities than sentences. They are imbedded in practice in a way that cannot be captured by sentences. Given his understanding of paradigms or language-games (i.e., as larger than sentences), Feyerabend continues to operate within the statement view of theories. His approach continues to differ from Kuhn's. I thank an anonymous reviewer for drawing my attention to Feyerabend's comment on 'theories'.

philosophers, just like Popper was,[11] for their "long-winded inquiries," which may be uninteresting, but may also hide methodological presuppositions (ibid., 61n71). He calls them scholastic, and had most probably in mind the ordinary language philosophers, mainly J. L. Austin, whom he had met at Berkeley in 1958.[12] But, at the same point in his paper, he mentions disapprovingly Wittgenstein and his remark in *Philosophical Investigations* (§124) that philosophy must not interfere with the actual use of language and "leave everything as it is" (ibid., 61). Feyerabend thinks that this attitude encourages conformism and suppresses criticism which is vital for the advancement of science. His whole paper aims to show that, despite the all- pervasiveness of theories which permeate even observation statements, theories can be criticized and tested.

Now, this particular understanding of Wittgenstein shows that Feyerabend thought that language games are closed systems with their own concepts and standards. That Feyerabend had this view is also shown by what he says in the same article about his "contextual theory of meaning". According to this theory, "the meaning of a term is not an intrinsic property of it but is dependent upon the way in which the term has been incorporated into a theory" (ibid., 68). The context, Feyerabend adds, extends beyond one specific theory to all the theories that are empirically adequate, are tested with it, and are incompatible with it (ibid., 69). Feyerabend credits Wittgenstein for "defending most forcefully" this view in the twentieth century, but criticizes him for "restrict[ing] this theory to the inside of his language games: Platonism of concepts is replaced by Platonism of (theories or) games." (ibid., 68n83). Platonism for Feyerabend means stability of meaning.[13] However, Wittgenstein did not hold this view. Language-games have a practical dimension, and are not mere sets of statements. The practice that makes possible and sustains each one of them extends indefinitely into the future. Language games are open, and their concepts, which are uses of words in particular circumstances,

---

[11] See, for instance, Popper (1992, 90):

> I may perhaps say here that what I regard as the ultimate cause of the dissolution of the Vienna Circle and of Logical Positivism is not its various grave mistakes of doctrine (many of which I had pointed out) but a decline of interest in the great problems: the concentration upon *minutiae* (upon "puzzles") and especially upon the meanings of words; in brief, its scholasticism. This was inherited by its successors, in England and in the United States.

[12] In several letters to Karl Popper from Berkeley, Feyerabend discusses Austin both positively and negatively. In one of these letters (Nov. 19, 1958), he says the following: "Whatever philosophical thesis Austin may hold—and he is very careful not to commit himself—is pure Wittgenstein, only a watered down Wittgenstein. I got this admission out of some very ardent admirers of his." (Collodel and Oberheim 2020, 304) Maybe Feyerabend got this idea about Austin from Stanley Cavell who was his colleague at Berkeley and a very ardent admirer of both Austin and Wittgenstein.

[13] Cf. Feyerabend (1962, 31): "it is one of the basic assumptions of Platonism that the key terms of sentences expressing knowledge (episteme) refer to unchangeable entities and must therefore possess a stable meaning."

are flexible and fluid (see Kindi 2012a).[14] Feyerabend thinks of a set context conferring meaning upon words, and confines context to theories, i.e., sets of statements. For Wittgenstein, on the other hand, language use is embedded in practice, and things are more dynamic. While there are established uses of words that can be taught, each new use contributes to the building and extension of the respective practice, which remains in a state of flux, encompassing a dense network of activities.

So far, I have tried to show that Feyerabend has held the so-called statement view of theories which brought him close to the logical positivists and Popper. He even said that his concept of incommensurability is connected to this view (Feyerabend 1991b, 526). However, Feyerabend has also said, without elaborating, that Wittgenstein's idea of meaning had a decisive influence on his concept of incommensurability (Feyerabend 1982, 115; 2020, 28). Given that Wittgenstein's later work cannot be assimilated to the statement view of theories, which of the two options, as regards incommensurability, should one opt for? Was Feyerabend's notion of incommensurability connected to the positivists' statement-view or originated in Wittgenstein's philosophy? There is no doubt that, Feyerabend's attack on the theory-observation distinction, intervened in the protocol-sentence debate which concerned the statement-view of theories that was held by the logical positivists and Popper. Observation statements, of critical significance for transmitting empirical content to theoretical terms and for testing theories, became Neurath's protocol statements and Popper's basic statements. Feyerabend, as already noted, challenged this distinction arguing that observation sentences are fully theoretical. This is where Wittgenstein's philosophy enters. Feyerabend had heard Wittgenstein's lecture at the Kraft Circle's seminar where Wittgenstein spoke about what one sees when looking through a microscope (Feyerabend 1995, 76). Feyerabend has said that "Wittgenstein was very impressive in his way of presenting concrete cases, such as amoebas under a microscope" (Feyerabend 1966, 4). The amoeba example is the first N. R. Hanson uses in the first chapter of his book *Patterns of Discovery* (1958, 4) in order to llustrate his view about the theory-ladenness of observation. According to Hanson, two microbiologists observing an amoeba, depending on the theories they have, see either a one-celled or a non-celled animal. So, Feyerabend used Wittgenstein's discussion of perception in order to undermine the role of observation statements in the statement-view of theories. If perception is

[14] Feyerabend refers in his footnote (1962, 68, n,83) to his review of N.R. Hanson's *Patterns of Discovery* for a criticism of "this attitude". It's not very clear whether "this attitude" refers to Wittgenstein's alleged Platonism or to Wittgenstein's alleged contextual theory of meaning. In Feyerabend's review of Hanson's book, we find several criticisms of Wittgenstein, but not of Wittgenstein's supposed Platonism. Actually, we find Feyerabend's criticism of the idea that "differences in use indicate differences of meaning," which may be taken to imply that Wittgenstein is criticized for not allowing for the stability of meaning. The other critical remarks against Wittgenstein in this review target Wittgenstein's use of examples and his many qualifications in stating a view.

theory-laden as Wittgenstein was taken to suggest,[15] then observation sentences cannot test the theories and their testing is left to be carried out by other, alternative incommensurable theories that may or may not be empirically adequate. Still, Feyerabend oversimplified and, thus, misinterpreted Wittgenstein's remarks about seeing. Contrary to Wittgenstein and Hanson who rejected a two-layer account of perception (observation plus interpretation), Feyerabend seems to uphold the view that observations are interpreted with the help of a theory (Feyerabend 1965, 204).[16] He may say that all observations are theoretical, but he seems to imply that there is a separate act of interpretation to make observations theoretical.[17]

### *10.2.2 Wittgenstein's Influence on Feyerabend's Later Work: Emphasis on Particulars, the Possibility of Alternatives, the Significance of Tradition*

If Wittgenstein's influence on Feyerabend was overshadowed by the statement-view of theories which brought him closer to the logical positivists and Popper, how did it resurface in Feyerabend's later work? We have noted above the ambivalent attitude Feyerabend had towards Wittgenstein's philosophy. Originally, though fascinated, he distanced himself from it, given his connections to Popper's milieu. Popper considered Wittgenstein his "eternal arch rival," (Collodel 2016, 37), and Joseph Agassi (2008, 170) has reported that Wittgenstein was "the most popular—the only really popular—leader of philosophy in England at the time, professional and amateur alike," and, as such, was Popper's "only bête noire": there could be no greater public expression of loyalty to him than to lunge at Wittgenstein." Feyerabend, in several letters in the 1950s, expressed his loyalty to Popper by keeping him posted

[15] For Wittgenstein's discussion of *seeing* and *seeing as* and their relation to the theory-ladenness of observation, see Kindi (2021).

[16] Feyerabend (1991b, 526) thinks that his thesis that observation sentences are fully theoretical is more radical than the theory-ladenness thesis of Hanson, Toulmin and Kuhn who were influenced by Wittgenstein. He, however, retains the two-layer account of perception. In his view, theories are used to explain what we observe (ibid.).

[17] An anonymous reviewer noted that Feyerabend, in his book *Against Method*, introduces the concept of 'natural interpretation' which puts observations and theories much closer together. Indeed, Feyerabend (2010, 57–60) compares 'natural interpretations" to Bacon's "mental operations which follow so closely upon the senses," and claims, *pace* Bacon, that it is nearly impossible to separate sensations from interpretations. He even says that, were we to remove the interpretations, we wouldn't be able to perceive (ibid., 60). Similarly, he says that "[t]here are not two acts - one, noticing a phenomenon; the other, expressing it with the help of the appropriate statement- but *only one* [...] for the speaker, ... statement and phenomenon are firmly glued together" (ibid., 57, his emphasis). Still, it seems that Feyerabend takes observation statements to involve two distinct elements: a clear sensation and a connection with theory and language (ibid., 57, 59). Feyerabend does not deny the distinctness of the two elements; he only points to the difficulty of separating them and arriving at the pure sensations.

as to how he [Feyerabend] dealt with Wittgenstein's philosophy and assured Popper that his attitude towards *Philosophical Investigations* remained critical and similar to Popper's critical attitude towards the *Tractatus* (Collodel and Oberheim 2020, 83, 94–96, 161).[18] Later, in the early 1960s, Feyerabend appropriated what he understood as Wittgenstein's account of perception to challenge the then dominant philosophy of science with its models of explanation and reduction which relied on the theory-observation dichotomy.[19] In his late writings, aspects of Wittgenstein's philosophy provided the backbone of Feyerabend's take on knowledge, science and the world. Wittgenstein's influence is not explicitly acknowledged and, in some cases, it need not be, as Feyerabend's views, although influenced by Wittgenstein's work, move in a different direction.

What does Wittgenstein's influence consist in? Feyerabend had always thought of Wittgenstein as a critic of fundamentalist doctrines (Feyerabend 1985b, 36n2), but this concern, i.e., combating monolithic, abstract theorizing, became more pronounced and central in Feyerabend's late work. Feyerabend could achieve this goal by relying on Popperian and Millian liberal ideas (and, to a certain degree, he did; see Feyerabend 1982), but Wittgenstein's philosophy offered a broader palette: among others, the emphasis on particulars, the possibility of alternatives, and the significance of tradition. As mentioned already, Feyerabend was relying originally on the statement-view of theories, but came to defend a richer approach that presupposed a practical dimension of language use, which results in the formation of traditions. This is very much in line with Wittgenstein's stress on the practical aspect of language games, "the practice of the use of language" (PI §7), "linguistic practice" (PI §21), "the practice of the language" (PI §51), and on the idea that following a rule (for instance, in the employment of words) is a practice (PI §202).

Feyerabend also advanced a pluralistic worldview, which dethroned science from its privileged position wherefrom it exclusively pronounced on knowledge. Feyerabend may have drawn, in this respect, on Wittgenstein's concern for the possibilities of phenomena (PI §90), and Wittgenstein's presentation of a variety of language games (PI §23), but Feyerabend's investigation is not grammatical, i.e., logical, as Wittgenstein's, but more political and polemical against the unique advantage that science enjoys on epistemic matters in contemporary societies, over other forms of knowledge. Feyerabend, in his late work, was not appreciative of purely logical investigations. He thought that logic makes crude distinctions and is far removed from the actual practice of scientists and human beings to be really helpful. Philosophers, he said, "deal with a never-never land that has hardly any point of contact with the actual lives of scientists, politicians, people like you and me."

---

[18] Feyerabend was even preparing a collection of Popper's remarks on Wittgenstein (Collodel and Oberheim 2020, 94).

[19] Cf.: "[Wittgenstein] saw that whenever we observe and report observations we make use of traditional elements ('forms of life')." (Feyerabend 1985a, 37) Wittgenstein did not really say that forms of life have a role in how we observe things. One may only presume that a certain training in a culture may shape our observations.

(Feyerabend 1991b, 492).[20] He was particularly frustrated with philosophers that used references to logic "not just to clarify but to evade scientific problems. "We are making a logical point," the philosophers would say when the distance between their principles and the real world became rather obvious." (Feyerabend 1995, 143) Feyerabend also contrasted his approach to Wittgenstein's in disputing Herbert Feigl's claim that there are "neo-Wittgensteinsteinian traces" in his reasoning. Feyerabend said that in Wittgensteinian analyses,

> [a]lternatives are brought in, not in order to arrive at a better theory through the criticism of the existing one, but rather *in order to get better insight into the existing theory*. [...] A Wittgensteinian uses alternatives with a *dogmatic*, or a *conservative*, purpose. Knowing where the argument is going to lead, he does not really take them seriously, although he may admit that they possess an important therapeutic function. Our own point of view is radically different. Alternatives are taken seriously. They are all candidates for a better theory in the future. Their function is critical and progressive, that is, they are used with the purpose of finding the weaknesses of the customary set of ideas. (Feyerabend 1965, 227n20, emphases in the original)

Here, Feyerabend criticizes the purely logical considerations in order to recommend practical and reformist solutions. Alternatives have for him practical value; they are not supposed to solely illustrate the diversity of language use in the effort to attack essentialism, as is the case with Wittgenstein who marshals, not just actual, but also imaginary examples of different language use. Wittgenstein is portrayed by Feyerabend as conservative, as unwilling to allow or explain how language games, traditions, or cultures change. For instance, he says that Wittgenstein "has failed to indicate how traditions can be improved and created the impression that such improvement is impossible" (Feyerabend 1985b, 37–38). He pauses at Wittgenstein's remark about philosophy leaving everything as it is (PI §124) and claims that, from such a point of view, criticism is quite impossible (Feyerabend 1965, 246n114; cf. ibid., 252n137). He also says that Wittgenstein does not allow revisions of conceptual schemes because, allegedly, Wittgenstein sticks with ordinary language and precludes its reform (Feyerabend 1985a, 159n29). But, this take on Wittgenstein's philosophy shows a misunderstanding. Wittgenstein may have been characterized as conservative,[21] but he was not complacent with the status quo, nor was he dogmatically attached to any particular ordinary language. Wittgenstein only thought that it was not philosophy's job to change whatever language people commonly use

---

[20] Cf. Feyerabend (1991a, 44): "a purely logical discussion is not only naive, it does not even make sense." He also said: "as a student I had mocked the intellectual tumors grown by philosophers." (Feyerabend 1995, 143) Note here the resemblance with Wittgenstein's reference to bumps: "The results of philosophy are the discovery of some piece of plain nonsense and the bumps that the understanding has got by running up against the limits of language. They—these bumps—make us see the value of that discovery" (PI §119).

[21] See the subchapter "Wittgensteinian Quietism and Conservatism" in *Virvidakis and Kindi (*2012*).*

as if it were defective. He did not believe that philosophy ought to dictate what is right or wrong to do, what to improve or what to leave unchanged.[22]

Feyerabend, unlike Wittgenstein, is more of an activist. He portrays science as a loose collection of heterogeneous subjects, with diverse approaches, with no firm structure, and suggests that other traditions and practices that are commonly deemed unscientific, should not be dogmatically dismissed (Feyerabend 1999, 193–196; 1987, 281). They can legitimately be considered as candidates for improving knowledge. Feyerabend castigates "monster science" which is understood as speaking with a single voice, delivering a coherent message that determines the nature of reality (1987, 27, 155; 2011, 55). With this metaphor, Feyerabend attacks a metaphysical understanding of science that erases differences in the name of oneness. Here, Feyerabend meets Wittgenstein on the significance of particulars and the differences they exhibit. In Feyerabend's view, universals are tyrants, the aggressors that annihilate the variegated landscape of particulars, and conquer the abundance of being (1999, 252–264). Yet, Feyerabend finds that in Homer they function in a different way. Not as tyrants, but as mediators between particulars, without superseding or annihilating them. Universals in that context, connect particulars as in lists. In that connection, Feyerabend criticizes Socrates's insistence on getting definitions instead of lists from his interlocutors in response to his questions. For instance, Socrates dismisses Theaetetus's answer to the question "what is knowledge?" which included a list of kinds of knowledge (geometry, astronomy, harmonics, arithmetic, cobblery). Similarly, in *Meno* where Socrates asks for a definition of virtue, Meno gives a list of kinds of virtue and is criticized by Socrates. Feyerabend writes that Socrates's quest reduces the complexity of things and of the world to simple formulas.

> The answers given by Theaetetus and Meno take this complexity into account. They provide *examples* and thereby invite us to consider more than one type of action; they use a *variety* of examples and thereby prevent a uniform interpretation of the material; and they are imprecise, for matters are in flux and cannot be nailed down. (Feyerabend 1999, 259)[23]

It should be noted here that Wittgenstein also comments on Socrates's quest for definitions in the Platonic dialogues and stresses the significance of giving examples.

> I cannot characterize my standpoint better than by saying that it is opposed to that which Socrates represents in the Platonic dialogues. For if asked what knowledge is I would list examples of knowledge, and add the words 'and the like'. No common element is to be found in them all. (cited in Hallett 1977, 33 from Wittgenstein's manuscripts)

---

[22] In his later years, Feyerabend came to appreciate this view:

> Scientists have their own way of doing things. In some of my earlier writings I wanted to interfere in their activity and I said: you won't make discoveries unless you proliferate. Now I would say that the only interference that counts is interference by the people on the spot. Why? Because they know the details including the details that are not written down but reside in their experience. (Feyerabend 2011, 126–127)

[23] Note, however, that in his review of Hanson's book *Patterns of Discovery*, Feyerabend expresses his disapproval for what he takes to be the Wittgensteinian practice of burying important points beneath concrete examples (Feyerabend 1960, 252).

Wittgenstein also says: "A main cause of philosophical disease -a one-sided diet: one nourishes one's thinking with only one kind of example." (PI 593) So, here, Feyerabend and Wittgenstein meet.

### *10.2.3 Look and See*

Feyerabend appeals to two more Wittgensteinian themes: Wittgenstein's injunction not to think but look (PI §66) and Wittgenstein's comments on secondary meaning. Let's consider, first, what use Feyerabend made of Wittgenstein's emphasis on looking rather than thinking. Feyerabend thought that Wittgenstein's injunction, not to think but look, is an invitation for empirical research. He says, for instance, that a paper by Michael Polanyi on the world view of the Azande. "was a concrete application of Wittgenstein's advice to look and see and not to wander off on an abstract tangent" (1991b, 492), and he avows that he himself will give preference to historical considerations over abstract models and theorizing.[24] This interpretation of Wittgenstein's remark is quite common,[25] but again does not chime well with Wittgenstein's philosophy. Wittgenstein was contrasting looking to theoretical, i.e., philosophical, thinking, but he was not inviting us to do empirical research.[26] He always insisted on distinguishing logic from the empirical and he explicitly denied that his philosophy was any kind of natural science (PPF §365). Regardless of Feyerabend's disapproval, Wittgenstein's investigations are grammatical, that is logical, not empirical, even when Wittgenstein asks us to look at how language is used.[27] His target in remark PI §66 is essentialism, and he asks us to look and see how various the use of the term 'game' is. He did not find this out by empirical research; the task of the philosopher, he says (PI §127), "consists in marshalling recollections," or, as it was put in the older translation of the first three editions of the *Philosophical Investigations*, in "assembling reminders".[28] Feyerabend (1982, 115) said that the Wittgensteinian invitation to look and see clashed with his own inclinations, which were theoretical, and, as a result, his papers that were influenced by Wittgenstein, are "mixtures of concrete examples" (Wittgenstein's influence

[24] Cf. Feyerabend (1991b, 515): "I look at history as an empiricist and I find that the empirically identifiable actions of individuals always break through even the most delicate theoretical scheme."

[25] See, for instance, Bloor (1983) and Lynch (1992), but also Jamie Shaw (2021, 467) who says that Wittgenstein's influence led Feyerabend to reject the conceptual analysis of statements and opt for a naturalistic approach that turns philosophy into an empirical investigation.

[26] Note also that Wittgenstein, in his *Remarks on Colour* (RC III.72) questions whether looking by itself can "teach us anything about the concepts of colour." My colleague Fay Zika has drawn my attention to this comment by Wittgenstein.

[27] For criticism of the view that Wittgenstein's philosophy can be interpreted as encouraging empirical investigations, see Kindi (2012b), Friedman (1998), and also Kuusela (2019), especially Chap. 6, "Non-Empiricist Naturalism: The Uses of Natural History in Logic".

[28] Wittgenstein also noted that he used imaginary examples and scenarios for his purposes (PPF§365): "we are not doing natural science; nor yet natural history a since we can also invent fictitious natural history for our purposes."

which he eventually rebuffed),[29] and "sweeping principles" (his own inclinations that he also eventually resisted). On the other hand, in his last book (1999, 84), Feyerabend criticizes the same Wittgensteinian injunction to "look and see" for not taking into account that "looking is not a simple matter," implying that it involves theoretical considerations. Feyerabend and Wittgenstein, however, are talking about different things. Wittgenstein criticizes the sublimation of logic and the inclination of philosophers to make idealizations and speak metaphysically. He urges them to look and see what they already know as users of language, to see what his philosophy draws attention to and tries to make visible. Feyerabend, on the other hand, is interested in challenging the logical positivists' reliance on pure observation statements for the testing of theories and for the conference of meaning, and is eager to point out that seeing needs interpretation and will not, by itself, achieve anything significant. Wittgenstein does not offer any theory about observation and, though he does not believe in pure observation, he actually reserves *seeing as* (which is often assimilated to the theory-ladenness of observation) for special occasions of seeing (see Kindi 2021, 172–174).

### 10.2.4 Secondary Meaning

Feyerabend makes an explicit reference to "secondary meaning" in order to combat the view that meaning is given to terms directly by the phenomena to which they apply. Feyerabend (1965, 205) thinks that "the idea that meanings are determined by phenomena, and the corresponding idea that the truth of observation statements is determined by attention to phenomena, must be rejected."[30] Phenomena alone, in his view "cannot even select interpretations," cannot select, that is, which sentence correctly describes the respective phenomenon (ibid.).[31] Accordingly, Feyerabend

[29] In his review of Hanson's book *Patterns of Discovery*, Feyerabend (1960, 252) expresses his disapproval for what he takes to be the Wittgensteinian practice of burying important points beneath concrete examples.

[30] The reason he gives is that, in the absence of a teacher to help,

> [t]he phenomena must speak for themselves. This means that there must be, in addition to *S* and *P*, a further phenomenon *R*, indicating that *S* fits *P*. Moreover, it is not enough that there is just another element in the visual (or in the mental) field. The element *R* must be relevant, it must fit the whole situation in such a manner that it enables us to say that *S* fits *P*. We therefore need still a further phenomenon, guaranteeing relevance *R'*; the same arguments apply to this phenomenon, hence we need still another phenomenon *R"*; and so on, *ad infinitum*. (Feyerabend 1965, 204)

The same discussion is repeated in Feyerabend (1985a, 26–27).

[31] Cf. Feyerabend 1985a, 27):

> Phenomena cannot determine meaning, although the fact that we have adopted a certain interpretation may (psychologically) determine phenomena. That is, the strict adherence to an interpretation and the rejection of all accounts which are different from it may lead to a situation where the relation between phenomena and propositions will be one to one. In such a situation a distinction between phenomena and interpretations on the one hand and phenomena and objective facts on the other cannot readily be drawn…

appeals to Wittgenstein's discussion of 'secondary meaning" to show that listening to the sound *e* does not determine the meaning of 'yellow' in the example that both Wittgenstein and Feyerabend use, namely, that of having the "strong inclination to call the vowel *e* "yellow." "Yellow" is not applicable to sounds and so, its meaning is not derived from the phenomenon it is supposed to describe. "Yellow" carries its usual meaning, which is fixed, according to Feyerabend, by conventions or by the conventions of the language that it belongs to and is chosen as a means of communication (ibid., 206–207). Wittgenstein does not speak of meaning in terms of conventions, but in terms of use. In that connection, he presents examples, such as calling the vowel *e* "yellow," in order to underline that, even in such extreme cases where meaning is supposedly derived from some idiosyncratic inner experience (namely, synaesthetic experience), inaccessible to others, that is not the case. The term "yellow" has still its primary meaning which is fixed by the practice of using color words.[32] So, Feyerabend's appeal to Wittgenstein in relation to secondary sense in order to point out that outer or inner phenomena cannot by themselves fix meaning, is not unsuitable, although Wittgenstein does not speak of interpretations of phenomena, as Feyerabend does, and does not think of meaning in terms of conventions.

### 10.2.5 Scientism

Finally, I will consider two other issues that have been noted in the literature as common between Feyerabend's and Wittgenstein's philosophy. The first is the opposition of both philosophers to scientism. Ian Kidd (2013, 92) writes that both Feyerabend and Wittgenstein shared a "deep and abiding anti-scientism," while.

Juliet Floyd (2005, 102) observes that the two thinkers believed that

> part of the task of philosophy … [is] to investigate and critique certain symbolic, dogmatic, and/or empty pronouncements that arise about and within science, but without preaching about "truth" from on high, dreaming of a purified scientific method or language, or attempting to speak in general terms about technological thinking or science as such.

Scientism is a certain ideology about science, one that inflates its powers and scope, and gives license to science's imperialistic tendencies and its pretensions to speak with authority about almost anything. It's true that both philosophers denounced this ideology: Feyerabend was critical, for instance, of the manifesto of 186 scientists, including Nobel Laureates, that condemned astrology (see Feyerabend 1982, 91–96),[33] and Wittgenstein was jestingly collecting newspaper clippings that were

---

[32] For more on Wittgenstein's discussion of secondary sense see Kindi (2009).

[33] Feyerabend (1982, 92, 95) said that the 186 scientists "neither know the subject they attack, astrology, nor those parts of their own science that undermine their attack. … [The Manifesto against astrology] shows the extent to which scientists are prepared to assert their authority even in areas in which they have no knowledge whatsoever."

about, among others, scientists meandering in areas they did not belong to. In Wittgenstein's "Collection of Nonsense," as it was called, there were cuttings featuring, for instance, Sir Oliver Lodge, a distinguished physicist, who believed that "the mind of Edison was directing work from beyond the grave" (2008, 195), and Albert Einstein, whom Wittgenstein called "a bloody journalist" (ibid, 201).[34] It's also true that Feyerabend's and Wittgenstein's criticism of scientism did not imply that they did not have respect for science.[35] Still, the two philosophers differed significantly. Feyerabend undertook an almost political task to undermine science's rising hegemony in contemporary societies by questioning the privileged trust we reserve for scientific claims. Wittgenstein, honoring his view to leave things as they are and not to meddle with particular practices, did not issue any recommendations as to how we should treat science, nor did he pronounce on its merits in comparison to other types of knowledge, as Feyerabend did.

### *10.2.6 Wonder and Mysticism*

A second theme that has been said to be common between Feyerabend's and Wittgenstein's philosophy has been noted by Ian Kidd who, drawing on David Cooper's work (2017), argues that both Feyerabend and Wittgenstein "affirm a mysterious background to human life," and value a sense of wonder at something mysterious about life and the world (Kidd 2017, 107–109). Kidd connects this appreciation of wonder, and even mysticism, by the two philosophers, with their anti-scientism. The idea is that the scientific world conception is blamed for being

[34] Brian McGuinness, editor of Wittgenstein's letters, notes that the clippings on Einstein were illustrating "Einstein's love for publicity and clowning, others his views on politics, the future of man, religion, and other general matters" (Wittgenstein 2008, 201). Wittgenstein's friend, the physicist W. H. Watson, apart from sending Wittgenstein cuttings, also recommended in his letters that Wittgenstein read *Nature* for gems of nonsense (ibid., 195).

[35] Cf. Feyerabend (1991a, b, 74–75):

> I have no special love for astrology and much that is written in this area bores me to tears. But astrology is an excellent example of the way scientists deal with phenomena outside their area of competence. They don't study them, they simply curse them, insinuating that their curses are based on strong and straightforward arguments.

It should also be taken into account that Feyerabend's theoretical pluralism was supposed to advance sound knowledge. In criticizing the 186 scientists, Feyerabend defended a certain scientific ethos.

Wittgenstein, who had studied mathematics and engineering and had great appreciation for the rigor of the scientific way of thinking, had dismissed Carnap's interest in parapsychological phenomena as not serious. Carnap remembered that Wittgenstein "was shocked that any reasonable man could have any interest in such rubbish" (Carnap, 1963, 26). We see here that Wittgenstein did not share Feyerabend's belief that the proliferation of theories of any kind, and their mutual criticism would advance knowledge.

hostile to, and for eroding the sense of wonder at the mysteriousness of the world (ibid., 109, 112, 113).

I see two problems with this view. The scientific world conception should not be identified and confused with scientism. Science, and versions of scientific philosophy, i.e., philosophy inspired by science, or philosophy that has science as a model, are responsible for a disenchanted world conception which has become dominant in contemporary times. It may be said that it is a defining characteristic of contemporary science to drive out of the world any kind of mystery.[36] Scientism, on the other hand, which usually puffs up whatever has to do with science, cannot really outdo science in this particular matter. Science's disenchanting of the world is so definitive, that no room is left for scientism to exceed it. So, if both Feyerabend and Wittgenstein embrace wonder and mysticism, i.e., some form of enchantment of the world, they should not just target scientism, but also science itself. But, we have already agreed that they both respect science and what it does as science. Their criticism of science is mostly restricted to its presentation as monolithic and unitary.

The second point of contention concerns the thesis that Feyerabend and Wittgenstein do indeed invite and welcome a sense of wonder at the mysteriousness of life and the world. Feyerabend says explicitly in his "Letter to the Reader," published posthumously in Ian Hacking's review of *The Conquest of Abundance* (Hacking 2000),[37] that if a name is to be given to his very late view, it could be "mysticism, though it is a mysticism that uses examples, arguments, tightly reasoned passages of text, scientific theories and experiments to raise itself into consciousness." Ever a pluralist and an effective opportunist, Feyerabend does not shy from using anything in his panoply of ways to tackle the issues that concern him. Wittgenstein, on the other hand, does not affirm "a mysterious background to human life," as Ian Kidd puts it (Kidd 2017, 108), at least in the same way that Feyerabend did. According to Kidd, human life for Wittgenstein "is ultimately mysterious because it consists of forms of life or ways of experiencing and describing the world that presuppose and so cannot explain their own conditions of possibility" (ibid.). It is true that Wittgenstein does not believe that our forms of life or our worldviews can be ultimately justified. But this view does not commit him to a metaphysical understanding of some ultimate reality or Being. He may say in *Culture and Value* (CV 16) that "[p]erhaps what is inexpressible (what I find mysterious and am not able to express) is the background against which whatever I could express has its meaning," but what he means by that phrase is that he is unable to go any further than saying "This is simply what I do"; he has exhausted the justifications and he

[36] According to Ian Kidd (2017, 110), Wittgenstein thinks that science's erosion of our capacity for wonder, is not an inevitable feature of the scientific world picture, but is rather connected to "what has come to be the 'spirit in which science is carried on nowadays' (CV 5)." Wittgenstein, however, in this remark from *Culture and Value,* does not talk generally about the essential or accidental characteristics of science. He makes a specific point, namely, that people nowadays are not likely to fear natural phenomena because of the way science is nowadays practiced.

[37] The letter has also been published in the 4th edition of *Against Method* (Feyerabend 2010).

has reached bedrock (PI 217).[38] Wittgenstein does not have a robust idea of what reality consists in/of, unlike Feyerabend who argues in favor of ontological pluralism, a view that he thinks "seems closer to facts and to human nature" (Feyerabend 1999, 215).

## 10.3 Conclusion

I have presented several areas where one can trace affinities between Feyerabend's and Wittgenstein's philosophy. The influence from Wittgenstein's work was not straight forward and involved inconsistencies, tensions and misreadings. Yet, even where it was resisted by Feyerabend, it was always present either as an inspiration or as a sounding board.

**Acknowledgments** I would like to thank the editors and, especially, Stefano Gattei for the invitation to contribute to this volume, and the two anonymous reviewers whose comments and suggestions helped me to significantly improve the paper.

## Bibliography

Agassi, J. 2008. *A Philosopher's Apprentice in Karl Popper's Workshop*. Amsterdam: Rodopi.
Bloor, D. 1983. *Wittgenstein. A Social Theory of Knowledge*. New York: Columbia University Press.
Carnap, R. 1963. Intellectual Autobiography. In *The Philosophy of Rudolf Carnap: The Library of Living Philosophers*, ed. P.A. Schilpp, vol. 11. La Salle: Open Court.
Collodel, M. 2016. Was Feyerabend a Popperian? Methodological Issues in the History of the Philosophy of Science. *Studies in History and Philosophy of Science* 57: 27–56.
Collodel, M., and E. Oberheim. 2020. *Feyerabend's Formative Years. Feyerabend and Popper*. Vol. I. Cham: Springer.
Cooper, D. 2017. Superstition, Science, and Life. In *Wittgenstein and Scientism*, ed. J. Beale and I.J. Kidd, 28–38. New York: Routledge.
Feigl, H. 1970. The Orthodox View of Theories. In *Minnesota Studies in the Philosophy of Science*, ed. M. Radner and S. Winokur, vol. IV, 3–16. Minneapolis: University of Minnesota Press.
Feyerabend, P. 1955. Wittgenstein's *Philosophical Investigations*. *The Philosophical Review* 64 (3): 449–483.
———. 1958. An Attempt at a Realistic Interpretation of Experience. *Proceedings of the Aristotelian Society* 58: 143–170.
———. 1960. Review: Patterns of Discovery. *Philosophical Review* 69: 247–252.
———. 1962. Explanation Reduction and Empiricism. In *Scientific Explanation, Space, and Time: Minnesota Studies in the Philosophy of Science*, ed. H. Feigl and G. Maxwell, vol. III, 28–97. Minneapolis: University of Minneapolis Press.

[38] It should be noted that this is an early remark, made in 1931, and it may reflect the spirit of Wittgenstein's early work, as expressed mainly in the *Tractatus*, where the efforts to state the conditions of possibility of language are characterized as nonsensical. Wonder at the conditions of possibility of language is not so prominent in Wittgenstein's later work.

———. 1965. Problems of Empiricism. In *Beyond the Edge of Certainty. Essays in Contemporary Science and Philosophy*, ed. R.G. Colodny, 145–260. Englewood, NJ: Prentice Hall.
———. 1966. Herbert Feigl: A Biographical Sketch. In *Mind, Matter, and Method: Essays in Philosophy and Science in Honor of Herbert Feigl*, ed. P. Feyerabend and G. Maxwell, 4–13. Minneapolis: University of Minnesota Press.
———. 1982/1978. *Science in a Free Society*. London: Verso Editions.
———. 1985a/1981. *Realism, Rationalism, and Scientific Method. Philosophical Papers Vol. 1.* Cambridge: Cambridge University Press.
———. 1985b/1981. *Problems of Empiricism. Philosophical Papers Vol. 2.* Cambridge: Cambridge University Press.
———. 1987. *Farewell to Reason*. London: Verso Editions.
———. 1991a. *Three Dialogues on Knowledge*. Oxford: Blackwell.
———. 1991b. Concluding Unphilosophical Conversation. In *Beyond Reason: Essays on the Philosophy of Paul Feyerabend*, ed. G. Munévar, vol. 132, 487–527. Dordrecht: Springer. (Boston Studies in the Philosophy of Science).
———. 1995. *Killing Time*. Chicago: The University of Chicago Press.
———. 1999. In *Conquest of Abundance. A Tale of Abstraction versus the Richness of Being*, ed. Bert Terpstra. Chicago: The University of Chicago Press.
———. 2010. *Against Method*. 4th ed. London: Verso Editions.
———. 2011. In *The Tyranny of Science*, ed. E. Oberheim. Cambridge: Polity Press.
———. 2020/1976. Autobiographical Remarks. In *Feyerabend's Formative Years. Feyerabend and Popper*, Vol. I., ed. M. Collodel and E. Oberheim, 19–44. Cham: Springer.
Floyd, J. 2005. Homage to Vienna: Feyerabend on Wittgenstein (and Austin and Quine). In *Paul Feyerabend (1924–1994) – Ein Philosoph aus Wien*, ed. F. Stadler and K.R. Fischer, 465–470. Berlin: Springer Verlag.
Friedman, M. 1998. On the Sociology of Scientific Knowledge and Its Philosophical Agenda. *Studies in the History and Philosophy of Science* 29: 239–271.
Glock, H.-J. 2008. The Influence of Wittgenstein on American Philosophy. In *The Oxford Handbook of American Philosophy*, ed. C.J. Misak, 375–402. Oxford: Oxford University Press.
———. 2013. Wittgenstein's Later Philosophy. In *The Oxford Handbook of Analytic Philosophy*, ed. M. Beaney, 572–593. Oxford: Oxford University Press.
Hacking, I. 2000. 'Screw You, I'm Going Home. Review of Paul Feyerabend's *The Conquest of Abundance*. *London Review of Books*, vol. 22: 12, 22 June 2000.
Hallett, G. 1977. *A Companion to Wittgenstein's Philosophical Investigations*. Ithaca: Cornell University Press.
Hoyningen-Huene, P. 1995. Two Letters of Paul Feyerabend to Thomas S. Kuhn on a Draft of *The Structure of Scientific Revolutions*. *Studies in History and Philosophy of Science* 26: 353–387.
Kidd, I.J. 2013. Review: Paul Feyerabend, *Against Method*, 4th ed., Paul Feyerabend, *the Tyranny of Science*, ed. Eric Oberheim. *Philosophical Investigations* 36 (1): 90–94.
———. 2017. Reawakening to Wonder: Wittgenstein, Feyerabend and Scientism. In *Wittgenstein and Scientism*, ed. J. Beale and I.J. Kidd, 101–115. New York: Routledge.
Kindi, V. 2009. Second Thoughts on Wittgenstein's Secondary Sense. In *Language and World*, ed. Volker A. Munz, Klaus Puhl, Joseph Wang, 202–204. 32nd International Wittgenstein Symposium, Kirchberg am Wechsel.
———. 2012a. Concept as Vessel and Concept as Rule. In *Scientific Concepts and Investigative Practice*, ed. U. Feest and F. Steinle, 23–46. Berlin: De Gruyter.
———. 2012b. The *Structure*'s Legacy: Not from Philosophy to Description. *Topoi* 32 (1): 81–89.
———. 2017. Wittgenstein and Philosophy of Science. In *A Companion to Wittgenstein*, ed. Hans-Johann Glock and John Hyman, 587–602. NY: Wiley Blackwell.
———. 2021. Kuhn, the Duck and the Rabbit – Perception, Theory-Ladenness and Creativity in Science. In *Interpreting Kuhn. Critical Essays*, ed. K. Brad Wray, 169–184. Cambridge: Cambridge University Press.

Kuby, D. 2018. Carnap, Feyerabend, and the Pragmatic Theory of Observation. *HOPOS* 8 (2): 432–470.

Kuusela, O. 2019. *Wittgenstein on Logic as the Method of Philosophy*. Oxford: Oxford University Press.

Lynch, M. 1992. Extending Wittgenstein: The Pivotal Move from Epistemology to the Sociology of Science. In *Science as Practice and Culture*, ed. A. Pickering, 215–265. Chicago: Chicago University Press.

Motterlini, M., ed. 1999. *For and Against Method*. Chicago: The University of Chicago Press.

Oberheim, E. 2006. *Feyerabend's Philosophy*. Berlin: De Gruyter.

Popper, K. 1992/1976. *Uneded Quest. An Intellectual Autobiography*. London: Routledge.

Preston. J. 2000. Science as Supermarket: 'Postmodern' Themes in Paul Feyerabend's Later Philosophy of Science. In *The Worst Enemy of Science. Essays in Memory of Paul Feyerabend*, ed. J. Preston, G. Munevar, and D. Lamb, 80–101. Oxford: Oxford University Press.

Shaw, J. 2021. Feyerabend on Naturalism, Pluralism, and the Foundations of Philosophy. *Logique et Analyse* 256: 467–486.

Virvidakis, S., and V. Kindi. 2012. Quietism. In *Oxford Bibliographies in Philosophy*, ed. Duncan Pritchard. New York: Oxford University Press. https://www.oxfordbibliographies.com/display/document/obo-9780195396577/obo-9780195396577-0184.xml?rskey=BzuyWk&result=1&q=quietism#firstMatch. Last accessed 16 Feb 2023.

Wittgenstein, L. 1980. In *Culture and Value*, ed. G.H. von Wright. Chicago: The University of Chicago Press. Abbreviated as CV page.

Wittgenstein, L. 1990/1977. *Remarks on Colour*, ed. G.E.M. Anscombe. Oxford: Blackwell. Abbreviated as RC paragraph.

Wittgenstein, L. 2008/1995. *Wittgenstein in Cambridge. Letters and Documents 1911–1951*, ed. B. McGuinness. Oxford: Blackwell.

Wittgenstein L. 2009/1953. *Philosophical Investigations* (G.E.M. Anscombe, P.M.S. Hacker and J. Schulte, Trans.). 4th edition. Blackwell, Oxford. References to the first part of the bool are abbreviated as PI paragraph, and to the second part, which in the 4th edition is given the title "Philosophy of Psychology-A Fragment," as PPF paragraph.

## Archival Material

Letter from Paul K. Feyerabend to Thomas S. Kuhn April 4, 1983. Thomas S. Kuhn Papers, MC 240, box 21, Folder 28. Massachusetts Institute of Technology, Department of Distinctive Collections, Cambridge, Massachusetts.

Letter from Paul K. Feyerabend to Thomas S. Kuhn February 24, 1988. Thomas S. Kuhn Papers, MC 240, box 21, Folder 28. Massachusetts Institute of Technology, Department of Distinctive Collections, Cambridge, Massachusetts.

# Chapter 11
# Feyerabend and Popper

**Roberta Corvi**

**Abstract** The relations between Popper and Feyerabend have been carefully reconstructed by a few scholars, but the question about the influence of the former on the latter has not yet found convergent answers. According to this paper, Feyerabend was a pupil of Popper's, because he deals with typically Popperian topics and develops themes that share a Popperian imprint. At the same time, however, Feyerabend was not a true disciple, for he took the central tenets of critical rationalism to their extreme consequences, distorting them to some degree, and making them alien to very nature of Popper's philosophy, if not altogether incompatible with it. In order to support this interpretation, the author examines two themes common to both philosophers: the reappraisal of metaphysics and the relationship between science and rationality. As far as metaphysics is concerned, Feyerabend, unlike Popper, believes that a demarcation criterion between science and metaphysics is not necessary because Feyerabend's conception of science and of rationality is different from Popper's one. Indeed, while for Popper science is the paradigm of rational discussion, for Feyerabend science, like any other human product, is the result of intuition and imagination no less than of reasoning.

**Keywords** Paul Feyerabend · Karl Popper · Metaphysics · Rationalism · Rationality

## 11.1 Was Feyerabend a Disciple of Popper's?

Many scholars have investigated the connection between Feyerabend and Popper, in the attempt to assess whether Feyerabend was a disciple of Popper's, or merely one of his pupils. To put it differently, the question is: was Feyerabend significantly

R. Corvi (✉)
Department of Philosophy, Catholic University of Milan, Milan, Italy
e-mail: roberta.corvi@unicatt.it

© The Author(s), under exclusive license to Springer Nature Switzerland AG 2024

S. Gattei, R. Corvi (eds.), *Feyerabend in Dialogue*, Boston Studies in the Philosophy and History of Science 346,
https://doi.org/10.1007/978-3-031-71938-7_11

influenced by Popper's teaching, or was their relation merely circumstantial, and did not significantly influence the author of *Against Method*?

The relations between the two Viennese philosophers have been carefully reconstructed by Matteo Collodel, who—based on their published works as well as on private correspondence—has traced Feyerabend's step by step path from his early contacts with Popper to his rejection of critical rationalism.[1] As a consequence, I do not intend to repeat a description of the events that accompanied Feyerabend and Popper from their first meeting, in August 1948, when young Paul agreed to attend the summer school in Alpach, Tyrol, in order to take note of the more important discussions, in exchange for travel and accommodation. On that occasion, as Feyerabend himself recounts in his autobiography, Popper treated him familiarly. Four years later, in the autumn of 1952, the young scholar opted for the advocate of critical rationalism when he had to choose a supervisor for the British Council scholarship he had been awarded (in fact, he had hopes of working with Wittgenstein, who had, however, died the previous year). During his time in London, Feyerabend worked closely with Popper and his pupils at the London School of Economics and Political Sciences, including Imre Lakatos and Joseph Agassi. In the summer of 1953, Feyerabend returned to Vienna, with the task of translating *The Open Society and Its Enemies*, and for one year he was assistant to Arthur Pap. Later, Popper (together with Schrödinger) helped him obtain a position at the University of Bristol. Eventually, Feyerabend's move to Berkeley, in 1958, marked his geographical departure from Popper's school, although his personal relations with some of its members, especially Imre Lakatos, did not decrease. However, it is more difficult to establish at what time and in what ways did Feyerabend's intellectual estrangement from Popper actually take place—an estrangement that turned into a deep rift, which led Feyerabend to employ scathing words to stress the distance from Popper's philosophical positions.

Opinions vary, of course: John Watkins has provided a psychological explanation for Feyerabend's parting arguing that his criticism stems from an attempt to mask his "unwanted indebtedness" to the old master.[2] John Preston has argued that Feyerabend was a disappointed Popperian since the 1970s.[3] By contrast, Robert Farrell has suggested that Feyerabend proposed a pluralist methodology, rejecting Popper's methodological proposal, but without giving up the essential values of critical method; accordingly, he has concluded that "Feyerabend was not a disappointed Popperian, but, in many respects, a die-hard pluralistic Popperian".[4] Finally, Eric Oberheim has even come to deny that Feyerabend ever belonged to the Popperian school.[5]

---

[1] Cf. Collodel (2016).

[2] Cf. Watkins (2000).

[3] Cf. Preston (1997).

[4] Cf. Farrell (2000), p. 257.

[5] Cf. Oberheim (2006), p. 78.

For my part, I think the question might be addressed through an analogy with two great philosophers of antiquity, Plato and Aristotle. Does it make sense to ask whether Aristotle was Plato's disciple or only his pupil? Undoubtedly, Aristotle did not follow his master by faithfully iterating his doctrines: in fact, he radically reformed them in many cases. But can we really deny that Aristotle's theories are akin, in theme, to those of his master? To make but one example: the two philosophers solve the problem of being in different ways, but it cannot be denied that the theme is common, and that the influence of the master on the pupil is clearly visible.

Similarly, I believe that Feyerabend was, indeed, a pupil of Popper's, in the sense that his thought addresses the central problems of Popper's thought and develops themes that share a Popperian imprint. At the same time, however, Feyerabend was not a true disciple, for he took the central tenets of critical rationalism to their extreme consequences, distorting them to some degree, and making them alien to very nature of Popper's philosophy, if not altogether incompatible with it.

I shall attempt to support this interpretation by examining two themes that are dear to both authors, in order to assess whether there is continuity or discontinuity between their two approaches. First, I will discuss the reappraisal of metaphysics that stems from their critique of empiricism; and then I will address the question of the relationship between science and rationality.

## 11.2 The Role of Metaphysics

### *11.2.1 Theory and Observation*

First, it should be noted—as it also appears from the very contributions to the present volume—that the young Feyerabend had a varied education, one in which scientific disciplines found their way together with forays into the arts, not to mention Viennese philosophy. The essays published between 1954 and 1958 show that he devoted himself to studying Wittgenstein, but also the works of well-known exponents of Positivism.[6] However, subsequent publications, dating from the 1960s, reveal that the critique of positivist empiricism he was developing was very much influenced by Popper.[7]

Feyerabend himself admitted that some of his discussions with Popper influenced his early writings, but he thought the same happened with Anscombe, Wittgenstein, Hollitscher, Bohr and even Dadaism, Expressionism and Nazism, thus downplaying the role of Popper's teachings in his education.[8]

[6] Cf. Feyerabend (1954a, b, 1955a, b, 1958).

[7] Cf. Feyerabend (1963, 1965, 1970).

[8] Feyerabend (1979) acknowledges that some of his discussions with Popper are reflected in his early writings, but the same happened with many people and even his "reading of Dadaism, Expressionism, Nazi-authorities has left trace here and there" (pp. 63–64). He states also that he "didn't know Popper had a philosophy": indeed, he claims, "Popper is not a philosopher, he is a schoolmaster" (p. 64).

Even in *Science in a Free Society* (1978), Feyerabend includes Popper among those who influenced him. Indeed, he acknowledges having admired him in 1948, among other things, for his humour and ability to state challenging problems in simple language; at that time, Popper "was a free mind, joyfully putting forth his ideas, unconcerned about the reaction of the 'professionals'".[9] Later, however, his admiration waned, as Popper turned into "the established Sir Karl", becoming "a mere propagandist".[10]

After all, the estrangement from Popper was already clear in 1970, with the publication of an essay that constituted the core of the 1975 volume bearing the same title, *Against Method*, in which Feyerabend did acknowledge that critical rationalism could be regarded as "the most liberal positivistic methodology in existence today",[11] but did not hesitate to qualify Popper as a positivist, a characterization that the old master could not accept willingly, believing that he had exposed its fundamental errors and was accordingly responsible for the death of Positivism, rather than its continuator.[12]

However, it is undeniable that Feyerabend's epistemological thought shares more than one aspect with Popper's. Feyerabend's effort to downplay Popper's influence notwithstanding, it must be acknowledged that his characterization of the relationship between observation and theory, which turns out to be much more complex than the one described by empiricism, owes much to Popper's teaching. For, according to empiricism, by observation we identify the facts that form the basis upon which theories are built; but Popper criticized this assumption, arguing for the primacy of the theoretical viewpoint over observation, since he believed that knowledge does not begin with observation, but with expectations that get disappointed. Indeed, he claims that "theory dominates the experimental work from its initial planning up to the finishing touches in the laboratory".[13] This statement is accompanied by a note in which Popper points out that "observations, and even more so observation statements and statements of experimental results, are always *interpretations* of the facts observed; [...] they are *interpretations in the light of theories*. This is one of the main reasons why it is always deceptively easy to find *verifications* of a theory, and why we have to adopt a *highly critical* attitude towards our theories if we do not wish to argue in circles: the attitude of trying to *refute* them".[14]

Popper famously believed that observation cannot be the source and the basis of knowledge, since "observation is always selective. It needs a chosen object, a definite task, an interest, a point of view, a problem".[15] Any observation presupposes

[9] Feyerabend (1978), p. 115.

[10] Ibid., pp. 114–115.

[11] Feyerabend (1975), p. 171.

[12] Cf. Popper (1974), p. 69, where he "admits responsibility" for the death of logical positivism.

[13] Popper (1935), p. 90.

[14] Ibid., p. 90, fn. *3.

[15] Popper (1963), p. 46.

"the adoption of a frame of reference", which consists of expectations or theories that are *a priori* with respect to observation itself. Of course, Popper does not advocate innatism, but claims that every organism possesses innate reactions or responses that are *a priori* with respect to experience. "These responses we may describe as 'expectations' without implying that these 'expectations' are conscious [...]. In view of the close relation between expectation and knowledge we may even speak in quite a reasonable sense of 'inborn knowledge'".[16] This knowledge, however, is not valid a *priori*, but it is psychologically or genetically *a priori*, namely, prior to all observational experience. In other words, "knowledge cannot start from nothing—from a *tabula rasa*—not yet from observation. The advance of knowledge consists, mainly, in the modification of earlier knowledge".[17] In short, from Popper's point of view, observation is always selective and depends on some prior knowledge or expectation, even if unconscious; so, it is not possible to isolate any fact that is objective and neutral with respect to the theory, of which it would allegedly form the foundation.

Feyerabend's production in the 1960s is in continuity with Popper's cue. Even in *Against Method*, he claims that "according to our present results, hardly any theory is *consistent with* the facts. The demand to admit only those theories which are consistent with the available and accepted facts again leaves us without any theory. (I repeat: *without any theory*, for there is not a single theory that is not in some trouble or other)".[18] Feyerabend intends to counter the assumption, deeply rooted in empiricism, that facts are obvious and unquestionable, since he believes, by contrast, that the empirical basis is always imbued with theory. In line with Popper, he notes that experience does not precede theory, but always comes with theoretical assumptions. Indeed, pure sensations cannot even be isolated, as sense knowledge is only possible because there is a system of interpretations of perceptions, which certainly cannot depend, in turn, on sense experience.

Consequently, Feyerabend rejects the fundamental principle of empiricism, according to which theories must be supported by facts. In this regard, he appeals to Hume, who showed that theories cannot be derived from facts—indeed, facts alone cannot even determine the rejection or acceptance of a theory because the latter will always say something more than what the facts suggest.[19] Therefore, empiricism cannot be the universal basis of all our knowledge, not even when accurate experiments are appealed to, because precisely in that case the influence of the theoretical apparatus is more powerful and more pervasive. Not only that: Feyerabend goes so far as to state that there are facts that cannot be discovered without the help of theories that contrast established ones, received by the scientific community.

[16] Ibid., p. 47.

[17] Ibid. p. 28.

[18] Feyerabend (1975), p. 65.

[19] "According to Hume, Theories cannot be *derived from* facts. The demand to admit only those theories which followfrom facts leaves us without any theory" (Feyerabend 1975, p. 65).

Hence, facts cannot be considered the testbed for theories, as empiricism would have it, as observations depend on the very theories that are to be assessed.

In order to subject a theory to scrutiny, then, all that remains is to compare it with other theories, proceeding counter-inductively: we have "to introduce and elaborate hypotheses which are inconsistent with well-established theories and/or well-established facts".[20] The proliferation of theories allows for an epistemological pluralism that is healthy for the progress of science, since even ideas that fail to prevail make a contribution that may prove decisive later on, as was the case with atomism or the heliocentric theory of Aristarchus of Samos. Favouring the proliferation of theories means, then, increasing the possibilities for criticism and thus the growth of scientific knowledge. Pluralism thus becomes "an *essential* feature of all knowledge that claims to be objective".[21]

This idea certainly comes from Popper. However, he was convinced that experience, or rather certain experiments, could invalidate the theory by falsifying it. Feyerabend, by contrast, denies this, arguing, on the one hand, that falsificationism is based on the unproven assumption that the observed object is independent of the observer and, on the other hand, that each theory creates the assumptions of its own experience, hence different and rival theories cannot be related to any neutral empirical data, which are supposed to judge impartially which theory is the best. From this assumption derives the thesis of the incommensurability of theories, according to which no theory can ever be checked by direct comparison with facts, and thus it becomes impossible to compare two theories by reference to common observation statements, since they have no point of contact between one another. This conclusion goes beyond Popper's theses, and thus Feyerabend, while taking his cue from the master to criticize empiricism, departs from him, following a different path that nevertheless crosses Popper's own with regard to a topic that is anything but secondary: the role of metaphysics.

### *11.2.2 Metaphysics for the Good Empiricist*

On this subject, too, Feyerabend takes up one of Popper's central ideas. Popper argued that, far from having to be rejected as meaningless, as positivism had claimed, metaphysics performs a very important function: it generates or can generate unusual ideas that prelude to new scientific theories. In the third volume of the *Postscript* (*Quantum Theory and the Schism in Physics*), Popper goes so far as to consider "a metaphysical theory as similar to a scientific one. It is vaguer, no doubt, and inferior in many other respects; and its irrefutability, or lack of testability, is its greatest vice. But, *as long as a metaphysical theory can be rationally criticized,*

[20] Feyerabend (1975), p. 29.

[21] Feyerabend (1963), p. 80.

I should be inclined to take seriously its implicit claim to be considered, tentatively, as true".[22] According to Popper, there is no need to demarcate between science and metaphysics; rather, we have to look for a criterion "*within metaphysics,* between rationally worthless metaphysical systems, and metaphysical systems that are worth discussing, and worth thinking about".[23] In so doing, Popper likens good metaphysics to science, since both pursue the aim "to gather all the true aspects of the world". What makes a picture worthy of consideration depends "upon its capacity to provoke rational criticism, and to inspire attempts to supersede it by something better".[24] In this way, Popper distinguishes the sphere of rational discussion from the sphere in which rational criticism cannot be exercised. Within the sphere subject to rational criticism, which includes both science and good metaphysics, one must, however, demarcate scientific theories that are falsifiable—that is, that can collide with experience—from those that are not, such as metaphysical theories.

Feyerabend agrees with Popper that a researcher free of prejudice, in order not to be entangled in the web of facts, will have to formulate hypotheses that are not conditioned by observation, and hence metaphysical. Once again, however, he follows the master only to surpass and oppose him. Indeed, he observes that if one forces a theory to forget its metaphysical past to be scientific, one remains in the realm of dogmatic rather than critical empiricism, as the empirical aspects still have the upper hand. He therefore proposes a very different evaluation of metaphysics, especially when it contradicts a well-corroborated theory, because in this case it becomes very useful for the purposes of criticism, offering a different point of view from which to reconsider the problem, thus favouring the discovery of new facts and avoiding dogmatically defending theories that might turn out to be inadequate. Feyerabend, in short, overturns logical empiricism's assumption that science must free itself from metaphysics, arguing that "a science that is free from metaphysics is on the best way to becoming a *dogmatic* metaphysical system".[25]

Accordingly, science, far from having defeated metaphysics, draws continuous nourishment from it to the extent that a science without metaphysics would end up becoming a dogmatic metaphysical system. Metaphysics stimulates creativity and can generate, as has already happened, brilliant ideas that can undermine the most widespread and deep-rooted prejudices, opening extraordinary perspectives for research. The indispensable function performed by metaphysics is, for Feyerabend, the confirmation that theory does not spring from experience—in fact, it is just the opposite, since every possible experience springs from theory. And every theory is always, to some extent, metaphysical, since it necessarily goes beyond empirical evidence:

---

[22] Popper (1982b), p. 199.

[23] Ibid., p. 211.

[24] Ibid.

[25] Feyerabend (1963), pp. 80–81. About relations between Feyerabend and logical positivism see also the chapter in the present collection.

> metaphysical systems which contradict observational results or well-confirmed theories *are most welcome* starting points of such criticism. Far from being misfired attempts at anticipating, or circumventing, empirical research which were deservedly exposed by a reference to experience, they are the only means at our disposal for examining those parts of our knowledge which have already become observational and which are therefore inaccessible to a criticism 'on the basis of observation'.[26]

Therefore, in Feyerabend's eyes, metaphysics provides the tools to criticise that knowledge which, having found empirical confirmation, points our observations in a precise direction and thus escapes possible refutation, and certainly cannot point to counterexamples. Consequently, "a good empiricist will not rest content with the theory that is in the centre of attention and with those tests of the theory which can be carried out in a direct manner. Knowing that the most fundamental and the most general criticism is the criticism produced with the help of alternatives, he will try to invent such alternatives".[27] The first step of a good empiricist will be the invention of a new metaphysics; then, he must elaborate it so that it is able to compete with the theory we want to try: "We may sum up both activities by saying that a good empiricist must be a critical metaphysician".[28] Feyerabend calls attention to the need to criticise the available alternatives in order to avoid turning theories into dogmas. Theoretical pluralism is adopted by Feyerabend as a qualifying factor of any good method, where a method is understood to be good insofar as it is non-dogmatic. Indeed, "*Unanimity of opinion may be fitting for a church, for the frightened victims of some (ancient, or modern) myth, or for the weak and willing followers of some tyrant; variety of opinion is a feature necessary for objective knowledge; and a method that encourages variety is also the only method that is compatible with a humanitarian outlook*".[29] One cannot fail to note that the appreciation of metaphysics, as well as the insistence on criticism, understood as a methodological tool, certainly demonstrates a Popperian inheritance, accentuated by a political colouring—of Popperian descent, as well—that condemns totalitarianism even in science in favour of humanitarian instances.

According to Feyerabend, metaphysics defies the claim to appeal to neutral and objective facts, since, in reality, there are no facts that are not embedded in a theoretical framework and are not the result of some form of interpretation. Such ideological ingredients, as we have seen, can only surface through counter-induction, that is, through that process that relies on the principle of proliferation and elaborates hypotheses that contradict well-confirmed theories and established facts. "Ideological ingredients of our knowledge and, more especially, of our observations, are discovered with the help of theories which are refuted by them. *They are*

[26] Ibid., p. 100.

[27] Ibid., p. 102.

[28] Ibid.

[29] Ibid, p. 97.

*discovered counterinductively*".[30] At this point, Feyerabend reminds us that the facts considered to be decisive for checking theories include points of view that are never explicitly formulated, of which we are unaware, thereby employing concepts similar to Popper's. The recourse to counter-induction also has an undeniably Popperian flavour, since, as is well known, the advocate of falsificationism criticised the inductive method ever since his first work, which begins precisely by declaring that the theory developed in the book is "opposed to all attempts to operate with the ideas of inductive logic".[31]

As is well known, Hume denied that induction has logical value, but considered it to have psychological value. Popper, while agreeing with the first part of Hume's argument, disagreed with the second, because he felt dissatisfied with Hume's psychological explanation, and so he suggested inverting it: "repetition-for-us as the result of our propensity to expect regularities and to search for them. [...] Without waiting, passively, for repetitions to impress or impose regularities upon us, we actively try to impose regularities upon the world. We try to discover similarities in it, and to interpret it in terms of laws invented by us".[32]

To sum up, Feyerabend's ideas about the usefulness of metaphysics and its counter-inductive consequences seem close to Popper's theoretical framework. That said, however, a further reason for contrast between Feyerabend and Popper must be pointed out, since the latter, as mentioned above, believed that there is a problem about the demarcation between what is science and what is not science, namely metaphysics, although he himself doubted that a sharp demarcation could be identified, because "science has at all times been profoundly influenced by metaphysical ideas; certain metaphysical ideas and problems (such as the problem of change, or the Cartesian programme of explaining all change by action at vanishing distances) have dominated the development of science for centuries, as regulative ideas; while others (such as atomism, another attempt to solve the problem of change) have by degrees turned into scientific theories».[33]

In other words, Popper believed that there is no clear demarcation line between science and metaphysics. However, he thought that the criterion of demarcation also has a practical function, related to the assessment of theories and their claims.[34] As is well known, the demarcation criterion advanced by Popper is that of "testability or refutability or falsifiability". Feyerabend agrees that it is not possible to identify a precise dividing line between science and non-science, between science and metaphysics, but differs from his teacher in that he believes that a demarcation criterion

---

[30] Feyerabend (1975), p. 77. Feyerabend argues that Galileo himself never offered any inductive support to Copernicanism, nor any refutations of geocentric theory, but praised Aristarchus and Copernicus for having proceeded conterinductively: see ibid., pp. 102–103. About counter-induction, see also ibid., Chaps. 2 and 3.

[31] Popper (1935), p. 6.

[32] Popper (1963), p. 46. Popper faced the problem of induction since the early 1930s, criticizing the inductive method: cf. Popper (1935), ch. 1, § 1, pp. 3–7.

[33] Popper (1983), pp. 159–160.

[34] Cf. ibidem, pp. 162 and 174.

is not necessary, not only because the history of science does not prove that in science the "most important developments are brought about by falsification",[35] but also, in my opinion, because Feyerabend has a conception of science that derives from his thoughts about rationality, which—once again—takes its cue from Popper's ideas, but—once again—continues in a divergent direction. Indeed, while for Popper science is the paradigm of rational discussion, for Feyerabend, science is not the paradigmatic expression of rationality, since science too, like any other human product, is the result of intuition and imagination no less than of reasoning.

## 11.3 Rationality and Science

### *11.3.1 Against Western Rationalism*

If it cannot be denied that Feyerabend's critique of empiricism has a clear Popperian imprint, the same is also true for his critique of rationalism, although in both cases the pupil has departed significantly from the master. In fact, for the author of *Against Method*, if there is no solid neutral basis upon which to build scientific theories, then one must conclude, paraphrasing Einstein, that the scientist is just "a type of unscrupulous opportunist".[36] This statement implies a critique of the canons of scientific rationality that in truth becomes a battle not only against rationality *tout court*, understood as the set of values and norms on which Western culture is based, but also against the social organisation that derives from it. In short, it is a radical critique of Western rationalism.[37] Feyerabend's attacks on epistemology, which are articulated in severe criticisms against empiricism, positivism, and falsificationism, all eventually converge against a double target: the idea of rationality, which has been elaborated by Western thought with its centuries-old history (of which Popper's critical rationalism is but the most recent version), and the related idea of objectivity, which underlies modern science.[38]

To support this point, Feyerabend criticises Western rationalism and its claim to identify the canons of rationality that can be consistently applied in the various fields of knowledge. He considers this claim to be completely baseless, for "*all*

[35] Feyerabend (1987), p. 171.

[36] Feyerabend (1975), p. 18. A few lines earlier, Feyerabend notes that the history of science, and especially the history of revolutions, shows the complexity of the changes and processes that lead to new knowledge, and therefore "successful *participation* in a process of this kind is possible only for a ruthless opportunist who is not tied to any particular philosophy and who adopts whatever procedure seems to fit the occasion" (ibid.).

[37] For a careful consideration on the reconstruction of Western rationalism in Feyerabend, see Preston (2016).

[38] Feyerabend himself declares that he will "criticize two ideas that have often been used to make Western expansion intellectually respectable—the idea of Reason and the idea of Objectivity": Feyerabend (1987), p. 5.

*methodologies, even the most obvious ones, have their limits*",[39] and none has absolute value. This is the core of the methodological anarchism that intends to demolish the very structure of the theory of knowledge: there can be no theory of knowledge, since at any level knowledge is a personal conquest, a creation and not a reproduction of objective contents through procedural protocols.

Since Xenophanes and Parmenides, philosophers have introduced abstract concepts, to which nothing corresponds in reality, and which therefore have no equivalent among the objects and events experienced in real life. Abstraction is based on a simplification of reality, which found a satisfactory application in mathematics and astronomy, but it impoverished the perception of reality and supported the idea that the only reliable form of knowledge is rational, objective knowledge, based on proofs and demonstrations, to the exclusion of everything that is an expression of individual creativity. With the emergence of modern science, the view that came to be received—also would be shared by Popper, as we have seen—was that rational knowledge identifies with science, and that rationality is the faculty of devising a universally valid method and sticking to it. The exaggerated esteem in which science is held rests precisely on this equation between science and rationality. Such equation is false, though, as the history of science—and, in particular, the work of Galileo, to which Feyerabend devoted much attention—shows.[40] Even contemporary epistemology, albeit in the version of critical rationalism, takes the reliability of science for granted, but the assertion that "there is no knowledge outside science – *extra scientiam nulla salus*—is nothing but another and most convenient fairy-tale",[41] because "everywhere science is enriched by unscientific methods and unscientific results, while procedures which have often been regarded as essential parts of science are quietly suspended or circumvented".[42]

True, Popper's falsificationism counters dogmatism and promotes criticism indefinitely, but this is not enough, since it assumes that facts can refute theory, but, as we have seen, if every observation is always conditioned by a theoretical point of view, there is no neutral observation capable of recording objective facts, so no refutation by empirical means will be possible. Moreover, critical rationalism does not regard science as a human fact, that is, as an expression of human freedom and creativity, but rather as something that imposes itself on human consensus due to its objectivity. In this regard, too, Feyerabend disagrees with Popper, although for the latter, objectivity is not the result of impartial observation, but rather the outcome of a critical effort on the part of the knowing subject.[43] Knowledge always derives from human actions that undoubtedly have a character of irrepressible subjectivity.

[39] Feyerabend (1975), p. 32.

[40] Cf. Feyerabend (1970) and (1975), Chaps. 6–12. See also the chapter on Feyerabend and Galileo in the present collection.

[41] Feyerabend (1975), p. 306. In *Dialogue on Method* Feyerabend says that "science is our religion. What happens inside religion, is Big news. What happens outside of it, in Pagan nonsense": Feyerabend (1979), p. 100.

[42] Feyerabend (1975), p. 305.

[43] Cf. Popper (1983), p. 48.

However, as soon as theories are controlled, discussed and criticised, one leaves the subjective sphere of individual psychology to enter the field of what is publicly testable for any knowing subject: objectivity coincides precisely with those aspects of knowledge that are public and on which reason can be critically exercised.[44] Scientific objectivity is only provided by critical method, which, if we are lucky, allows us to rout out errors and prejudices that may creep into our theories.

Feyerabend objects that Popper "identifies the problem of (the objectivity of) *knowledge* with the problem of the truth and/or objectivity of *theories*. These may work for parts of physics (though there is 'tacit knowledge' even here) but it is much too narrow an approach for history, psychology and for the wide domain of common sense".[45] Feyerabend rejects both the classical idea of tradition-independent truth, which he identifies with the material notion of objectivity, and the more recent idea of tradition-independent ways of finding truth, which he identifies with the formal notion of objectivity. The ancient idea of tradition-independent truths (the material notion of objectivity) had to come to terms with the presence of different cultural perspectives and so it "was replaced by the somewhat less ancient idea of tradition-independent *ways of finding truth* (the formal notion of objectivity). And being rational or using reason now meant using these ways and accepting their results".[46] Later, with the progress of science, the formal notion of objectivity was appealed to not only to produce knowledge, but also to legitimise an already existing body of knowledge. This formal notion of objectivity raises problems, because there exists no finite set of rules that enforces or forbids well defined procedures and is compatible with all events we find in the history of science.[47] Such a procedure only works in a stable world, which rarely offers counterexamples, but it is ineffective when theories deal with phenomena presenting many anomalies, such as most social problems. Even assuming that being rational means following a standard procedure, it must be acknowledged that there is often disagreement as to which procedure is correct: "hardnosed empiricists regard it as irrational to retain views plainly in conflict with experiment while hardnosed theoreticians smile at the irrationality of those who revise basic principles at every flicker of the evidence".[48]

In fact, it should be noted that Popper had already offered a weaker notion of rationality, denying that reason has a demonstrative function and merely acknowledging its critical function. In *The Open Society and its Enemies*, the author says that the adoption of a rational attitude expresses "an irrational *faith in reason* [...] the fundamental rationalist attitude results from an (at least tentative) act of faith—from faith in reason".[49] This conviction was reiterated later, and remained firmly in

---

[44] Cf. Popper (1935), p. 22. Referring to Kant, Popper uses the word 'objective' "to indicate that scientific knowledge should be *justifiable,* independently of anybody's whim: a justification is 'objective' if in principle it can be tested and understood by anybody" (ibid.).

[45] Feyerabend (1987), p. 80.

[46] Ibid., p. 8.

[47] Ibid., pp. 8–9.

[48] Ibid. p. 10.

[49] Popper (1945), Vol. II, p. 231. The appeal for a faith in reason is repeated on p. 258.

Popper's thought until the end: "My rationalism is not dogmatic. I fully admit I cannot rationally prove it. I frankly confess that I choose rationalism because I hate violence, and I do not deceive myself into believing that this hatred has any rational grounds. Or to put it in another way, my rationalism is not self-contained, but rests on an irrational faith in the attitude of reasonableness. I do not see that we can do beyond this".[50]

Feyerabend radicalises this position and draws the consequences: rationalism is unfounded, abstract rationalism is contradicted by the practice of science. As has already been said, Feyerabend acknowledges that critical rationalism is "the most liberal positivistic methodology in existence today", yet he cannot forgive its levity of unquestioningly assuming modern science as the criterion of all criteria: "Science is only *one* of the many instruments man has invented to cope with his surroundings. It is not the only one, it is not infallible, and it has become too powerful, too pushy, and too dangerous to be left on its own".[51] Feyerabend "does not deny the effectiveness of science as a provider of technologies and basic myths", nor does he assert "that we can do without the sciences".[52] He does not doubt that science is a reservoir of knowledge, but the same can be said for "myths, fairytales, tragedies, epics and many other creations of non-scientific traditions".[53]

### 11.3.2 Science as Art

According to Feyerabend, the typical aim of any cultural expression is not to reflect an objective reality that we cannot know, but rather to bring together dissimilar heterogeneous elements into a comprehensible and coherent image, using different tools such as numbers, words or canvases and colours (depending on circumstances), and generating different results. In no case, however, will it be possible to remove from these products the indelible, subjective personal imprint left by those who made them. That is why science is very similar to art, in that both offer interpretations of the world, develop different styles, but use similar modes of communication: for, in order to be followed, the scientist must offer an argument that is not only correct, but also persuasive. Consequently, "*good sciences are arts, or humanities, not sciences in the textbook sense*".[54]Here too, the affinity with Popper, at least at the beginning, is evident, as the author of the *Postcript*, too, said that "science is not only, like art and literature, an adventure of the human spirit, but it is among the

[50] Popper (1963), p. 357.

[51] Feyerabend (1975), p. 217.

[52] Feyerabend (1987), p. 126.

[53] Feyerabend (1979), p. 123.

[54] Feyerabend (1987), p. 295. On the function of art in relation to epistemological issues, cf. Ambrosio (2021).

creative arts perhaps the most human".[55] Popper claims that reason works by trial and error both when we invent myths and when we elaborate theories.[56] Indeed, even scientific theories, like myths, are human inventions, "nets designed by us to catch the world. To be sure, these differ from the inventions of the poets and even from the inventions of the technicians. Theories are not *only* instruments. What we aim at is truth: we test our theories in the hope of eliminating those which are not true. In this way we may succeed in improving our theories—even as instruments: in making nets which are better and better adapted to catch our fish, the real world".[57]

Once again, the differences between Feyerabend and Popper must be emphasised regarding both the role of reason and the purpose of knowledge. For Popper, reason can criticise theories either through logic or observations; these tools, although they cannot prove the validity of our hypotheses, can refute them. It is no coincidence that for Popper critical is the best synonym for rational, at least in the context of scientific theories.[58] On the contrary, Feyerabend does not wish to assign any precise or even privileged role to reason, since much of knowledge is "tacit", and cannot be expressed in rational terms.[59]

Furthermore, for the advocate of critical rationalism, rational discussion, unlike myths and poetic inventions, always aims at truth.[60] He insists that critical discussion is dominated by the idea of finding a true theory and "*we do justify our preferences by an appeal to the idea of truth*: truth plays the role of a regulative idea. *We test for truth*, by eliminating falsehood".[61] For Feyerabend, however, truth is an "abstract monster", and he wishes it could vanish alike "other abstract monsters such as Obligation, Duty, Morality".[62] According to him, we must forget "an ideal philosophical heaven"—Popper's third world, that is –and turn instead to the material universe, relying on our meagre resources: a fallible brain, imperfect measuring instruments and erroneous theories.[63]

Different interpretations have also been offered about Feyerabend's assessment of rationalism: some have argued that from his first publication in 1948, until his work in the late 1970s, he developed a critical stance towards rationalism, taking a position that has certainly evolved over time, but which has essentially remained consistent.[64] By contrast, Luca Tambolo noted that not only did Feyerabend change his opinion towards Popper's falsificationism, but that Popper himself presented different variants of falsificationism, implying different conceptions of progress. In

[55] Popper (1983), p. 249.

[56] Cf. Popper (1963), p. 192.

[57] Popper (1982a), p. 42.

[58] Popper (1972), p. 82.

[59] Cf. Feyerabend (1987), pp. 106ff.

[60] At least, after *The Logic of Scientific Discovery*, in which truth plays no significant role.

[61] Popper (1972), pp. 29–30.

[62] Feyerabend (1975), p. 180.

[63] Cf. Ibid., p. 159.

[64] Cf. Shaw (2020).

the first variant, offered in *The Logic of Scientific Discovery*, progress is given by corroborated theories that have greater explanatory and predictive power, while in *Conjectures and Refutations* progress is indicated as a better approximation to the truth.[65] Consequently, Feyerabend, who could accept the first version of Popperian methodology, could not accept the second, which implied the recognition of truth as the ultimate goal of knowledge, and of an objective truth, irreconcilable with relativism that would lead him to develop a substantial rift with his master.

Feyerabend believes that Popper's rationalism was still conditioned by Neopositivism, as it defines rationality in terms of scientific rationality, taking it for granted that it is an unsurpassed and unsurpassable paradigm. We must note, however, that Feyerabend also falls into the same trap, in that his sharp critique of scientific rationality ends up by foregrounding this mode of thinking, believing that the only form of rationality is scientific rationality. Thus, Feyerabend would not have succeeded in freeing himself completely from scientism, since he would continue to hold that the paradigm of rationality is that which is derived from science, even if he draws different conclusions, without assuming the possibility of defining rationality itself differently.

Feyerabend rejects the notion of a rationality, the paradigm of which is science, in the name of creativity, intuition, tacit inference. But he does not ask whether there is an alternative notion of it, according to which these abilities are not irrational at all, but may be part of a rationality broadly conceived, extending well beyond logical-mathematical reason.[66] In fact, Feyerabend seems to have accepted the idea that rationality is either of the type theorised by Western rationalism, especially since the modern age, or it is not altogether. He opted for the second option, as, unlike Popper, he could not accept a reason founded "on an irrational faith in the attitude of reasonableness".

## References

Ambrosio, Chiara. 2021. Feyerabend on Art and Science. In *Interpreting Feyerabend. Critical Essays*, ed. K. Bschir and J. Shaw, 11–39. Cambridge: Cambridge University Press.

Collodel, Matteo. 2016. Was Feyerabend a Popperian? Methodological Issues in the History of the Philosophy of Science. *Studies in History and Philosophy of Science* 57: 27–56.

Corvi, Roberta. 2024. Addio alla ragione? Il contributo di Feyerabend alla teoria della razionalità. *aut aut* 402: 8–21.

Farrel, Robert P. 2000. Will the Popperian Feyerabend Please Step Forward: Pluralistic, Popperian Themes in the Philosophy of Paul Feyerabend. *International Studies in the Philosophy of Science* 14: 257.

Feyerabend, Paul K. 1954a. Ludwig Wittgenstein. *Merkur* 8: 1021–1038.

———. 1954b. Wittgenstein und die Philosophie. *Wissenschaft und Weltbild* 7: 212–220. (Part I) and 283-287 (Part II).

[65] Cf. Tambolo (2015), p. 34.

[66] About critical remarks on Feyerabend's theory of rationality, cf. Corvi (2024).

———. 1955a. Carnaps Theorie der Interpretation theoretischer Systeme. *Theoria* 21: 55–62.
———. 1955b. Wittgenstein's *Philosophical Investigations*. *Philosophical Review* 64: 449–483.
Feyerabend, Paul K. 1958. Reichenbach's Interpretation of Quantum Mechanics. *Philosophical Studies*, 9, pp. 47–59; reprinted in Feyerabend (1981), pp. 236–246.
Feyerabend, Paul K. 1963. How to be a Good Empiricist: A Plea for Tolerance in Matters Epistemological. In *Philosophy of Science: The Delaware Seminar*, Vol. II, ed. Bernard Baumrin. New York: Interscience Publishers, pp. 3–39; reprinted in Feyerabend (1999), *Knowledge, Science and Relativism: Philosophical Papers*, Vol. 3, edited by John M. Preston, Cambridge: Cambridge University Press, pp. 78–103.
———. 1965. Problems of Empiricism. In *Beyond the Edge of Certainty: Essays in Contemporary Science and Philosophy*, ed. Robert G. Colodny, vol. 2, 145–260. Englewood Cliffs: Prentice Hall.
———. 1970. Problems of Empiricism, Part II. In *The Nature and Function of Scientific Theories: Essays in Contemporary Science and Philosophy*, ed. Robert G. Colodny, 275–353. Pittsburgh: University of Pittsburgh Press.
———. 1975. *Against Method: Outline of an Anarchistic Theory of Knowledge*. London: New Left Books.
———. 1978. *Science in a Free Society*. London: New Left Books.
———. 1979. Dialogue on Method. In *The Structure and Development of Science*, ed. Gerard Radnitzky and Gunnar Andersson, 63–131. Dordrecht-Boston: D. Reidel.
———. 1987. *Farewell to Reason*. London: Verso.
Oberheim, Erich. 2006. *Feyerabend's Philosophy*. Berlin: Walter de Gruyter.
Popper, Karl R. (1935). *Logik der Forschung*, Vienna: Springer; English translation, *The Logic of Scientific Discovery*, London: Hutchinson, 1959.
———. 1945. *The Open Society and Its Enemies*, Vol. 1: *The Spell of Plato*, and Vol. 2: *The High Tide of Prophecy: Hegel, Marx, and the Aftermath*, 4th ed. London: Routledge & Kegan Paul, 1962.
———. 1963. *Conjectures and Refutations: The Growth of Scientific Knowledge*. London: Routledge & Kegan Paul.
———. 1972. *Objective Knowledge: An Evolutionary Approach*. Oxford: Clarendon Press.
———. 1974. Intellectual Autobiography. In *The Philosophy of Karl Popper*, ed. Paul A. Schilpp, vol. I, 3–181. La Salle: Open Court.
———. 1982a. In *The Open Universe: An Argument for Indeterminism, from the Postscript to The Logic of Scientific Discovery*, ed. William W. Bartley III. London: Hutchinson.
———. 1982b. In *Quantum Theory and the Schism in Physics, from the Postscript to The Logic of Scientific Discovery*, ed. William W. Bartley III. London: Hutchinson.
———. 1983. In *Realism and the Aim of Science, from the Postscript to The Logic of Scientific Discovery*, ed. William W. Bartley III. London: Hutchinson.
Preston, John M. 1997. *Feyerabend: Philosophy, Science and Society*. Cambridge: Polity Press.
———. 2016. The Rise of Western Rationalism: Paul Feyerabend's Story. *Studies in History and Philosophy of Science, Part A* 57: 79–86.
Shaw, Jamie. 2020. The Revolt against Rationalism: Feyerabend's Critical Philosophy. *Studies in History and Philosophy of Science, Part A* 80: 110–122.
Tambolo, Luca. 2015. A Tale of Three Theories: Feyerabend and Popper on progress and the Aim of Science. *Studies in History and Philosophy of Science, Part A* 51: 33–51.
Watkins, John W.N. 2000. Feyerabend Among Popperians, 1948-1978. In *The Worst Enemy of Science? Essays in Memory of Paul Feyerabend*, ed. John Preston, Gonzalo Munévar, and David Lamb, 47–57. Oxford-New York: Oxford University Press.

# Chapter 12
# Feyerabend and Kuhn

**Paul Hoyningen-Huene**

**Abstract** I begin the chapter with biographical remarks, especially, when and how Feyerabend and Kuhn met. Then, I discuss the concept of incommensurability. Perhaps surprisingly, Feyerabend and Kuhn did disagree about several aspects of this concept, which has often not been seen appropriately in the literature. I first discuss their agreements, and then the main differences in their understanding of incommensurability. Then I discuss Feyerabend's general criticism of Kuhn's *Structure of Scientific Revolutions*. Feyerabend's criticism is rather harsh, explicitly accusing Kuhn of propaganda. Feyerabend is especially enraged about Kuhn's positive evaluation of Kuhn's normal science. However, a closer analysis of Feyerabend's arguments against normal science reveal that they are not very strong and somewhat biased. The chapter ends with some remarks about Feyerabend's and Kuhn's reconciliation in the late 1980s.

**Keywords** Thomas Kuhn · Paul Feyerabend · Incommensurability · Normal science · Ideology · Scientific progress

## 12.1 Introduction

In this paper, I want to discuss several aspects of the relationship between Paul Feyerabend and Thomas Kuhn. There are seven sections. The next section will contain biographical information about the two, especially when, where, and how these two thinkers came together. Then, in the third section, I'm going to discuss aspects of Kuhn and Feyerabend's theories that have become trademarks. By discussing these issues one can see the greatest parallels between their theories. In the fourth section, I'm going to show the large differences that existed in the sixties between

P. Hoyningen-Huene (✉)
Institute of Philosophy, Leibniz University, Hannover, Germany
e-mail: hoyningen@ww.uni-hannover.de

© The Author(s), under exclusive license to Springer Nature Switzerland AG 2024

S. Gattei, R. Corvi (eds.), *Feyerabend in Dialogue*, Boston Studies in the Philosophy and History of Science 346,
https://doi.org/10.1007/978-3-031-71938-7_12

these two thinkers which can be seen by Feyerabend's general criticisms of Kuhn's *Structure of Scientific Revolutions*. It is important to show these differences because Feyerabend and Kuhn are often seen as sitting in the same boat, as if their opinions in the philosophy of science were identical.[1] This, however, is not the case, because in the sixties they could not put aside their massive ideological differences. In the fifth section, I'm going to analyze Feyerabend's direct criticism of Kuhn—specifically of Kuhn's account of normal science. In the sixth section, I'll critically discuss Feyerabend's argument against Kuhn's account. Finally, I'm going to briefly show how, in the late eighties, Feyerabend changed his opinion about his philosophical relationship to Kuhn.[2]

## 12.2 Biographical Remarks

Paul Feyerabend, born 1924, went as a guest professor to the Philosophy Department of the University of California at Berkeley in 1958.[3] The year after he received a permanent position. Prior to that he worked at Bristol. Thomas Kuhn, born 1922, went as assistant professor to the University of California at Berkeley in 1956.[4] Later he became a full professor, primarily in the Department of the History of Science. Prior to that he was at Harvard. According to Feyerabend himself, he first read Kuhn's ideas in 1959.[5] By the latest they were well acquainted by 1960, and there most intensive interaction was in 1960 and 61.[6] Their positions at the same university ended in 1964, when Kuhn went to Princeton. Since then, they only met a few times. However, what I had not known until after Feyerabend's death in 1994, Kuhn and Feyerabend were not really on speaking terms anymore, at least since

[1] Sometimes, however, Feyerabend and Kuhn kindled this impression themselves, see, for instance (Feyerabend 1987), p. 156. Some philosophers have been well-aware of their differences, for instance (Suppe 1977, p. 636; Burian 1984, pp. 10–12; Shaw and Bschir 2021, p. 2).

[2] This chapter is a substantial elaboration of (Hoyningen-Huene 2000a). I shall not deal with Feyerabend's and Kuhn's different relations to relativism. Readers interested in this topic may consult (Sankey 2019).

[3] Information about Feyerabend's life can be found, for instance, in Hoyningen-Huene (1994, 2000b, 2005). Feyerabends autobiography (Feyerabend 1995) is not always reliable. More information can be drawn from Feyerabend's published letter exchanges with Imre Lakatos (Lakatoset al. 1999), Thomas Kuhn (Hoyningen-Huene 1995, 2006), (Collodel and Oberheim 2024), Hans Peter Duerr (Feyerabend and Duerr 1995), Paul Hoyningen-Huene (Augustin 2010), Hans Albert (Feyerabend et al. 1997, 2008), and Karl Popper (Feyerabend and Popper 2020).

[4] Information about Kuhn's life can be found, for instance in Bird (2022), Hoyningen-Huene (1997), and Hoyningen-Huene (2005).

[5] See (Feyerabend 1983), p. 375. I have not found a similar statement in the English editions of *Against Method*.

[6] Feyerabend (1970, pp. 197, 219, 1978b, p. 117 fn. 49, 1979, p. 205 fn. 125).

1962.[7] The reason was that Kuhn could not stand anymore Feyerabend's criticism of his position, which he found unfair and ungrounded (see Sections 4–6, below). When I went to M.I.T. in August 1984 to work with Kuhn, Feyerabend had asked me to invite Kuhn to Zürich, to which invitation Kuhn first reacted slightly reservedly (which I could not understand at the time). However, in May and June of 1985 Kuhn spent a month in Paris and then accepted Feyerabend's invitation, spending 3 days in Zurich in June 1985.[8] They were days of intensive academic and personal discussions. Kuhn gave a talk in the ETH Zurich to a full auditorium maximum. The following day, there was a pleasant cruise on Lake Zurich with a dozen friends and acquaintances. Feyerabend called this trip "the intellectual cruise."

Feyerabend became known noticeably earlier than Kuhn. He made a name for himself in the late fifties, predominately as a knowledgeable philosopher of physics, who had some provocative opinions that were clearly formulated and full of sharp arguments.[9] He travelled often, gave talks, and had many discussions with many people. Kuhn, at this time, was only known to the inner circle of historians of science. This circle was small because in the fifties, the historiography of science had just begun to establish itself professionally in North America.[10] In the early sixties, Feyerabend was noticeably instrumental in helping Kuhn become more well-known among philosophers. In several talks from 1960 to 61, Feyerabend mentioned that a book from Kuhn was forthcoming: a book that would enforce and support Feyerabend's own theoretically developed opinions by way of concrete examples drawn from the history of science.[11] Even so, some time went by before Kuhn became generally well-known, even after the publication of Kuhn's *Structure of Scientific Revolutions* in 1962. In the well-known *Encyclopedia of Philosophy* published in eight volumes in 1967 (Edwards 1967), which in the following decades was a standard reference work in English-speaking philosophy, there is not one hint of Thomas Kuhn. Furthermore, in over 4200 large, two-column pages there is not a single article about the history of science. Paul Feyerabend, by contrast, was mentioned and discussed in an article about the philosophical consequences of quantum mechanics (Hanson 1967, pp. 46–47) and he was the author of four articles about the physicists Boltzmann (Feyerabend 1967a), Heisenberg (Feyerabend 1967b), Planck (Feyerabend 1967c), and Schrödinger (Feyerabend 1967d).

What brought Feyerabend and Kuhn together, so that they were so often mentioned in the same breath? In these days, both had reservations about the dominant philosophical tradition in the Anglo-Saxon world—logical empiricism. Both had a solid scientific background. Feyerabend had a master's degree in Astronomy. Kuhn

[7] Kuhn related this to me in private conversations after Feyerabend's death, after having read a draft of mine on their philosophical relationship.

[8] Feyerabend mentions this meeting in (Feyerabend 1995), p. 162.

[9] Clifford Hooker once related to me that Feyerabend was seen at this time as the guy who will shake up philosophy of science.

[10] Kuhn was an active participant in the institutionalization of historiography of science in the U.S., see, e.g., (Hoyningen-Huene 2001).

[11] See, e.g., (Feyerabend 1999 [1961]), p. 61.

had a Ph.D. in theoretical physics (his supervisor was John Van Vleck, who won the Nobel prize in 1977). Feyerabend developed his critical attitude towards the neo-positivist tradition not primarily from the history of science, but from the discussion of the empirical basis of science, the so-called protocol sentence discussion, and from his intensive analysis of Popper. Kuhn's skepticism toward the empiricist philosophical tradition had its roots in his analysis of the history of science, which he began in 1947. It appeared to him that the *actual* history of science did not fit well with the *normative* image of science that philosophers had developed. Most important to their philosophical similarity, in the eyes of their peers anyway, was the fact that they both simultaneously introduced a new concept into the philosophy of science in their extra-ordinarily influential works from 1962: Feyerabend in his paper "Explanation, Reduction and Empiricism" (Feyerabend 1962) and Kuhn in his book *The Structure of Scientific Revolutions* (Kuhn 1962). This concept is incommensurability; it took center stage in the discussion about Kuhn and Feyerabend from the early sixties on.

## 12.3 Incommensurability

The concept of incommensurability turned out to be an extra-ordinarily difficult and controversial concept. Feyerabend once expressed it like this: "Apparently everyone who enters the morass of this problem comes up with mud on his head."[12] In the following, I cannot discuss the concept of incommensurability in all its details, but only give a somewhat schematic overview.[13] First, I want to present what is meant by the concept of incommensurability and why it was so provocative for the preceding tradition in the philosophy of science. Then, I will talk about the main difference between Feyerabend's concept of incommensurability in comparison to Kuhn's.

The central point of incommensurability is that theories that are separated by a scientific revolution and replace one another, do not work with exactly the same concepts. On the one hand, a new theory introduces new concepts and throws away certain concepts of the old theory. For example, in Newton's theory of gravitation, the new concept of a gravitational force is introduced and the concept of the center of the universe, which was essential for the older theory, was rejected as baseless. This case is very straightforward. The more interesting instantiations of incommensurability, however, are those in which concepts of the old theory are used in the new theory as well, but with a somewhat different meaning. For example, the

[12] Feyerabend (1977), p. 363.

[13] The literature on incommensurability is endless. For incommensurability in general, see (Oberheim and Hoyningen-Huene 2024). For an analysis of Kuhn's concept of incommensurability see Hoyningen-Huene (1990, 1993), Sect. 6.3 (with additional references), and (Hoyningen-Huene 2005). For Feyerabend's concept of incommensurability, see (Hoyningen-Huene 2005; Oberheim 2005, 2007), part II. The discussion of the relationship between Kuhn's and Feyerabend's concepts of incommensurability is somewhat more extensive in (Hoyningen-Huene 2005).

concept of mass is used in Newton's mechanics and in relativity theory, but it does not mean exactly the same thing in both theories. These changes of concepts, often called 'conceptual shifts,' lead to a special relationship between the pre-revolutionary and post-revolutionary theory. It is this relationship that incommensurability is all about. As I said, Feyerabend and Kuhn differ in several points about how the concept of incommensurability is exactly characterized, to which I will turn in a minute. But first, I would like to mention four central characteristics of the relationship between incommensurable theories that are valid for both Feyerabend and Kuhn's conceptions.

1. Incommensurable theories are incompatible, but their incompatibility cannot be transformed into a logical contradiction. Today, this may not be extraordinarily provocative, but for the received philosophical programs of neo-positivism and logical empiricism it was. These programs contended that philosophical insight can only be accomplished by logical analysis.[14] But according to Feyerabend and Kuhn, incommensurability cannot be completely characterized with the tools of logic—at least not with the standard tools of existing logic.[15] Feyerabend is quite explicit about this point: incommensurable concepts "bear no logical relation to each other" (Feyerabend 1962, p. 89). This led them to criticize logical analysis as the exclusive tool of the philosophy of science, and to promote hermeneutic or anthropological methods for the philosophy of science as well.[16]
2. Incommensurable theories make different claims about what exists in the world, or more sharply, what the world is. Correspondingly, one can also say, as Feyerabend and Kuhn did, that the world changes with a scientific revolution. What this exactly means is another question.[17]
3. Incommensurable theories are not literally translatable into each other. That means that with the concepts of the one theory, one cannot formulate the other theory completely. Correspondingly, one has to learn a new language, more precisely, a cluster of related concepts, to understand the new theory. This cluster of concepts partly consists of new concepts and partly of concepts that seem to resemble established concepts but are, in fact, strangely and weirdly altered.

[14] See, e.g., (Carnap et al. 1929).

[15] This remark aims at hinting that Kuhn's "no overlap principle" by which he later characterized the conceptual relation between incommensurable theories (see Kuhn 1991, p. 4), can indeed be seen as a part of logic, namely of the logic of kind terms.

[16] In Feyerabend, this becomes only later explicit, for instance in (Feyerabend 1988), pp. 196–226. In Kuhn, the underlying hermeneutical approach is more indirect. His philosophy of science is based on the new internal historiography of science as he declares very early on in *Structure*: this book's "aim is a sketch of the quite different concept of science that can emerge from the historical record of the research activity itself," (Kuhn 1970 [1962]), p. 1). It is the new internal historiography of science that investigates this research activity mainly by a close, i.e. hermeneutic, reading of the original sources. For more context, see (Hoyningen-Huene 1993), Chap. 1.

[17] For Feyerabend, see (Feyerabend 1983), pp. 370–371 and (Feyerabend 1988), chapter 16; for Kuhn, see Hoyningen-Huene (1993), Chaps. 2, 3, and the epilogue; (Hoyningen-Huene 2021, 2022, 2023).

In order to have a good command of both theories, one must, in a certain sense, become bilingual. Even if one has accomplished this, one is not able to literally translate one theory fully into the language of the other theory because being bilingual does not imply being able to literally translate.

4. The comparison of two incommensurable theories with respect to their capabilities is substantially more complicated than comparing commensurable theories. With commensurable theories that employ substantially the same apparatus of concepts, one compares the predictions of the two theories with respect to their empirical precision. For every individual prediction, there is a corresponding prediction of the other theory. Difficulties can only arise if one theory is better than the other theory in a certain area, while the other theory is better in another area. This case does not arise if one theory is a special case of the other—and this is how theory succession in modern science was widely understood in philosophy of science. With incommensurable theories, however, one cannot compare theories by super-imposing single predictions of the theories because of the difference in the concepts and the untranslatability of both theories. Incommensurable theories conceptualize the more or less same domain differently. That is why some claims of one theory do not find counterparts in the other theory, and what is from the perspective of the one theory evaluated as a substantial achievement may, from the perspective of the other theory, not be understandable.

Here is an illustration.[18] In chemistry in the first half of the eighteenth century, combustion was understood as a procedure in which a substance called phlogiston was set free. That is not so implausible if one observes a candle flame with gas apparently emanating upwards, or if one notices that the ash of wood is much lighter than the wood that was burned. The question arose, how heavy is phlogiston, and different methods of measuring were suggested. After the so-called Chemical Revolution, combustion was understood as oxidation, as a process of binding with oxygen, and phlogiston was declared not to exist. Obviously, the question about the weight of phlogiston does not make sense any longer and quantitative results of the older chemistry about it find no direct counterpart in the new chemistry. From the perspective of the new chemistry, all data about the weight of phlogiston are simply irrelevant, no matter how precise and well reproducible they were; they had be radically reinterpreted. Similarly, many claims in the new chemistry have no counterpart in the old. Although such claims must play a role in the assessment of a theory, they cannot compete directly with the corresponding claims of the other theory. This is why the comparison of the theories becomes substantially more complex.

A certain part of the controversial discussion about the conception of incommensurability can be traced back to the fact that Feyerabend and Kuhn were often understood as wanting to deny the possibility of an achievement-oriented comparison of theories altogether.[19] But if that were true, the development of science would be

[18] Compare (Hoyningen-Huene 2008).

[19] See e.g., among many others, (Sankey 1994), Chap. 1; for a critique of Sankey's book, see Hoyningen-Huene et al. (1996).

arbitrary as it would lack any rational theory choice procedure. This interpretation does not fit with the intentions of Feyerabend and Kuhn (even if Feyerabend in particular often sounded this way). They primarily wanted to argue against the received, in their opinion, over-simplified understanding of the relationship of replacing theories, where the choice of theories by comparison of their capabilities exclusively was made by the comparative evaluation of individual predictions.

Now to the main differences between Feyerabend's conception of incommensurability and Kuhn's.[20] Kuhn's conception of incommensurability has a larger scope (or extension) than Feyerabend's. The reason is that for Feyerabend only *comprehensive* theories can be incommensurable if they are interpreted in a *certain way*. Examples of pairs of incommensurable theories are, for Feyerabend, quantum mechanics and classical mechanics or relativist and classical physics—all in a certain interpretation. Such comprehensive theories are part of the constitution of their objects. If one uses a theory that is incompatible with another comprehensive theory, one gets different objects and the claims of both theories can no longer be directly compared with each other. As a consequence of the restriction to comprehensive theories of Feyerabend's incommensurability thesis, there are some theory pairs that are incommensurable for Kuhn which are not for Feyerabend. For example, the Ptolemaic and Copernican theories of planets, which is for Kuhn one of the standard examples of incommensurability, are for Feyerabend, on the other hand, explicitly rejected as candidates for incommensurability.[21]

Now I shall turn to the controversy between Feyerabend and Kuhn about Kuhn's *Structure of Scientific Revolutions.*

## 12.4 Feyerabend's General Criticism of Kuhn's *Structure of Scientific Revolutions*

Feyerabend respected Kuhn a great deal. Modest, as he actually was, contrary to the impression he could also give, he always thought that Kuhn was more important than himself. He began his famous 1970 article on Kuhn entitled "Consolations for the Specialist" that is contained in the volume edited by Lakatos and Musgrave with the following words:

> In the years 1960 and 1961 when Kuhn was a member of the philosophy department at the University of California in Berkeley I had the good fortune of being able to discuss with

[20] Feyerabend discusses the relationship of his conception of incommensurability to Kuhn's in several places: (Feyerabend 1970, pp. 219–222, 1976, pp. 389–391) (the first English edition of *Against Method* (Feyerabend 1975a, b) does not contain this dispute with Kuhn); (Feyerabend 1978b, pp. 65–70, 1981b), pp. x–xi (identically in Feyerabend 1981a, p. viii–ix, 1981a, p. 154 fn. 54, Feyerabend 1983, p. 375, Feyerabend 1987, p. 156, Feyerabend 1988, pp. 229–230). Kuhn discusses the relationship between the two concepts of incommensurability only in (Kuhn 1970b), pp. 232, 266–267, 268, 269 fn. 3.

[21] Feyerabend (1983), p. 172.

> him various aspects of science. I have profited enormously from these discussions and I have looked at science in a new way ever since.[22]

Despite this, Feyerabend had deep reservations about Kuhn's *Structure of Scientific Revolutions*. Although Feyerabend acknowledged the problems Kuhn struggled with (especially the problem of the omnipresence of anomalies in science), he often disagreed with Kuhn's theoretical treatment of these problems. I cannot deal with the many detailed critical remarks that Feyerabend formulated in his four letters to Kuhn. Instead, I will concentrate on the global point of criticism with which Feyerabend began the first letter to Kuhn and which is the central point of his article from 1970. Feyerabend suspects that underlying Kuhn's thinking is an ideology that "could only give comfort to the most narrow-minded and most conceited kind of specialism." For Feyerabend, this ideology "would tend to inhibit the advancement of knowledge" and "is bound to increase the anti-humanitarian tendencies [...] of [...] science."[23] Such an ideology is, for Feyerabend, naturally a flagrant target open for attack. Feyerabend, on the one hand, sees himself as a strict opponent of all dogmatic tendencies—in science, in philosophy, and later in politics as well. On the other hand, he sees himself as a champion of humanitarian values such as liberty and the right to develop one's individuality. Thus, he is an opponent to all forms of oppression of minorities and other cultures. With respect to science, for Feyerabend it is "the most important question of all [...] to what extent the happiness of individual human beings, and to what extent their freedom, has been increased."[24] This question is a result of his conviction that the happiness and the full development of an individual human being is now as ever the highest possible value.[25] How does it come that Feyerabend suspects Kuhn of a hidden agenda, an ideology with anti-humanitarian tendencies?

Feyerabend's accusation of Kuhn has, on closer inspection, three components that I will summarize by way of introduction before we consider them in detail. First, Kuhn formulates, as is well-known, a certain picture of how science in the basic disciplines typically develops historically. Kuhn's book about the "structure" of scientific revolutions formulates a general model of the stages of the development of science that should be valid for basic research.[26] There are, in other words, certain regularities in the succession of different phases of science. Feyerabend first doubts the historical adequacy of some aspects of Kuhn's (in this sense) schematic description of the development of science. Thus, he contradicts Kuhn in historical (or descriptive) respect. Second, beyond these descriptions, Kuhn discusses the

[22] Feyerabend (1970), p. 197.

[23] Feyerabend (1970), pp. 197–198.

[24] Feyerabend (1970), p. 209. Similarly in his third letter to Kuhn: "I judge the importance of a topic from the influence a specific solution of it may have upon the well being [sic] of mankind. And by 'well being of mankind' I do not mean physical well being, but the well being which derives, among other thing, from the exercise of one's imagination, from the full development of human faculties, and from spiritual happiness", (Hoyningen-Huene 2006), p. 613–614.

[25] Feyerabend (1970), p. 210.

[26] Compare (Hoyningen-Huene 1993), Sect. 1.3.

functional role that different elements of science have for scientific progress. For example, Kuhn characterizes what he calls "normal science" as containing a certain dogmatic tendency and explains why this dogmatic tendency is, paradoxically, not an obstacle to the development of science, but rather conducive to it. Kuhn, thus, assesses certain aspects of science positively by explaining their function for further development. Feyerabend doubts secondly Kuhn's assessment of these elements of science. Thus, he contradicts Kuhn in methodological (or evaluative or normative) respects. Third, in Kuhn's presentation of his theory in the *Structure of Scientific Revolutions*, the descriptive and evaluative aspects are not cleanly separated, more often descriptive and evaluative elements alternate. The reason for this, from Kuhn's perspective (at least in retrospect from 1969 and in reaction especially to Feyerabend), is that under certain assumptions, the descriptive claims about the development of science indirectly imply normative claims for a successful operation of science.[27] Feyerabend, on the other hand, sees in Kuhn's mode of presentation an insidious way of taking the reader in with his own ideology without the possibility of critical distance; thus, he criticizes Kuhn with respect to his mode of presentation.

So, now to the concrete details. At heart of these criticisms is the so-called normal science' which Kuhn brought to the discussion.[28] Normal science, after Kuhn, is a phase of the development of science that is marked by the broad consensus that the scientific community shares with respect to the basic questions within their discipline. This consensus is based on concrete scientific achievements which are so convincing, and which have so much heuristic potential, that further research in the corresponding discipline can model itself upon them and choose problems and solutions in analogy to them. Such exemplary concrete scientific problem solutions are called paradigms by Kuhn. The concrete research activity in normal science has certain similarities with a very different activity, namely, with puzzle-solving. The given characterization of normal science has as an assumption that there is no doubt about the exemplary character of the paradigmatic problem solutions—more strongly put, that the paradigms are effective dogmatically. They are effectively dogmatic in the sense that during normal science they cannot be called into question, thus they do not have to be justified or continually tested. Instead, in the practice of normal science one assumes that they are valid. But for Kuhn, normal science leads always to scientific revolutions, where the until then valid exemplary problem solutions are replaced by new ones. This happens in phases of extra-ordinary science where several different theories, the old theory, improved versions of it, and totally new ones are tried out and compared.

First, Feyerabend doubts that the existence of normal science is a historical fact.[29] Instead, he believes that there is no temporal differentiation of, as he calls it, periods of theory proliferation and monism, but that those are ways of science which coexist simultaneously. Feyerabend's reasoning for this assertion is not very convinc-

---

[27] Kuhn (1970a, p. 207, b), pp. 233, 237.

[28] Compare for the following (Hoyningen-Huene 1993), Chap. 5 with references.

[29] Feyerabend (1970), pp. 207–208, 212.

ing: he just gives one example which he thinks proves the co-existence of theory proliferation and the work inside one tradition. But this example, as so often in the history of science, could also be interpreted differently. It does not seem to me to be a specific weakness of argumentation by Feyerabend, but rather an expression of the difficulty of justifying or refuting basically statistical assertions about regularities in the history of science by historical material.[30] I will not follow this criticism of Kuhn and its inherent problems here because in comparison to the following second and third point of criticism by Feyerabend, it is marginal.

Feyerabend's second criticism concerns the evaluation of normal science. For Feyerabend, normal science is, to put it simply, a horror, just as it is for the other critical rationalists of the 1960s—especially Popper and Watkins.[31] I group Feyerabend with the critical rationalists because in 1960–61, when he wrote the two letters to Thomas Kuhn, an essential part of his arsenal of arguments was indeed taken from the critical rationalist camp.[32] The reason for his aversion against Kuhn's normal science is its dogmatic or quasi-dogmatic element, as Kuhn himself explicitly developed it.[33] For Kuhn, this dogmatic element is not at all tied to a devaluation of normal science but is functional for scientific progress.[34] On the other hand, for a critical rationalist, a dogmatic element inside science is absolutely unacceptable. The reason is that for a critical rationalist science is at heart critical and must be so. Science consists of producing highly falsifiable hypotheses and testing these hypotheses as rigorously as possible. One of the central tenets of critical rationalism is that if one decides not to test certain hypotheses, then one ceases to do science.[35] If Kuhn evaluates the dogmatic element of normal science positively, he shows, in the eyes of the critical rationalist, a fundamental violation of the scientific ethos, namely, to be critical and undogmatic. And for the enlightener Feyerabend, this violation is also anti-humanitarian, because,

> Progress has always been achieved by probing well-entrenched and well-founded forms of life with unpopular and unfounded values. This is how man gradually freed himself from fear and from the tyranny of unexamined systems.[36]

[30] Compare on these problems (Meehl 1992; Faust and Meehl 1992).

[31] See Popper (1970), Watkins (1970).

[32] This is particularly evident in Feyerabend's defense of falsificationism in his first letter to Kuhn, see, e.g., (Hoyningen-Huene 1995), p. 363. On the same page, there is a sentence that shows how far away the Feyerabend of 1960 is from the Feyerabend from the seventies: "In short, the rules according to which the game of science is played are simply the rules of scientific method."

[33] See Kuhn (1963b).

[34] For this reason, the word "dogmatic" is not quite appropriate in the given context as it has in its everyday use a negative value connotation. Kuhn therefore used "quasi-dogmatic" instead, (Kuhn 1963b), p. 347 n. 1 and p. 349. Finally, because of its misleading connotation, Kuhn gave up the qualification of normal science as "dogmatic" or "quasi-dogmatic" altogether: see (Kuhn 1963a), pp. 390 and 392.

[35] See Popper (1959 [1934]), Section 85.

[36] Feyerabend (1970, pp. 209–210). I wonder whether Feyerabend would have accepted this utterly abstract and general (and historically quite questionable) statement some 20 years later.

That is why Feyerabend identifies Kuhn's normal' element of normal science as a conservative and anti-humanitarian element.[37] I will return to more details in the next section.

But what enraged Feyerabend even more is Kuhn's mode of presentation, as Feyerabend understood it, and this is his third main criticism.[38] Feyerabend grants that Kuhn needs a point of view, or, as he also puts it, "an ideology," which provides the background for his account in the sense that it influences the interpretation of the historical facts. Without such an interpretation generating point of view, any historical account "would be the most drab and uninteresting affair imaginable."[39] In other words, Feyerabend allows a "methodological", that is evaluative point of view, according to which Kuhn evaluates certain elements of science as rational and other elements as irrational. Feyerabend has especially Kuhn's positive evaluation of normal science in mind. However, Feyerabend accuses Kuhn of not making this evaluative point of view explicit, so that the readers can take notice that there are also alternative points of view, which can lead to other evaluations. Instead, with Kuhn it looks as if the evaluation of historical facts follows immediately from these facts themselves. Feyerabend summarizes this aspect of his criticism very concisely in the first letter to Kuhn:

> What you are writing is not just history. It is *ideology covered up as history*.[40]

And,

> It is this bewitching way of representation to which I object most, the fact that you take your readers in rather than trying to persuade them. This manner of presentation you share with Hegel and Wittgenstein.[41]

Feyerabend calls this aspect of Kuhn's account its "ambiguity": a vacillation between description and prescription. Feyerabend goes as far as to claim openly and in print that Kuhn's ambiguity is intended:

> I venture to guess that the ambiguity is *intended* and that Kuhn wants to fully exploit its propagandistic potentialities. He wants on the one side to give solid, objective, historical support to value judgements which he just as many other people seem to regard as arbitrary and subjective. On the other side he wants to leave himself a safe second line of retreat: those who dislike the implied derivation of values from facts can always be told that no such derivation is made and that the presentation is purely descriptive.[42]

[37] Feyerabend (1970), p. 212.

[38] See Feyerabend (1970, pp. 198–199) and the first paragraph of Feyerabend's first letter to Kuhn: (Hoyningen-Huene 1995), pp. 354–355.

[39] Hoyningen-Huene (1995), p. 355.

[40] Hoyningen-Huene (1995), p. 355, Feyerabend's own italics.

[41] Hoyningen-Huene (1995), p. 355.

[42] Feyerabend (1970), p. 199, Feyerabend's own italics.

I do not know if Feyerabend felt good about this massive and in fact slandering accusation.[43] In any case, in his first letter to Kuhn, he asks at the end of a similar paragraph:

> Or have I perhaps completely misunderstood you?[44]

And in his printed paper from which the above quote was taken, he pleads for the support for his own interpretation because he is fortified in his opinion

> by the fact that almost every reader of Kuhn's *Structure of Scientific Revolutions* interprets him as I do, and that certain tendencies in modern sociology and modern psychology are the result of exactly this kind of interpretation.[45]

This passage is rather surprising as Feyerabend is not known as a philosopher who is very impressed by majority opinions, not to mention of using them to legitimize his own position.[46] A bit later in the same article, Feyerabend then asks if those readers misinterpreted Kuhn[47]—and I do not think that for Feyerabend this was a purely rhetorical question.

## 12.5 Feyerabend's Criticism of Kuhn's Evaluation of Normal Science

In any case, at the end Feyerabend leaves open the question about the mode of presentation of Kuhn's book. But their deepest point of difference is the assessment of normal science, to which I now turn again. How deep this difference is, can be seen by the fact that Feyerabend is obviously unable even to describe normal science in neutral, that is non-evaluative terms. In passages where he intends to present Kuhn's view about normal science, he describes normal science as "the most boring and most pedestrian part of the scientific enterprise"[48] which deals with tiny puzzles"

---

[43] Also in later publications, Feyerabend makes strongly derogatory remarks about Kuhn, for example in (Feyerabend 1999 [1975a, b]), p. 185: "Kuhn's ideas are interesting but, alas, they are much too vague to give rise to anything but lots of hot air. [...] Never before has the literature on the philosophy of science been invaded by so many creeps and incompetents. Kuhn encourages people who have no idea why a stone falls to the ground to talk with assurance about scientific method." Or in (Feyerabend 1999 [1980]), p. 215: "Note how clearly and simply Mill describes a phenomenon which Kuhn much later used to support a rather cumbersome theory of 'revolutions' and 'normal science'."

[44] Hoyningen-Huene (1995), p. 355

[45] Feyerabend (1970), p. 198.

[46] An anonymous referee suggested that the cited passage should not be read "as an appeal to popularity, but to show that Kuhn has been successful in his 'propaganda' and this makes it more important to be clear about his normative commitments (in other words, Feyerabend's worry about the mode of presentation is not a mere idle worry)." This is certainly also a legitimate way of reading the passage.

[47] Feyerabend (1970), p. 199.

[48] Feyerabend (1970), p. 209.

and which is based on the monomaniac concern with only one single point of view."[49] Briefly put, normal science is "professional stupidity."[50] I remark, by the way, that ironically these characterizations invite the same criticism of Feyerabend that Feyerabend made of Kuhn—the mixing of descriptions with evaluations.[51] Such characterizations strongly inhibit Feyerabend's ability to handle the positive function for the development of science that Kuhn attributes to normal science. But apart from these reactions that can probably be qualified as emotional, he tries to disprove Kuhn's argument for the functionality of normal science for scientific development by a sophisticated counter-argument. He formulates this argument by stating "three difficulties" with which Kuhn's functional argument for normal science is confronted.[52]

Feyerabend's argument against Kuhn's positive assessment of normal science has the following structure. First, he describes Kuhn's functional argument for normal science. Then, he shows that Kuhn's functional argument is based on two presuppositions that are both untenable. Finally, he denies the historical existence of normal science. Feyerabend's argument against Kuhn's positive assessment of normal science can thus be reconstructed in four steps.

Step 1: Feyerabend's description of Kuhn's functional argument for normal science: "Normal science, [Kuhn] says, is a *necessary presupposition of revolutions.*"[53] This is a functional characterization of normal science, which aims at explaining why normal science is a good thing: because it has the function to lead to revolutions. The persuasiveness of this functional characterization of normal science rests on two presuppositions, that "revolutions are desirable" and that "the particular way in which normal science leads to revolutions is desirable also."[54] Steps 2 and 3 analyze these two presuppositions.

Step 2: Analysis of the first presupposition of Kuhn's functional argument, that revolutions are desirable. The "first difficulty of [Kuhn's] functional argument" which Feyerabend identifies consists in his contention that the desirability of revolutions cannot be founded in Kuhn's theory.[55] According to Feyerabend, it is impossible for Kuhn to assess the changes that a revolution leads to as improvements because of the incommensurability that Kuhn himself recognizes. If a scientific revolution brings only a change but not an improvement, then there is no argument for the desirability of revolutions.

[49] Feyerabend (1970), p. 201.

[50] Feyerabend (1970), p. 208 fn. 2.

[51] This holds especially for Sect. 12.4 of (Feyerabend 1970) that is the source of the citations given.

[52] Feyerabend (1970), pp. 202–208. Despite the title of Sect. 12.5, "Three difficulties of [the] functional argument," the third difficulty is only discussed in Sect. 12.6, see the end of Sect. 12.5, p. 207. Yes, Feyerabend can be very sloppy.

[53] Feyerabend (1970), p. 201, Feyerabend's italics.

[54] Feyerabend (1970), p. 202.

[55] Feyerabend (1970), p. 202.

Step 3: Analysis of the second presupposition of Kuhn's functional argument, that the particular way in which normal science leads to revolutions is desirable. The "second difficulty of [Kuhn's] functional argument" that Feyerabend identifies consists in his contention that there is a better route to revolutions than normal science. Feyerabend thinks that according to Kuhn,

> scientists would stick to their paradigms to the bitter end until disgust, frustration and boredom makes it quite impossible from them to go on;[56]

they would "suddenly [give] up when the problems get too big."[57] Feyerabend confronts this route to revolutions with an alternative. According to this alternative route, revolutions should be caused by following the principle of the proliferation of theories; thus, by creating competing theories to a given theory. In light of the competing theories, the difficulties of the former theory will be emphasized and simultaneously the means for repairing or getting rid of the difficulties can be seen.[58] For Feyerabend, this procedure also leads to revolutions, but as opposed to Kuhn, in a rational way. Thus, Kuhn's normal science route to revolutions is not desirable.

Step 4: The third difficulty of Kuhn's functional argument is that the historical existence of normal science is more than doubtful.[59] And this is the end of Feyerabend's counter-argument against Kuhn's positive evaluation of normal science.

## 12.6 Critical Discussion of Feyerabend's Argument

Let me now assess Feyerabend's argument with respect to how convincing it is. I will discuss the argument's four steps in turn.

Step 1: First of all, and most importantly, the presentation of Kuhn's functional argument for normal science is far from appropriate. The positive assessment of normal science by Kuhn is not at all derived from the fact that normal science leads to revolutions, as Feyerabend alleges. Rather, its leading to revolutions makes it only *acceptable* that normal science contains a certain quasi-dogmatic element. The main function of normal science is rather that it produces, in an extra-ordinarily efficient way, scientific knowledge of which a certain part also survives the next revolution.[60]

Step 2: Feyerabend's "first difficulty" is an immanent critical argument based on a certain interpretation of Kuhn's concept of incommensurability. According to this interpretation, incommensurability implies that one can no longer talk of

---

56 Feyerabend (1970), p. 205.

57 Feyerabend (1970), p. 207.

58 Feyerabend (1970), p. 205.

59 Feyerabend (1970), pp. 207–208.

60 See Hoyningen-Huene (1993), Sects. 5.6 and 6.3.

scientific progress through revolutions. But in truth, this implication is not part of Kuhn's theory: for Kuhn, incommensurability does not exclude scientific progress at all, but only precludes certain conceptions of scientific progress, to which the conception of the cumulative progress of knowledge as approaching truth belongs in particular.[61] The first difficulty that Feyerabend mentions is thus based on a misunderstanding of Kuhn's conception of incommensurability and is thus irrelevant. I point out, by the way, the deep irony that one of the inventors of incommensurability misunderstands the other inventor's conception of incommensurability with respect to a very substantial aspect!

Step 3. The "second difficulty" that Feyerabend mentions is that he sees a much better way to trigger scientific revolutions than Kuhn's normal science—the permanent proliferation of theories. At first, let there be no doubt that step 3, if it is correct, is not an argument that hits Kuhn directly because of the misdescription of Kuhn's functional argument. But even independently, Feyerabend's argument is problematic. The background of this argument is Feyerabend's conviction that sometimes substantial anomalies of a theory can only be discovered from the perspective of a competing theory. From the perspective of the original theory, these anomalies are invisible, according to Feyerabend.[62] I call this thesis of Feyerabend the *anomaly importation thesis*. But this thesis has its own difficulties.[63] They become especially clearly visible when one asks in exactly what relationship are the two theories involved in the anomaly importation thesis. Let us suppose that the two theories are *commensurable*. In this case, one theory can be articulated basically with the vocabulary of the other. Under these circumstances, it is difficult to see why a substantial anomaly of one theory is invisible from its own perspective, but visible from the perspective of the other theory. Let us therefore suppose things are the other way around, that the theories are *incommensurable*. Then one has to explain how it is possible that the defenders of the second theory who see a substantial anomaly for the first theory can ever convince the defenders of the first theory of this circumstance. Because of the incommensurability of the two theories and the resulting ontological disparities between them, the proponents of the first theory always have the possibility to claim that the difficulties of their own theory which are diagnosed from the perspective of the second theory are irrelevant. I do not claim that these difficulties with Feyerabend's argument are insurmountable, but in any case, they present a challenge for his position.

Step 4. The "third difficulty" of the functional argument concerns the question of whether normal science really exists historically.[64] But that is not truly a *difficulty*

[61] Compare (Hoyningen-Huene 1993), Sect. 7.6.

[62] See Hoyningen-Huene (1995, pp. 364–365), Feyerabend (1962, pp. 65–66, 1999 [1972]), p. 141–142, Bschir (2015). It is remarkable that Brownian motion is the only well worked-out historical example that Feyerabend produces for the anomaly importation thesis.

[63] These difficulties have already been discussed in (Laymon 1977) and (Couvalis 1988), but not with recourse to incommensurability.

[64] Feyerabend (1970), pp. 207–208.

of the functional argument (however it is formulated). Rather, the historical reality of normal science is a *presupposition* of the functional argument. It only makes sense to ask the question of whether normal science is functional for scientific progress if normal science really exists. The functional argument would not be problematic by the non-existence of normal science, but it would become superfluous.[65] The "third difficulty" according to Feyerabend, is thus no *difficulty* for Kuhn's functional argument, but at most a difficulty with this entire theory. But I mentioned already earlier that Feyerabend's historical argument against the existence of normal science is not very convincing.

Let me briefly add a word about Kuhn's own defense of the existence of normal science against Feyerabend and the other Popperians. He did not mainly try to prove the historical existence of normal science[66] but attempted to show that the existence of normal science necessarily follows from the existence of scientific revolutions. This argument assumes that scientific revolutions exist, and this assumption Kuhn shares with his Popperian critics and can thus be used unproblematically in the given context.[67] Revolutions presuppose, according to Kuhn, normal science in the sense that there must be stages *between* revolutions, thus non-revolutionary stages of science—if the talk about scientific revolutions or upheavals should make sense.

Howsoever one assesses the strength of Feyerabend's arguments against normal science, Kuhn felt misunderstood by Feyerabend in the decisive points, and he found Feyerabend's arguments against normal science entirely insufficient. First, Feyerabend himself reports this candidly at the beginning of his 1970 article. Kuhn had, according to Feyerabend, often said this to him in their discussion in the sixties:

> On all these points my discussions with Kuhn remained inconclusive. More than once he interrupted a lengthy sermon of mine, pointing out that I had misunderstood him, or that our views are closer than I had made them appear.[68]

Second, in many places of Kuhn's published reaction to Feyerabend's article from 1970 Kuhn's disappointment with the arguments against his theory is obvious. There Kuhn goes so far as to say that the reaction of his critics leads him to postulate the existence of two Thomas Kuhn's.[69] One Kuhn had published a book with the title *The Structure of Scientific Revolutions* with certain theses. The second Kuhn wrote a totally different book, but with the same title. And this book contains, according to Kuhn, many theses which are incompatible with those from the first book. It is from this second book that his critics, Feyerabend included, often cite.

---

[65] Kuhn has expressed this weakness in the argument quite nicely by summarizing the criticism of normal science in the following way: "Normal science does not exist *and* it is uninteresting", (Kuhn 1970b), p. 233.

[66] See Kuhn (1970b), pp. 233, 249–259.

[67] Kuhn (1970b), pp. 233, 241–249.

[68] Feyerabend (1970), p. 198.

[69] Feyerabend (1970), p. 231.

## 12.7 Feyerabend's Reconciliation to Kuhn in the Late Eighties

In the sixties, when Feyerabend wrote his article, he did not allow his opinion to be changed that he understood Kuhn properly. Also in 1977, he reinforced his conviction that he contested the idea of normal science from the very beginning.[70] But another 10 years later, his tone changed. In the revised English edition of *Against Method* from 1988 he wrote:

> But having read Paul Hoyningen-Huene's admirable account of Kuhn's philosophy (*Die Wissenschaftstheorie Thomas S. Kuhns*, Thesis, ETH Zürich, 1988) and thus having been made aware of the great complexity of Kuhn's thought I am not at all sure that our differences are as great as I often thought they were.[71]

He reinforced this opinion in the third English edition of *Against Method* of 1993 by writing that "our views [...] by now seem to be almost identical."[72] In a 1994 posthumous review, Feyerabend discussed my reconstruction of Kuhn's theory (Hoyningen-Huene 1993), an expanded English version of my above-mentioned habilitation thesis, as a "surprisingly coherent and powerful system of thought."[73] Thus, it seems that Feyerabend reconciled with Kuhn in the last years of life. Kuhn reapproached Feyerabend in his late years, too. In a letter on the occasion of Feyerabend's 65th birthday, Kuhn wrote to him with respect to incommensurability:

> But I've come to recognize that what we mainly differed about was not substance, as it then seemed, but goals: I wanted a theory of the way things were; you, I think, wanted to change them.[74]

In this somewhat ironic way, the somewhat superficial but wide-spread opinion that these two extremely influential thinkers of the second half of the twentieth century were really close in thoughts, is justified after all.

## References

Augustin, C. 2010. Aber ein Paul hilft doch dem anderen. Paul Feyerabend – Paul Hoyningen-Huene, Briefwechsel 1983–1994. In. Wien: Passagen Verlag.

---

[70] Postscript 1977 to the expanded German version of Feyerabend (1970) in Feyerabend (1978a), p. 204.

[71] Feyerabend (1988), p. 230. A modified version of the habilitation thesis mentioned in the quote was published as (Hoyningen-Huene 1989). Feyerabend writes something very similar also in (Feyerabend 1989), p. 405 fn. 26.

[72] Feyerabend (1993), p. 123.

[73] Feyerabend (1994), p. 173.

[74] Kuhn (1988).

Bird, A. 2022. Thomas Kuhn. In *The Stanford Encyclopedia of Philosophy*, ed. E. N. Zalta *(Spring 2022 Edition)*: https://plato.stanford.edu/archives/spr2022/entries/thomas-kuhn/.

Bschir, K. 2015. Feyerabend and Popper on Theory Proliferation and Anomaly Import: On the Compatibility of Theoretical Pluralism and Critical Rationalism. *HOPOS: The Journal of the International Society for the History of Philosophy of Science* 5: 24–55.

Burian, R.M. 1984. Scientific Realism and Incommensurability: Some Criticisms of Kuhn and Feyerabend. In *Methodology, Metaphysics and the History of Science: In Memory of Benjamin Nelson*, ed. R.S. Cohen and M.W. Wartofsky, 1–31. Dordrecht: Reidel.

Carnap, R., H. Hahn, and O. Neurath. (1929). *Wissenschaftliche Weltauffassung – Der Wiener Kreis*. Wien: Artur Wolf. Engl. transl.: The Scientific Conception of the World: The Vienna Circle, in O. Neurath: *Empiricism and Sociology*, ed. M. Neurath and R.S. Cohen, 299–318. Dordrecht: Reidel, 1973.

Collodel, M., and E. Oberheim. 2024. *Feyerabends Formative Years. Volume 2. Feyerabend on Logical Empiricism*, ed. Bohm and Kuhn: Correspondence and Unpublished Papers. Cham: Springer.

Couvalis, S.G. 1988. Discussion: Feyerabend and Laymon on Brownian Motion. *Philosophy of Science* 55: 415–421.

Edwards, P. 1967. *The Encyclopedia of Philosophy*. New York: Macmillan.

Faust, D., and P.E. Meehl. 1992. Using Scientific Methods to Resolve Questions in the History and Philosophy of Science: Some Illustrations. *Behavior Therapy* 23: 195–211.

Feyerabend, P.K. 1999 [1961]. Knowledge Without Foundations. In *Knowledge, Science and Relativism: Philosophical papers*, ed. J. Preston, vol. 3, 50–77. Cambridge: Cambridge University Press.

———. 1962. Explanation, Reduction and Empiricism. In *Scientific Explanation, Space, and Time. Minnesota Studies in the Philosophy of Science, Vol. III*, ed. H. Feigl and G. Maxwell, pp. 28–97; reprinted (modified) in *Philosophical Papers*, vol. 21, pp. 44–96. Minneapolis: University of Minnesota Press.

———. 1967a. Boltzmann, Ludwig. In *The Encyclopedia of Philosophy*, ed. P. Edwards, vol. 1, 334–337. New York: Macmillan.

———. 1967b. Heisenberg, Werner. In *The Encyclopedia of Philosophy*, ed. P. Edwards, vol. 3, 466–467. New York: Macmillan.

———. 1967c. Planck, Max. In *The Encyclopedia of Philosophy*, ed. P. Edwards, vol. 6, 312–314. New York: Macmillan.

———. 1967d. Schrödinger, Erwin. In *The Encyclopedia of Philosophy*, ed. P. Edwards, vol. 7, 332–333. New York: Macmillan.

———. 1970. Consolations for the Specialist. In *Criticism and the Growth of Knowledge*, ed. I. Lakatos and A. Musgrave, 197–230. Cambridge: Cambridge University Press.

———. 1972 [1999]. On the Limited Validity of Methodological Rules. In *Knowledge, Science and Relativism: Philosophical papers*, ed. J. Preston, vol. 3, 138–180. Cambridge: Cambridge University Press.

———. 1975a. *Against Method: Outline of an Anarchistic Theory of Knowledge*. London: New Left Books.

———. 1975b [1999]. How to Defend Society Against Science. In *Knowledge, Science and Relativism: Philosophical papers*, ed. J. Preston, vol. 3, 181–191. Cambridge: Cambridge University Press.

———. 1976. *Wider den Methodenzwang. Skizze einer anarchistischen Erkenntnistheorie*. Frankfurt: Suhrkamp.

———. 1977. Changing Patterns of Reconstruction. *British Journal for the Philosophy of Science* 28: 351–382.

———. 1978a. *Der wissenschaftstheoretische Realismus und die Autorität der Wissenschaften. Ausgewählte Schriften*. Vol. 1. Braunschweig: Vieweg.

———. 1978b. *Science in a Free Society*. London: NLB.

———. 1979. *Erkenntnis für freie Menschen*. Frankfurt: Suhrkamp.

———. 1980 [1999]. Democracy, Élitism and Scientfic Method. In *Knowledge, Science and Relativism: Philosophical papers*, ed. J. Preston, vol. 3, 212–226. Cambridge: Cambridge University Press.
———. 1981a. *Problems of Empiricism. Philosophical Papers*. Vol. 2. Cambridge: Cambridge University Press.
———. 1981b. *Realism, Rationalism and Scientific Method: Philosophical Papers*. Vol. 1. Cambridge: Cambridge University Press.
———. 1983. *Wider den Methodenzwang*. 2nd ed. Frankfurt: Suhrkamp.
———. 1987. *Farewell to Reason*. London: Verso.
———. 1988. *Against Method:* Revised Edition. London: Verso.
———. 1989. Realism and the Historicity of Knowledge. *Journal of Philosophy* 86: 393–406.
———. 1993. *Against Method*. 3rd ed. London: Verso.
———. 1994. Little Review of M. Biagoli: *Galileo Courtier: The Practice of Science in the Culture of Absolutism* (University of Chicago Press, 1993) and of P. Hoyningen-Huene: *Reconstructing Scientific Revolutions* (University of Chicago Press, 1993). *Common Knowledge, 3*, 173.
———. 1995. *Killing Time: The Autobiography of Paul Feyerabend*. Chicago: University of Chicago Press.
Feyerabend, P., and H.P. Duerr. 1995. *Briefe an einen Freund*. Frankfurt am Main: Suhrkamp.
Feyerabend, P.K., and K.R. Popper. 2020. The Feyerabend-Popper Correspondence (1948–1967). In *Feyerabend's Formative Years. Volume 1. Feyerabend and Popper: Correspondence and Unpublished Papers*, ed. M. Collodel and E. Oberheim, 59–420. Cham: Springer International Publishing.
Feyerabend, P., H. Albert, and W. Baum. 1997. *Paul Feyerabend-Hans Albert Briefwechsel, herausgegeben von Wilhelm Baum*. Frankfurt: Fischer.
Feyerabend, P., W. Baum, M. Mühlmann, and H. Albert. 2008. *Briefwechsel*. Klagenfurt: Kitab.
Hanson, N.R. 1967. Quantum Mechanics, Philosophical Implications of. In *The Encyclopedia of Philosophy*, ed. P. Edwards, vol. 7, 41–49. New York: Macmillan.
Hoyningen-Huene, P. 1989. *Die Wissenschaftsphilosophie Thomas S. Kuhns. Rekonstruktion und Grundlagenprobleme*. Wiesbaden: Vieweg (2. Aufl. Wiesbaden: Springer Fachmedien 2014).
———. 1990. Kuhn's Conception of Incommensurability. *Studies in History and Philosophy of Science* 21: 481–492.
———. 1993. *Reconstructing Scientific Revolutions: Thomas S. Kuhn's Philosophy of Science*. Chicago: University of Chicago Press.
———. 1994. Obituary of Paul K. Feyerabend (1924–1994). *Erkenntnis* 40: 289–292.
———. 1995. Two Letters of Paul Feyerabend to Thomas S. Kuhn on a Draft of *The Structure of Scientific Revolutions*. *Studies in History and Philosophy of Science* 26: 353–387.
———. 1997. Thomas S. Kuhn. *Journal for General Philosophy of Science* 28: 235–256.
———. 2000a. Paul Feyerabend and Thomas Kuhn. In *The Worst Enemy of Science? Essays in Memory of Paul Feyerabend*, ed. J. Preston, G. Munevar, and D. Lamb, 102–114. New York: Oxford University Press.
———. 2000b. Paul K. Feyerabend: An Obituary. In *The Worst Enemy of Science? Essays in Memory of Paul Feyerabend*, ed. J. Preston, G. Munévar, and D. Lamb, 1–15. Oxford: Oxford University Press.
———. 2001. Thomas Kuhn und die Wissenschaftsgeschichte. *Berichte zur Wissenschaftsgeschichte* 24: 1–12.
———. 2005. Three Biographies: Kuhn, Feyerabend, and Incommensurability. In *Rhetoric and Incommensurability*, ed. R. Harris, 150–175. West Lafayette: Parlor Press.
———. 2006. More Letters by Paul Feyerabend to Thomas S. Kuhn on *Proto-Structure*. *Studies in History and Philosophy of Science* 37: 610–632.
———. 2008. Thomas Kuhn and the Chemical Revolution. *Foundations of Chemistry* 10: 101–115.
———. 2021. The Genealogy of Kuhn's Metaphysics. In *Interpreting Kuhn*, ed. K.B. Wray, 9–26. Cambridge: Cambridge University Press.

———. 2022. Is Kuhn's "World Change Through Revolutions" Comprehensible? *Epistemology and Philosophy of Science* 59: 55–72.

———. 2023. The Plausibility of Kuhn's Metaphysics. In *Perspectives on Kuhn: Contemporary Approaches to the Philosophy of Thomas Kuhn*, ed. P. Melogno, H. Miguel, and L. Giri, 139–154. Springer.

Hoyningen-Huene, P., E. Oberheim, and H. Andersen. 1996. On Incommensurability. *Studies in History and Philosophy of Science* 27: 131–141.

Kuhn, T.S. 1962. *The Structure of Scientific Revolutions*. 2nd ed. Chicago: University of Chicago Press.

———. 1963a. Discussion. In *Scientific Change: Historical Studies in the Intellectual, Social and Technical Conditions for Scientific Discovery and Technical Invention, from Antiquity to the Present*, ed. A.C. Crombie, 381–395. London: Heinemann.

———. 1963b. The Function of Dogma in Scientific Research. In *Scientific Change: Historical Studies in the Intellectual, Social and Technical Conditions for Scientific Discovery and Technical Invention, from Antiquity to the Present*, ed. A.C. Crombie, 347–369. London: Heinemann.

———. 1970a. Postscript – 1969. In *The Structure of Scientific Revolutions*, 2nd ed., 174–210. Chicago: University of Chicago Press.

———. 1970b. Reflections on My Critics. In *Criticism and the Growth of Knowledge*, ed. I. Lakatos and A. Musgrave, 231–278. Cambridge: Cambridge University Press.

———. 1988. Letter to Paul Feyerabend on Occasion of His 65th Birthday. https://www.pkfcentennial.org/memoirs/thomas-s-kuhn.

———. 1991. The Road Since Structure. In *PSA 1990, Proceedings of the 1990 Biennial Meeting of the Philosophy of Science Association*, ed. A. Fine, M. Forbes, and L. Wessels, vol. 2, 3–13. East Lansing: Philosophy of Science Association.

Lakatos, I., P. Feyerabend, and M. Motterlini. 1999. *For and Against Method: Including Lakatos's Lectures on Scientific Method and the Lakatos-Feyerabend Correspondence*. Chicago: University of Chicago Press.

Laymon, R. 1977. Feyerabend, Brownian Motion, and the Hiddenness of Refuting Facts. *Philosophy of Science* 44: 225–247.

Meehl, P.E. 1992. Cliometric Metatheory: The Actuarian Approach to Empirical, History-Based Philosophy of Science. *Psychological Reports* 71: 339–467.

Oberheim, E. 2005. On the Historical Origins of the Contemporary Notion of Incommensurability: Paul Feyerabend's Assault on Conceptual Conservativism. *Studies in History and Philosophy of Science* 36: 363–390.

———. 2007. *Feyerabend's Philosophy*. Berlin: de Gruyter.

Oberheim, E., and Hoyningen-Huene, P. 2024. The Incommensurability of Scientific Theories. In *The Stanford Encyclopedia of Philosophy*, ed. E. N. Zalta. https://plato.stanford.edu/entries/incommensurability/

Popper, K.R. 1959 [1934]. *The Logic of Scientific Discovery*. London: Hutchinson.

———. 1970. Normal Science and Its Dangers. In *Criticism and the Growth of Knowledge*, ed. I. Lakatos and A. Musgrave, 51–58. Cambridge: Cambridge University Press.

Sankey, H. 1994. *The Incommensurability Thesis*. Aldershot: Avebury.

———. 2019. The Relativistic Legacy of Kuhn and Feyerabend. In *The Routledge Handbook of Philosophy of Relativism*, ed. M. Kusch, 379–387. London: Routledge.

Shaw, J., and K. Bschir. 2021. Introduction: Paul Feyerabend's Philosophy in the Twenty-First Century. In *Interpreting Feyerabend: Critical Essays*, 1–10. New York: Cambridge University Press.

Suppe, F. 1977. *The Structure of Scientific Theories*. 2nd ed. Urbana: University of Illinois Press.

Watkins, J.W. 1970. Against 'Normal Science'. In *Criticism and the Growth of Knowledge*, ed. I. Lakatos and A. Musgrave, 25–37. Cambridge: Cambridge University Press.

# Chapter 13
# Feyerabend and Lakatos

**John Kadvany**

**Abstract** Paul Feyerabend and Imre Lakatos were the closest of friends and disputing intellectual comrades until Lakatos' early death in 1974. *Against Method* (1975) was to have been co-written as *For and Against Method*, with Lakatos delivering his own historicist and rationalist attack on Feyerabend's skepticism about universal scientific method. This chapter describes their complementary and competing ideas on some major episodes from the history of science, including a shared critical appraisal of Newton's philosophy of method compared to his scientific practice. Feyerabend's views in *Against Method* on the emergence of Western rationalism in the Presocratics is deepened by Lakatos' anti-foundationalism in mathematics, as developed in the posthumously published *Proofs and Refutations: The Logic of Mathematical Discovery* (1976). In the philosophy of science, Feyerabend argued that Lakatos' methodology of scientific research programmes was an 'anarchism' in disguise, with that appraisal placed here in the context of Lakatos' Hegelian historiography and its Hungarian intellectual sources.

**Keywords** Research programmes · Rationalism · Presocratics · Historiography · Hegel · Lakatos

## 13.1 1975

Following Imre Lakatos' death in February 1974 at age 51, Paul Feyerabend wrote an appreciation of Lakatos' work for the *British Journal for the Philosophy of Science* where Lakatos had been editor. The essay is a deeply felt encomium to Lakatos, but mainly a coda to Feyerabend's book *Against Method.* That book was meant to be co-authored with Lakatos who likely would have defended his theories

J. Kadvany (✉)
Policy & Decision Science, Menlo Park, CA, USA

© The Author(s), under exclusive license to Springer Nature Switzerland AG 2024
S. Gattei, R. Corvi (eds.), *Feyerabend in Dialogue*, Boston Studies in the Philosophy and History of Science 346,
https://doi.org/10.1007/978-3-031-71938-7_13

of modern scientific method through his methodology of scientific research programmes.

*Against Method* is written as a 'letter' to Lakatos, and so is a closing of the continuous communication by post and telephone between the two great friends up to Lakatos' death. Feyerabend's memorial essay is mostly a restatement of *his*, Feyerabend's, ideas in *Against Method*, including his account of Archaic Greece as the historical setting antedating the emergence of Greek rationalism through Presocratic philosophers and then Plato. That world, as understood through the Homeric epics, is put forward as something of a large-scale cultural counterexample to rationalist methods introduced in Greece through several centuries following Homer. That includes in particular roles for abstractions about nature and inchoate logical-deductive means for their use.

Feyerabend enlists Lakatos' philosophical history of nineteenth century mathematics[1] in support of his, Feyerabend's, intended contrast:

> ...Lakatos shows that informal mathematics, which is that part of mathematics that is responsible for its growth, can be incorporated into the fallibilistic scheme. He removes the last Aristotelian element, the element of necessity, from modern science. And he corrects the idea, originated by Xenophanes and Parmenides that knowledge deals with unchanging entities and must be presented in the form of principles describing such entities. He reverses a trend that has dominated western thought for more than two millennia.[2]

How can this be? Because *P&R* historicized nineteenth century mathematical proof. *P&R* describes how mathematical theorems and the entities involved—the polyhedra featured in *P&R*, but also mathematical functions, sets, series, kinds of numbers, vector spaces, etc.—became inextricably interdependent through theorem reformulations and concept formation. *P&R* is about that nineteenth century achievement as a history of the mathematical present.

The coordination of theorems and concept-formation is Lakatos' 'method of proofs and refutations'. This style of mathematical reasoning only gradually became prevalent through the nineteenth century. *P&R* organizes multiple often incorrect or partial proofs for Euler's theorem stating a basic relation for a polyhedron's **v**ertices, **e**dges and **f**aces, that $V—E + F = 2$, The formula is easily checked for cubes $(8 - 12 + 6)$, tetrahedra $(4 - 6 + 4)$ and many other convex (non-indented) polyhedra. But which ones, exactly, and which ones leading to 'answers' different from 2? Ongoing changes in theorem formulation address possible counterexamples, including cylinders and hollow picture frames. The reconstructed series of proofs and refutations stand for evolving ideas about mathematical rigor and creativity. But at a cost. Improved formulations of theorems and their domains, like which continuous functions are differentiable, or have integrals, or the possible varieties of the mathematical infinite were enabled, Lakatos argues, because of considerable 'concept-stretching' applied to how theorems are set up and the mathematical

---

[1] Lakatos' 1971 Cambridge PhD thesis appeared as three long papers in *BJPS* 1963–4, posthumously combined with thesis material as Lakatos (1976), cited here as *P&R*.

[2] Feyerabend (1975a), p. 14.

'objects' to which they may apply. This new way of mathematical thinking was in principle unconstrained, and is the strongest reason for seeing Feyerabend's point that Lakatos unwinds conceptual prerequisites for mathematical 'necessity', including its proto-forms in Greek philosophy. Lakatos' philosophical history is written as a dialogue among a Teacher and 18 students, named with Greek letters Alpha, Beta,...et al. The character Kappa, a radical skeptic, says that unconstrained 'concept-stretching will refute any statement, and will leave no true statement whatsoever.'(*P&R* 99) What happened historically, says Lakatos commenting on the dialogue, is that

> nineteenth-century mathematical criticism stretched more and more concepts, and shifted the meaning-load of more and more terms onto the logical form of the propositions and onto the meaning of the few (as yet) unstretched terms. In the 1930s this process seemed to slow down and the demarcation line between unstretchable ('logical') terms and stretchable ('descriptive') terms seemed to become stable. A list, containing a small number of logical terms, came to be widely agreed upon, so that a general definition of logical truth became possible; logical truth was no longer 'with respect to' an ad hoc list of constituents. [Alfred] Tarski [in 1935] was, however, puzzled about this demarcation and wondered whether, after all, he would have to return to a relativized concept of counterexample, and consequently, of logical truth—like Bolzano's. (*P&R* 103n2)

So *P&R* is meant to include modern logic, with Feyerabend rightly framing Lakatos' historical and skeptical account of mathematical provability and truth. Mathematical knowledge is presented, as Feyerabend alludes, as of *changing* theorems, (counter-) examples and techniques organized through the history and evolution of the method of proofs and refutations. In *P&R* that common ground appears through modern mathematics and its history, rather than the modern science appearing in *Against Method*. The arch-rationalist Lakatos is adroitly enlisted to support Feyerabend's own project in *Against Method*, principally its concluding chapter on the Homeric Greeks as illustrative of what modern rationality is *not*.

More can be said already about Lakatos and Feyerabend on scientific method. The confluence of skepticism and modern scientific reason would be an ongoing interest for both of them. In spite of the mathematical content of Lakatos' remarkable dialogue-history, there's also a direct connection to Lakatos' later philosophy of science. That would be through the remarkably intricate, insightful, sui generis and amusing historiography of *P&R*. Each of Lakatos' dialogue characters represents stereotypes of important nineteenth century mathematical methods, with a running commentary in the footnotes which identifies the 'actual history' to which the dialogue corresponds. That same intricate historiographic style, devised initially by Lakatos as a superannuated Cambridge graduate student in the late 1950s, reappears as the central expressive means for his philosophy of science and its historical reconstructions of science past.

Before exploring that, we can take advantage of Feyerabend's supportive role for Lakatos to gauge, now some five decades later, the force of the archaic world Feyerabend used prominently as reason's counter-culture.

## 13.2 Back to the Pre-Presocratics

The Archaic period lasted from the beginning of the eighth century BCE until the beginning of the fifth, with the Classical period in Greece being most of the fifth and fourth centuries. From Homer to the death of Socrates in 399 is some 350 years. Toward the middle of the Archaic era, early Greek philosophy includes new *kinds* of concepts and their use by means of new *kinds* of explanation or argument. For concepts, those are idealized substances, or abstractions, from which all else is composed, or ultimately reliant upon, such as *air* (Anaximenes), *fire* (Heraclitus), *water* (Thales), *apeiron* (Anaximander's 'unbounded'), *atoms* (Democritus), *number* (Pythagoras). These abstractions are non-mythological and de-personified alternatives for describing the natural world. Most important for Feyerabend is the novelty of the new 'abstractions' compared to what came before. In *Against Method*, the historical interpretation is addressed to Lakatos' philosophy of science, and then reappears for decades in Feyerabend's work. Hence we continue the dialogue between the two philosophers combined with a brief assessment of Feyerabend's approach to an important episode in the history of ideas.

That includes, for the Presocratics, explanations including proof-like reasoning completely unlike the gods-and-heroes accounts in Homer's *Iliad* and *Odyssey*. Arithmetic had augmented practical calculation and geometrical measurement with the strange new technique of proof by contradiction, famously used as early as Pythagoras to show that the length of a square's diagonal and the length of the square's sides were *not* expressible as a proportion of whole numbers *a:b*. Such a relation would make the lengths directly commensurate. To prove that's not possible, one assumes there *can* be such a common measure, and from that deduce a contradiction, and so reversing the false assumption. This revolutionary method is still today one of the most useful proof techniques used throughout mathematics.

Overlapping the new mathematics and natural philosophy are conceptually timeless notions not found in earlier eras. 'Being' or the 'One' are concepts to which the earliest expressions of proof-like methods are applied by Parmenides and his student Zeno.[3] Their abstractions simply *are*, outside of past, present or future.[4] The proof that a square's diagonal is not commensurate with its side, proved by contradiction, is similarly non-empirical. Feyerabend's interest lay in how this post-Homeric conceptual world of early Greek philosophers came to be. Opposition to Homer reaches a peak in Plato, who wants the earlier conceptual world and the practices supporting it to be abandoned for his timeless Forms with their mathematical-like character. Feyerabend makes two lessons from this conflict.

---

[3] See Szabó (1978), dedicated to Lakatos, with Part III on proof concepts and their origins in dialectic. See also Feyerabend (1999b), pp. 55–59.

[4] See Robb (1983). Havelock (1963) explores abstract language in Plato and literate Greece against Homeric predecessors. On parataxis, orality and writing as relevant to mathematics and logic see my (1991).

First, these earliest manifestations of proto-scientific reasoning are championed, whether by Plato or earlier philosophers, not because they are shown, or argued, to be better than what came before, but because they are valued and preferred. The confrontation is of two historical traditions. Lakatos and many others do not *justify* scientific methods as better at approaching truth in some transcendental sense, or as determining unchanging essences behind appearances. If mid-Archaic and classical Greece are seen as originating these methods, then, as Feyerabend puts it, the 'Presocratics only rarely argued with their predecessors, they started a new form of life and measure of excellence. For example the older thinkers are criticized for their polymathy but nobody explains why a unified account is preferable to a list'.[5]

Second, 'polymathy' and 'lists' bring up a more significant role for the Homeric world. Its conceptual organization differs so much from its rationalist successors that preferences, or comparative advantages, between the two are virtually incomparable, unless what comes later is assumed as better to start. Why indeed is an abstraction better than a list, or a unified account better than Homeric polymathy? One may 'choose' the new rationalist methods, but these are not self-justifying, nor underwritten as a better approximation to Truth when first proposed in the ancient world. Lakatos' conception of scientific method was normative, with theories or research programmes doing *better or worse*, and possibly to be continued or dumped. If such advice is not grounded with more than pragmatic or instrumental preferences, that considerably deflates the conception of the natural world which modern science is about. If indeed, Greek rationalist prototypes are introduced by fiat compared to what's before, just what does that contrast or trade-off involve? How significant is the historical role granted by Feyerabend to the distant form of life understood through Homer's epic poems?

The difference Feyerabend emphasizes in the Homeric world, in a word, is its *paratactic* character. He took up this insight and much supporting detail from Bruno Snell's *The Discovery of the Mind* and its Homeric philology.[6] As a prominent example, there's no language for an integrated body in Homer. Instead there are assemblages of head, torso, joints and limb segments. There's no holistic, coordinated body 'as such'. Vision is similar. There are many *particular* ways of looking, like carefully, with fear, staring, glancing, or some other specific visual attitude. But there's no vision as a neutral 'operation' common to all kinds of seeing, much as we freely use today. Pre-Homeric, 'polymathic' Greeks might 'list' kinds of seeing, and that's what the concept amounts to for them.[7] This paratactic world has no place for a unified self or soul, or at least no single word or expression having that sense.

[5] Feyerabend (1975a), p. 3n1.

[6] Snell (1953). In addition to *AM* on the Homeric world view see Feyerabend (1999a, b, 2016). Williams (1993) Ch. 2 criticizes Snell's realism, but oriented to decision-making and the self, rather than parataxis and polymathy as described by Plato and earlier Greek philosophers. The latter is Feyerabend's focus.

[7] Just 'who' is relevant here in the Archaic world is difficult to pin down. Homer post-dates Greek writing, while an idealized paratactic world should precede that, and is still known only through the epics and Greek philosophers.

Without knowing selves to invoke, there's knowledge, but, apparently, no separation of knowers and what's known, in the style of later natural world abstractions and reasoning about them. Our language games facilitate abstract talk in ways Homeric or pre-philosophical Greek thinking apparently did not. Feyerabend augments Snell with an account of archaic figurative art which, at least to us, makes people, chariots, animals, weapons and such appear as paratactic aggregates. Combined with the linguistic analysis, Feyerabend concludes that paratactic language and pictorial representations are indicative of Archaic perceptions and experience. Archaic language and art are interpreted *realistically*, another favorite concept of Feyerabend's. That's a considerable step beyond conceptual differences between the early Archaic world and what came later. Let's quickly bring the Homeric world, as 'counter-example', up to date, and back to the modern mathematical reasoning of *P&R* with which we began.

Feyerabend justifies his strong interpretations of Homeric language and imagery through Benjamin Lee Whorf's famous linguistic relativity, which has two very different versions. One kind is Feyerabend's, that (**A**) Homer's paratactic language, plus Archaic age figuration, represents how that world is perceived and experienced as non-unified aggregates and lists. The world literally *seems* quite different as suggested by paratactic language. A second (**B**) kind of relativity, also Whorf's, is that different languages *grammaticalize* different concepts or processes: time as tense, spatial relations used as prepositions, attitudes toward knowledge through modal verbs and much more across the world's languages.[8] Grammatical roles need not significantly influence perceptions or experience as in (**A**). For the Homeric world, Feyerabend takes the absence of later abstractions and a pervasive presence for paratactic representations to support his (**A**) interpretation. Feyerabend cautions against that speculative jump,[9] but makes it anyway. Whorf discusses (**A**) and (**B**), with these often confounded, especially by later hand-wringing critics of 'linguistic relativity'.

The difference can be clarified in Feyerabend by substituting *cognition* for *perception,* with a (**B**) version of an idealized Homeric worldview quite defensible and still philosophically provocative. Unqualified 'perception' and 'experience' are themselves post-Homeric abstractions, and something of an exaggerated misdirection. Changes introduced by the Presocratics remain, but are more like (**B**)-style differences seen in linguistically distinctive cultures today.[10] Lakatos could agree on a (**B**) interpretation for paratactic language, and its ('incommensurable') difference

[8] See Gentner and Goldin-Meadow (2003) and Whorf (1956), e.g. pp. 51–56 on Hopi grammaticalization.

[9] Feyerabend (1975b), p. 237 first para., Feyerabend (2016), p. 64.

[10] Linguistic relativity has been the focus of a cognitive linguistic research programme attending to (**A**) and (**B**) versions. There are (**A**)-positive results, though subtle, e.g. sensitivity to color differences. A dramatic example of cognitive influence is of languages (e.g. Mesoamerican Tzeltal) using *absolute* grammatical markers for location. Speakers must always know true directions—'a rabbit is north of my foot'—while an English location-*relative* ('right of', 'front of') speaker lacking a compass is quickly disoriented in a dense forest or homogeneous terrain. See Levinson (2003).

with early Greek philosophy, but falling short of (**A**). In response to the Homeric world of *Against Method*, Lakatos can side with the early natural philosophers—'back to the Presocratics', as endorsed by Popper. Novel language games enable novel conceptualization, so agreeing with (**B**), but suspending judgment on (**A**). This (**B**)-type language change is also explicit in *P&R*. The 'necessity' of modern mathematics is still neutralized by *P&R*, as Feyerabend suggests, but because of options created though language change and diversity. As put by the character Pi, Lakatos' philosopher of language,

> Progress indeed replaces naive classification by theoretical classification, that is, by theory-generated (proof-generated, or if you like, explanation-generated) classification. Conjectures and concepts both have to pass through the purgatory of proofs and refutations. Naive conjectures and naive concepts are superseded by improved conjectures (theorems) and concepts (proof-generated or theoretical concepts) growing out of the method of proofs and refutations. And as theoretical ideas and concepts supersede naive ideas and concepts, theoretical language supersedes naive language. (*P&R* 91)

What is triumphant in *P&R* for nineteenth century mathematics is likewise triumphant for the Presocratics and their rejection of the Homeric world. Motivated choices for abstract language games begun in the Archaic age are principally of new and enhanced theoretical language. Faced with the Homeric 'counterexample', Lakatos might gladly accept Whorf-type (**B**) language change, with Presocratic innovations exploiting opportunities for grammaticalizing timeless and anonymous concepts and reasoning with them. So indeed, Feyerabend can invoke *P&R* to historicize the 'necessity' of scientific and mathematical abstractions, but only because the underlying possibilities for language change are much the same, especially as the Presocratics and moderns all benefit from *written* language unavailable in the pre-Homeric Greek world. That's also the kind of anthropological history favored by Feyerabend, but with greater attention to linguistically-enabled cognition and the perceived autonomy and permanence of written texts.

One might not *prefer* intellectualist abstractions begun with the Presocratics as (part of) a form of life, but linguistic innovation and change is intrinsic to the abundance of languages themselves. For Lakatos, language is critical infrastructure for rationality. A value-laden choice *against* the abstractions initiated by early Greeks comes close to a choice against the value of languages themselves for conceptualizing the world's abundance. Lakatos polemicized at length against just that, with language the essential infrastructure for rational discourse, and tacit knowledge its enemy.[11] The relevant choices are not of perception, language and culture, but of language, theoretical-philosophical culture and more language. Karl Popper's 'objective knowledge' is the driving cultural value here, of knowledge separated from its knowing subjects through public discourse. Feyerabend notably replaced Popper's wooly 'third world' of knowledge with a Wittgensteinian account of

---

[11] '…[A]rticulated knowledge' is 'exactly this small tip of the human enterprise wherein rationality resides.' Lakatos (1978c) (*MSE*), p.226. For Popper too, that 'you can kill my words but not me' Is a foundational principle for an open society.

epistemic objectivity grounded in discursive practices implicit in rationalist forms of life.[12,13]

## 13.3 Reason in History

We can now pivot from *P&R* to the philosophy of science proper via Lakatos' novel and uniquely Hegelian historiographic method, integrating a 'rationally reconstructed' history of nineteenth century mathematics with its 'actual history' in the footnotes. This section provides some philosophical background followed by historical case studies relevant to Feyerabend in the next.

The dialogue form of *P&R* is misleading. The characters Alpha, Beta, . . and Teacher are involved in a fictional discussion, but as a deliberate historical narrative, with figures entering and others receding as the dialogue progresses. Except for Teacher, the characters represent historical stereotypes of nineteenth century mathematicians, characteristic proof patterns, or philosophical positions. Alpha constantly creates counterexamples; Beta and Sigma represent Augustin-Louis Cauchy's and Niels Abel's play-it-safe 'exception-barring'; Delta contracts and redefines definitions to answer Alpha's counterexamples; Lambda introduces 'lemma-incorporation' as the major change in mid-nineteenth-century mathematics, by which theorems are reformulated and qualitatively improved by incorporating counterexample conditions as supporting lemmas or definitions. Lambda's technique is re-baptized as the method of *proofs and refutations*.

This historiographic style adapts and improves on Hegel's *Phenomenology of Spirit*.[14] Hegel's successive 'gestalts' of consciousness are stereotyped versions of philosophies and historical eras from the Greeks through the Enlightenment to the *Phenomenology* itself. Lakatos' improvement is to document, separate, and comment on the 'actual history' of mathematics underwriting the reconstruction of the text. That 'rationally reconstructed' history of nineteenth century mathematics organizes changes in mathematical practice—proving and improving theorems—with proof patterns also becoming forward-looking, positive heuristics for individual proofs, exemplified by ongoing changes to, or variants of, proofs of Euler's theorem. The rational reconstruction is of an individual theorem ($V - E + F = 2$) set in an evolving historical background. That history, in caricature, is internalized as features of contemporary proof methods and individual proofs. *P&R* is deeply pedagogical, extending George Pólya's heuristics for problem-solving to theorem-proving. Lakatos labelled his presentation 'biogenetic', also Friedrich Engel's term for the *Phenomenology,* and meaning the text is explicitly presented as a combined

[12] *PE* Ch. 9, 'Popper's "Objective Knowledge", with reference to Lakatos p. 182.

[13] A systematic treatment of topics here could include: abstraction and content; proofs and reasoning; language and experience; realism and forms of life; cross-language representation; linguistic autonomy; immersive learning. See also the editors' Introductions in Feyerabend (1999a, b).

[14] Hegel (1977/1807).

pedagogy-history.[15] Where Lakatos follows a single theorem through its historical development, Hegel follows a single philosophical 'consciousness'. Lakatos represents the mathematical present as history using Hegel's same historiographic method. Reconstructed caricatures of actual history are Hegel's and Lakatos' shared approach for understanding reason in history. Historical 'determinism' is not involved.

With Lakatos' 'research programmes' in the philosophy of science, there's no 'individual' history as in *P&R* and proofs of Euler's theorem. Given that, actual history is also normatively 'corrected' by the rational reconstruction, whose falsifications of history may (or may not) be identified, say in the footnotes below.[16] 'Methodology is wedded to history', Lakatos says, 'since methodology is nothing but a rational reconstruction of history, of the growth of knowledge.'(*MSE* 178n3) As Hegel puts it, 'Whoever looks at the world rationally will find that it in turn assumes a rational aspect; the two exist in a reciprocal relationship.'[17]

What are research programmes?[18] Lakatos builds on Popperian falsification as confrontations of theories or models with theory-laden observations. Progress is non-inductive through 'negative' learning, as Popper called it, and as did Hegel.[19] Falsifiability for Popper is a normative 'demarcation criterion' distinguishing science from non-science, and Lakatos similarly wants research programmes to have normative force. While falsifiability is often noted positively as a feature of scientific method, problems quickly arise. It may not make sense to reject or abandon a theory when no adequate alternative is in place. If, as Lakatos argued, 'all theories are born refuted', meaning theories always face anomalies or contradictory facts, falsification alone is too strict. Theories can also be saved by ad hoc changes which may or may not be significant. If falsification is followed rigorously, theories could move in and out of 'refuted' status, with 'final' refutations requiring a maze of supporting qualifications. That's even more difficult if, as Popper asks, conditions of falsification need to be specified in advance.

Lakatos' lesson is that an isolated theory confronting (even) theory-laden observations makes for a poor 'unit' of progress. Instead, series of theories should be assessed, particularly with respect to ongoing predictions, their verifications or

---

[15] 'Biogenetic' refers to Ernst Haeckel's 'ontogeny recapitulating phylogeny' in biological development; see *P&R* p. 4n2. See the *Phenomenology's* Preface on *Bildung,* or constructive historical learning, as the book's organizing idea. The Introduction clearly states Hegel's historiographic method.

[16] '...Rational reconstruction is not just a *selection* of methodologically interpreted facts: it may be, on occasions, their *radically* improved version...One way to indicate discrepancies between history and its rational reconstructions is to relate the internal history in the text, and indicate *in the footnotes* how actual history misbehaved in the light of its rational reconstruction.' Lakatos (1978b) (HS) pp. 119–20.

[17] Hegel (1975/1822), p. 29.

[18] The long essay Lakatos (1970) (FM) develops research programme concepts as a reconstructed series of Popper variants plus case studies; some are summarized below. FM and the historiographic HS essay form an integrated whole.

[19] On learning through error for Mill, Boltzmann, Popper and Lakatos see Feyerabend (1975a), p. 9.

failures, changes in modeling or experimental methods, and especially the status of competing theories addressing similar problems. It's almost tautological that Popper's 'conjectures and refutations' lead to a temporalized science and changing theories. Lakatos's next step is to give Popper's conjectures and refutations an 'historical character': both as series of theories and as 'tested' by the history of science.[20] Historical materials are introduced as data to assess Popper's or any other methodology. The historical turn is introduced by Lakatos deliberately and with precision.

Finally, theories don't accumulate willy-nilly. In a research programme they share some 'hard core' of assumptions, like Newton's laws of motion plus universal gravitation; evolution by natural selection for Darwin; the central dogma that 'DNA makes RNA makes proteins' in molecular biology; the wave (or particle) theory of light. Theories and models develop as successive testable versions of the hard core, and may progress or not. Programmes further organize theories through a 'positive heuristic' for generating models, predictions, or experimental tests; and a 'negative heuristic' defending against competing research programmes, *their* theories and other confirmed (novel) predictions. There are anomalies or contradictions, and reinterpretations of observational data; 'touchstone' theories, such as chemistry and physics for biology, or optics for astronomy; and observational theories for apparatus operation and measurement methods. Progress occurs with additional theoretical predictions, additional confirmations, reduction of ad hoc hypotheses, new modeling concepts and applications. Programmes not showing relevant progress or over-using ad hoc hypotheses are 'degenerating' or just non-progressive.

As Feyerabend put it, Lakatos' concepts explicate the 'constantly expanding horizon of facts in modern science'.[21] Knowledge *grows*. Temporal 'problemshifts' have marks of scientific rationality or not. Research programme concepts are used to characterize an 'internal' history of science about theories and observation statements, or language. There are no people or scientists for such 'internal' histories, not unlike those found in many science textbooks. Whether theories are *true* or even *believed* are questions for other types of history or philosophy.[22] 'External' history is of actual scientists, groups, and social or personal events, influencing theory interpretation or change. Both kinds of history are needed, with their relative importance a nexus for philosophical and historical debate. Hegel made the same internal-external distinction for his world history.[23] Being 'scientific' means to be part of a research programme, regardless of progress, for which one often needs patience. There are no time limits nor, as Lakatos argues through case studies, any absolutely

[20] Marx 'measured Hegel's philosophy by the [historical] yardstick he himself discovered and systematically elaborated, and he found it wanting.'; Lukács (1971/1923), p. 17. That is, Marx compared actual political and social conditions with Hegel's rational reconstructions.

[21] Feyerabend (1975a), p. 6.

[22] '[T]here is no need to believe (rationally or non-rationally) in the hard core of the program on which one is working….'; examples include Newton, Maxwell and Planck. (*MSE* 220, 219n1) and see n11 *above*.

[23] Hegel (1830), p. 28.

crucial experiments. There is science and non-science, and research programmes which can be progressive or not.

Such in brief is Lakatos' historiographic and critical toolkit. Feyerabend saw much of value here by encouraging the importance of historical materials and the social or other contexts in which scientific knowledge was produced. Research programmes are meant to be mostly 'backward-looking'[24] to assess science past, but they are *also* used by Lakatos *normatively* to assess programmes today. We'll see below that Lakatos' attempt to *change* the world in addition to *interpreting* it was a bridge too far.

## 13.4 Methodology in Dialogue

The following sketches highlight historical accounts developed by Lakatos or his students intersecting with Feyerabend's philosophy of science.

### *13.4.1 Contradiction*

Lakatos reconstructs Niel Bohr's early quantum theory as a research programme whose hard core consists of Bohr's five postulates for his atomic model, articulated in an epochal 1913 paper. Bohr's famous idea was that electrons traveled only in orbits of discrete, not continuous, radii, and that within orbits, electrons radiate no energy. Instead, electrons radiate when jumping between orbits and only by quantum multiples.

Lakatos sees Bohr's programme elaborated through a sequence of theories and their models. First, a hydrogen model has the electron moving in a circular orbit; then a fixed-plane elliptical orbit; then removal of the constraint that the orbital plane and nucleus must be fixed. Successive models explained some new or existing emission spectral series, all the while confronted with anomalous results not explained until later models.

Electromagnetic theory might be replaced eventually, but Bohr's 'refutable variants' left the inconsistency with the Maxwell-Lorentz electromagnetic theory unchanged. Bohr's 'positive heuristic, even if it had been completely successful, would have left the inconsistency with the Maxwell-Lorentz theory unresolved'. The history of the programme shows that one can countenance 'growth-on inconsistent foundations'; 'how a progressive shift may lend credibility—and a rationale—to an inconsistent program'; how there can be 'marvelously fast progress—on inconsistent foundations'; and that 'some of the most important research programs

[24] That contrasts with the 'forward-looking' and creative heuristics of *P&R;* see Hacking (1983), Ch. 8.

in the history of science were grafted on to older programs with which they were blatantly inconsistent'. Such liberalism is not a 'crime against reason,' as Popper would have it, but a rational means for exploiting the heuristic power of the grafted programme 'without resigning [one]self to the fundamental chaos on which it is growing'.[25] For Lakatos, no further extra-scientific explanation is needed to account for such progress, formally logical contradictions notwithstanding. That tolerance for theory proliferation and contradictions is in line with Feyerabend's recommendation to 'Invent, and elaborate theories which are inconsistent with the accepted point of view, even if the latter should happen to be highly confirmed and generally accepted.'[26] A shared perspective on contradiction is one of many salvos directed by Feyerabend and Lakatos at a misplaced emphasis on formal logic by Popper and the positivists.

Lakatos' role for contradiction is related to other senses of contradiction in the philosophy of science. Feyerabend throughout his career was often supportive of Bohr's varied and conflicting approaches to quantum theory, while Lakatos hoped the new physics would avoid wave-particle duality as an ongoing and real contradiction; and to similarly avoid the problematic 'collapse' of quantum states from their 'superposition' into single expected values created by measurement. Popper, in support of one of Einstein's views, argued that quantum probabilities could be interpreted like traditional statistical ensembles, an approach which few believed and Feyerabend showed faces many technical difficulties. Lakatos sides with Popper in not accepting an instrumentalist role for quantum calculations whose interpretation to this day remains conceptually fragile.

Lakatos had no real arguments using the physics, and never did reconcile the ongoing progress of quantum-theoretic methods lacking a plausible realist interpretation. Lakatos may have seen Bohr, and later David Bohm, as too close to the 'dialectics of nature', whether in Hegel or the dialectical materialism Lakatos saw in Hungary before leaving for England after the failed 1956 revolution. Hegel famously made contradictions real in his *Naturphilosophie.*[27] Lakatos kept contradictions out of nature, instead making them prominent in *theories* of nature, but not rejected by logic alone. Feyerabend, for his part, was open-minded about philosophies of nature, and he praised Hegel and Goethe for astute comments on Newton.[28] Being supportive of Bohr and Bohm, Feyerabend might have seen 'contradictions in nature' as a misplaced perspective on quantum programme challenges. His critique of Popper's reinterpretation of quantum probabilities showed that conceptual

[25] FM 56, 64, 67, 90.

[26] Feyerabend (1981a) (*RR*), p. 105, orig. emph.

[27] Kant's 'idea that the contradiction introduced into the world of Reason by the categories of Understanding is inevitable and essential was to make one of the most important steps in the progress of Modern Philosophy…Its only motive was an excess of tenderness for the things of the world. The blemish of contradiction, it seems, could not be allowed to mar the essence of the world; but there could be no objection to attach it to the thinking Reason, to the essence of mind.' Hegel (1975/1830), p. 77.

[28] Feyerabend (1981b) (*PE*) 44n9, 46n13.

contradictions of the new physics were not easily resolved. Lakatos's account of the earlier, 'old quantum theory', nonetheless remains a compelling rationale for the positive management of theoretical or empirical contradictions as intrinsic to scientific reason. It's Hegelian contradictory reason without tears.

### *13.4.2 Hindsight*

A second prominent Hegelian feature of research programmes relevant to Feyerabend is the reinterpretation of existing results through novel theories or concepts, or with 'hindsight'. To start, Lakatos revives Pierre Duhem's 1906 arguments, in *The Aim and Structure of Physical Theory,* that there are no absolutely crucial experiments or refutations. What can appear to be a refutation or 'crucial experiment' can be the product of considerable staging and coordination. 'The problem should not be put in terms of whether a 'refutation' is real or not. The problem is how to repair an inconsistency between the 'explanatory theory' under test and the—explicit or hidden—'interpretative' theories; or, if you wish, the problem is which theory to consider as the interpretative one which provides the 'hard' facts and which the explanatory one which 'tentatively' explains them.'(FM 44).

That process need not be immediate, being only apparent through later ideas making sense of what came before. Lakatos' account of the Michelson-Morley (and later, Miller) experiments on the speed of light is about the role for 'hindsight'. Performed in the decades surrounding Einstein's 1905 special theory of relativity, these experiments were cited into the 1960s as refuting the need for, or existence of, a 'luminiferous ether', long thought to be an essential medium for the propagation of electromagnetic waves. The Michelson-Morley results were also described as 'leading to', or otherwise motivating, early relativity theory. Popper cites the Michelson-Morley experiments in *The Logic of Scientific Discovery* as a paradigm case of his falsification criterion at work.

The Michelson-Morley experiments were motivated by the competing ideas that either: the ether could be left slightly behind a moving earth, creating a virtual 'ether wind' (Augustin Fresnel); or, the ether could be dragged along with the earth with no wind (George Stokes). An ether wind would cause light to travel at slightly different speeds relative to the earth's motion, with Michelson's interferometer designed to measure such minute differences with the needed high precision. The wind/no wind alternatives are reasonably cast as ether programme competitors.

Michelson's 1881 experiment, meant originally to measure the velocity of the earth in absolute space rather than the speed of light, detected no ether wind. From then forward, Michelson was caught up in a series of confounding interpretations for this result and others *vis à vis* Stokes' and Fresnel's alternatives. The 1887 experiment was meant as sensitive enough for a slight ether wind, but still none detected. A year before, another experiment showing no ether wind had been taken as consistent with Fresnel's no-wind proposal. An 1897 test included a fifty-foot high measurement frame meant to better detect a faint wind, but again no result.

Before that, Hendrik Lorentz and George FitzGerald had both proposed the idea of explaining the null results assuming a material contraction of the measurement apparatus in the direction of motion. The contraction hypothesis was not intended by Lorentz to eliminate the ether, but to remove the contradiction between Fresnel's theory and Michelson's result. The contraction factor, applying to the arms of Michelson's interferometer, was the famous $1 / (\sqrt{(1—v^2/c^2)})$ later rederived by Einstein. At this point, Lorentz and Michelson were still well within ether programmes but distinguishing Fresnel and Stokes. The 1897 elevated height test again showed no significant interference fringes, and exasperating Michelson who thought they should be found. Nothing had happened by 1897 to suggest a problematic status for the ether, and that is Lakatos's historical point. Even Michelson's first 1881 interferometer experiment was meant to measure the earth's velocity in absolute space rather than the speed of light. Michelson took his 1897 result to mean that Stokes's ether theory could be in effect at even higher altitudes. Tests continued as late as Dayton Miller's 1921 high-altitude project showing a positive effect, but repeated by Michelson in 1928 with new altitude controls and again no positive result.

Whatever one believes about the role of the Michelson-Morley experiments before 1905, their subsequent role was only gradually and confusedly understood, and never as clearly refuting the existence of the ether. Lakatos makes the Michelson-Morley and Miller experiments a grand example of hindsight for interpreting experimental results in terms of what's refuted or counts as evidence. His moral is that the experiments status

> could not be seen instantly. Even if the experiment was negative, it was not clear, negative exactly to what? Moreover, Michelson in 1881 thought it was also positive: he held that he had refuted Fresnel's but had verified Stokes' theory. Michelson himself and then Fitzgerald and Lorentz explained the result also *positively* within the ether programme. As it is with all experimental results, its negativity for the old programme was established *only later,* by the slow accumulation of ad hoc attempts to account for it within the degenerating old programme and by the gradual establishment of a new progressive victorious programme in which it has become a positive instance. But the possibility of the rehabilitation of some part of the 'degenerating' old program could never be rationally excluded... [T]he status of an experiment as 'crucial' depends on the status of the theoretical competition in which it is embedded. As the fortunes of the competing camps wax or wane, the interpretation and appraisal of the experiment may change. Our scientific folklore however is impregnated with theories of instant rationality. (FM 76, 85)

For Feyerabend, the sciences all rely on their histories, so that 'All sciences are *Geisteswissenschaften,*' and 'the history of ideas is an essential part of scientific method'.[29] Lakatos takes that a further step, with hindsight relevant at the level of experimental interpretation. As Lakatos wrote elsewhere, 'Falsification can thus be said to have a 'historical character' and that 'Crucial counterevidence' or 'crucial experiments' can be recognized as such among the scores of anomalies only with hindsight, in the light of some superseding theory.'(FM 35, 36) This means that

[29] *PE* 12, *RR* 140. On Lakatos and Michelson-Morley, see the extensive historical criticism of Hacking (1983), pp. 256ff, directed at Lakatos as an overly theory-biased philosopher of science.

'rationality works much slower than most people tend to think, and, even then, fallibly. Minerva's owl flies at dusk.'(FM 87) That's Hegel's owl from the *Philosophy of Right* standing for the backward-looking, interpretative role for philosophy. Lakatos places the owl's retrospective wisdom in science itself. Feyerabend's perspective on hindsight or 'historical character' went further by including participants' sui generis understanding missed by generic methodological rules. The next two episodes provide examples.

### *13.4.3 Atomic and Phenomenological Theories of Heat*

Another perspective on late nineteenth century physics is provided by Peter Clark's research programme study on the atomic-kinetic and phenomenological theories of heat.[30] The former theory is basically that moving atoms and molecules *are* 'the kind of motion we call heat', as put by Rudolph Clausius in 1857. Kinetic models relied on mechanical laws constituting a powerful positive heuristic for the atomic-kinetic 'hard core'. The phenomenological theory as its competitor forbid, in *its* 'negative heuristic', specifics about the internal nature of matter. Instead, descriptive laws organize macroscopic quantities (temperature, pressure, work, viscosity, etc.) as phenomena to be derived and explained, suspending judgment on internal structure. Atomic theory was controversial until after 1900, so the phenomenological theory could progress while suspending judgment on molecular reality. Clark's research programme account is relevant to Feyerabend's career-long interest in realism as a nexus of scientific conflict and growth.

It's initially odd to consider a research programme with no series of predictive models. The hard core of the phenomenological theory is just that a definite relation always existed between a quantity of heat and its potential work, with Josiah Gibbs and Max Planck in the late 1870s stating their two laws, of conservation of energy and entropy increase. The two laws were meant as all-inclusive empirical generalizations applying to all processes. The programme's positive heuristic was to derive new consequences from other empirical regularities. That procedure, different from kinetic 'model building', was empirically or fact driven: start with a regularity like 'pressure x volume is proportional to temperature' for gases, and apply the two laws with boundary conditions.

The 'experience' of the phenomenological theory was still then theory-laden, and that is consistent with the programme's motivating philosophy. Ernst Mach criticized presentations of Newton's laws as *a priori*, arguing instead that they were generalizations of ordinary experience with forces and moving objects. That kind of 'positivism' is not philosophically objectionable, with Einstein also influenced by Mach's critique of Newtonian absolute space and time.

[30] Clark (1976). On positivist philosophy in the phenomenological theory see Morrison (1995) and Clark p. 44.

The contrasting ontologies and heuristics for creating testable theories are included in the programmes' hard cores with relevant 'philosophies' intrinsic to the programmes. Feyerabend for his part rails against accounts of Mach as equating observation to sense-data-like perception, with Mach instead analyzing sensation as a physical process dependent on physiology and its functional role for the human body (*PE* 95–96). The phenomenological programme, and especially Mach, were later associated with the philosophy of logical positivism and experiential 'sense-data' as grounding physical knowledge. Clark argues that 'degeneration of the kinetic programme compared with the empirical progress of [phenomenological] thermodynamics which accounts for the rise of scientific positivism,' not vice versa. Mach even supported the kinetic theory early on, but turned against it when its heuristic power ran out, and not just for 'philosophical' reasons. An oft-mentioned positivist philosophical bias is this way a misinterpretation.

The kinetic and phenomenological programmes both led to remarkable predictions and problematic anomalies before 1900. By then, the kinetic theory was stalling on predictions, versus phenomenological progress with a weak heuristic for moving beyond macroscopic phenomena. A needed 'creative shift' in heuristics would have to wait for quantum ideas of Planck and Einstein.

What finally decided for the kinetic programme was Einstein's 1905 derivation showing the possibility of Brownian motion, soon supported by Jean Perrin's quantitative analysis. Einstein's problem-shift identified molecular 'average velocity' as ill-defined, with probabilistic Brownian paths almost everywhere 'jagged,' or non-differentiable and fractal-like, and so making relevant averages impossible. Used earlier in failed attempts to refute the phenomenological law of entropy increase, Brownian motion combined with Einstein's stochastic paths supported a realist interpretation of atoms.[31]

Feyerabend objects to the implied importance of comparative programme progress, largely represented by novel predictions and confirmations. That's because there are also preferences here for how to do physics, whether as a kinetic-realist, or as a phenomenologist. While research programme concepts help sort out the role of heuristics in both programmes, missing is how, for some, very different theoretical styles for physics were themselves at issue. Einstein himself didn't see the kinetic victory as a complete rejection of the phenomenological programme. He took an entirely positive view of a 'principled' approach to physics, as with the two phenomenological laws, over kinetic models. That philosophy for physical theory was for Einstein correct, or at least appealing, notwithstanding overwhelming new support for atomism. In addition, kinetic-atomic predictions did not yet clearly rely on molecular processes in themselves, and 'not merely while changing, or acting in concert with other entities'.(*RR* 6). Details of *how* the kinetic programme made progress was as, or more, important than comparative research programme predictive progress. Research programmes, for Feyerabend, appear too coarse to capture

[31] Relevant to Lakatos and hindsight, 'Perhaps the phenomenon of Brownian motion seems to find its niche only by hindsight, especially considering the state of its comprehension as indicated in investigations prior to the middle of the nineteenth century.' Nye (1972), p. 21.

conceptual understanding of programme elements motivating theory choice or rejection, and not necessarily reflected in competing predictions and confirmations. That difference doesn't always matter, but was relevant to changes occurring with the new physics.

### *13.4.4 Special Theory of Relativity and Copernicus*

With the special theory of relativity Feyerabend again identifies orthodox predictive progress as not reflecting more fundamental changes in physical theory. As for the theory of heat, there's a constructive, model-building programme, being Hendrik Lorentz's electromagnetic theory; and Einstein's 'principled' *STR*. The question for Lakatos' student Elie Zahar is a 'rational' justification for pursuing *STR* circa 1905, especially as both Lorentz and Einstein wanted to unify electromagnetism and mechanics, seen as nearly separate branches of physics.[32]

For many at the time Lorentz's approach and *STR* were effectively equivalent. Circa 1905 there are no novel predictions or other research programme justifications for *STR* preference early on. Even after 1905 mass contraction and local time were variously combined as the 'Lorentz-Einstein,' 'relativity,' or 'Lorentz' theories. Nonetheless, *STR* was the choice for leading physicists including Max Planck and Hermann Minkowski. Research programme comparisons would prima facie make such choices for *STR* unjustified, or even 'irrational'. Difficulty is again created by confounding the usefulness of research programmes as a historiographic toolkit with normative implications for theory or research programme choice.

Given the predictive equivalence, and even shared goals, of Lorentz and *STR*, how can a research programme proceed? Zahar compares the implicit heuristic power of *STR* to Lorentz's constructive approach, with Einstein deriving relativistic versions of classical laws like $f = ma$, or $E = mc^2$ as a mass-energy conservation principle. The 1905 'progress' is supposed to be in that heuristic but not predictive power. As Feyerabend points, out, while Minkowski or Planck might see how to exploit the new approach, that capability was not shared. Some did not support Einstein, but it's unclear why those negative views, or others supportive of Lorentz, are not then 'irrational' alongside Planck and Minkowski's acceptance. For Feyerabend it becomes arbitrary who 'acted rationally' or not.

The non-specific methodological perspective also misses the point of reformulating physical laws with respect to relative motion, and unlike a 'constructive' theory like Lorentz's at all. Physical laws can themselves be first revised, but not for building and using predictive models. The 'principled' framework for the phenomenological theory of heat lives on even as it was set aside for atomism. The dualism of mechanical and electromagnetic explanations was to be overcome through relativity as a universal principle. Revised physical laws, like relativistic conservation of

[32] Zahar (1976). For Feyerabend on Zahar and *STR* see *PE* pp. 227–230.

momentum or mass-energy, would make dynamic and electromagnetic concepts interdependent with neither having a privileged status. In Lakatos' favor, 'positive heuristic' usefully frames part of the *STR* attraction by its novel derivations. But the reformulation of physical laws and measurement is poorly captured. Feyerabend saw Einstein's unification project as concerned 'with the form, rather than the content, of theory.'[33] *That* progress is poorly captured by research programme comparisons.

Feyerabend similarly criticized Lakatos' and Zahar's attempt to justify Copernican heliocentric planetary orbits as a progressive programme compared to Ptolemy's geocentric model, even as both had no established novel predictive differences.[34] Copernicus more easily derived *known* phenomena (e.g. stations and retrogressions of planets), and these 'novel deductions' of known facts are presented as research programme 'progress'. Troubling for Feyerabend is that 'novelty' is *not* valued in the Aristotelian physics from which Copernicus was turning away but still acting within. For Aristotelian physics, 'saving the phenomena' involved no new principles, say of motion or the nature of celestial bodies. Copernicus' calculations on his own account describe planetary orbits, and often more 'harmoniously' as circles around the sun. Lacking a dynamic theory from which planetary motions are derived Copernicus's system is strictly mathematical.

Extending dynamical arguments to astronomy would be a later major innovation of Kepler and Galileo. Copernicus has one foot in the Aristotelian project of saving the phenomena, and the other in an emerging science of correcting the senses and generating novel predictions. For Feyerabend, what is degeneration or stagnation, in failing to produce new facts, can be 'saving the appearances' in a late-medieval Aristotelian view, for which the universe of facts is, so to speak, *finite*. There is not an arbitrarily large inventory of facts to explain, only those accessible through ordinary perceptual resources, like naked-eye astronomy before Galileo. The background metaphysics doesn't fit modern ideas of progress and a 'constantly expanding horizon of facts' to which research programmes are better suited. Feyerabend liked to say Aristotle was the true empiricist, an idea he said he learned from the physicist Phillip Frank.

If so, then as Feyerabend points out, 'The revolution that was started by Copernicus's hypothesis was much more dramatic than a transition, *within the methodology of research programmes,* from one research programme to another. It brought in new standards and thus constituted a true paradigm change in the sense of Kuhn.'[35] The difference lay in the transition from the cosmological assumption of an Aristotelian 'finite' world of appearances, with no need of improving research programmes, to one for which hidden realities are possible with a constantly expanding horizon of facts. As with Einstein and the *STR*, research programme

---

[33] Hirosige (1976), p. 41.

[34] Lakatos (1978a) Ch. 4 written with E. Zahar. Kuhn in 1957 argued for this and more in *The Copernican Revolution*.

[35] *PE* 225 orig. emph.

interpretations break down when faced with such major conceptual changes or their precursors.

### *13.4.5 Newton*

Leaving aside research programme applications, in this final comparison Feyerabend and Lakatos complement each other's accounts of Isaac Newton's influential inductive philosophy of method, principally as codified in 'Rule IV' of the *Principia.* For both Lakatos and Feyerabend, that statement of method is to be sharply distinguished from Newton's scientific practice and how the two relate. A paraphrased Rule IV says:

> Propositions inferred from phenomena by induction are to be held true or nearly true, notwithstanding contrary hypotheses that may be imagined, and until other phenomena allow propositions to be made more accurate or subject to exceptions. Propositions (also 'theories') induced from phenomena may not be evaded by hypotheses.[36]

Rule IV authorizes theories induced from phenomena, in contrast to deducing them from a priori principles or concepts, notably René Descartes' system of mechanical vortices. Cartesian 'clear and distinct' concepts, say of space or material objects filling it, deductively explain specific processes and laws of nature.[37] 'Contrary' was crossed out by Newton and restored by editors for a somewhat weaker statement.

*Phenomena* act as premises for induction, in reverse from Aristotelian-style proofs like Descartes'. Induction allows Newton to introduce and mathematically justify propositions ('theories'), like Kepler's laws for elliptical planetary orbits, even if not derived from Cartesian-like first principles. Rule IV implies phenomena should closely match observations, say from astronomy or experiments with light. But if not, Newton may require his own (auxiliary-like) hypotheses, in spite of their apparent prohibition in Rule IV. With recalcitrant observations, propositions can be defended by listing exceptions.

For Lakatos, Rule IV compactly 'demands the prohibition of almost all possible criticism and by implication is a plea for concentration instead on the development' of Newton's physics. Feyerabend agrees that Rule IV implies a 'theoretical monism'. Notwithstanding the limitations of Rule IV, Newton finds ways to use his own theories, or unproven hypotheses, to redescribe phenomena which will *then* amount to inductive proof. In this way, as Feyerabend puts it, Newton's *practice* is *critical* 'in the sense that it allows for the revision of any part [of science], however close to 'experience'. Newton is the greatest practitioner of 'classical empiricism'—as Feyerabend ironically labels this 'fascinating, tortuous, schizophrenic combination

---

[36] On Rule IV text and 'contrary' edits see Lakatos (1978a), p. 204.

[37] Descartes' mechanistic vortices, made up of ultrafine corpuscles, were an alternative to Newtonian gravitation as an 'occult' force. Vortex theory supported heliocentrism with Earth at rest within a rotating vortex band, hopefully mollifying Church concerns about its movement.

of a conservative ideology and a progressive practice.'(*PE* 34) And for Lakatos, '.. .one may say that, while Newton's method created modern science, Newton's theory of method created modern philosophy of science.'[38] The flexible language of Rule IV combines theories, phenomena and inductive inference, with these inconsistently deployed by Newton to defend his own work or repel competitors. Lakatos and Feyerabend agree that Newton's problematic ideas were influential for centuries, with Newton many times progressing only by violating his own inductive precepts.

For example, Newton understood that Kepler's laws didn't correctly describe orbits or velocities of Saturn and Jupiter, but wanted nonetheless to utilize them inductively. In place of Kepler's real planets and moons, Newton assumes a fixed sun acting by gravitation on planets, but not reciprocally as they should be by gravitation. Planets also do not act on one another, notably Jupiter and Saturn, whose changing relative positions are thereby neutralized. The qualified phenomena are contradicted by Newton's own universal gravitation, but then support Kepler's laws. The transfer of empirical conditions to theoretical/propositional status allows phenomena to function like laws closer in style to Cartesian-like principles. As Feyerabend says, 'This separation of *phenomenon* and *actual fact* and the definition of phenomena with the help of theory' reverses what's stated in Rule IV.[39] Lakatos notes it's not revised phenomena used inductively, but a *negation* of Newtonian gravitation meant to apply to *all* objects.

Outside of the *Principia,* Newton explained to John Flamsteed, the first Astronomer Royal, a better theory of atmospheric refraction, with which Flamsteed revised his data, along with other corrections consistent with Newton's theories. Newton's hypotheses acted as 'touchstone' theories for revised and newly theory-laden observations. For Lakatos, this is the sense in which Newton and his followers 'established *in practice* the basic features of scientific methodology...Newtonians never let the authority of observations prevail against their research programme; with the help of their positive heuristic they produced one theory after the other to accommodate counterexamples...'[40]

Feyerabend examines Rule IV in the context of Newton's *ray theory* of light and its phenomena. The difficulty is to identify light as phenomena. An object's 'color' tells us about that body's disposition to propagate rays' 'motions', but that's not light itself. Colors as they seem to us as sensations are also not useful, as colors 'seen' in a dream have nothing to do with light. We perceive only a limited set of colors while refrangibility implies a continuum of mostly unseen hues. Rays of various colors for Newton are also mixed in our senses, but not in themselves. These are perfectly valid observations, but Newton still needs substantially abstracted phenomena for any propositions at all.

---

[38] Lakatos (1978a), p. 221.

[39] *PE* 46n13 orig. emph. See *PE* Ch. 2 for Feyerabend on Newton and Feyerabend (1999a) Ch. 7.

[40] *Ibid.* Lakatos uses research programme concepts without a rational reconstruction.

The situation is quite different from observing planetary orbits. Pure, simple colors as posited by Newton are, Feyerabend notes, 'strictly speaking *unobservable*'. 'To present the results of concrete experiments with all their imperfections in a unique form and as *supporting* (and not merely *illustrating*) the ray theory therefore involves idealizing assumption which bend the concrete and very complex experimental result in the direction of Newton's theory.' Such transitions are unaccounted for. 'And in this redescription [Newton] introduced the machinery of the very same theory he wanted to prove.' Christiaan Huygens' wave theory could better explain diffraction and mirror images, both of which Newton dealt with using ad hoc hypotheses of no lasting value. Mirror reflections were to depend on a smoothing ether or similar 'power' to even out surface imperfections otherwise creating incoherent scattering.[41] Rule IV effectively *forbids* alternatives like the wave theory, so in principle it isn't available, even if only to describe 'exceptions'.

The theories of light contrast with Newton's successful dynamical theories, where overwhelming progress could neutralize problems of justification or evidence, including an adaptable Rule IV, its phenomena and inductive proof. Lakatos positively sees Newton's practice in mechanics and astronomy as initiating the first major scientific research programme, complete with anomalies, reinterpreted observations, observational theories, predictions, confirmations and certainly auxiliary and even ad hoc hypotheses. But then, Feyerabend writes,

> It is a marvelous accident, worthy of an equally marvelous explanation that this way of building up the theory of gravitation did not create a miserable patchwork but a coherent system that astounds by its simplicity and its effectiveness in the mastery of concrete phenomena.. .. [L]ight did not submit to this procedure quite as readily. What we obtain here is a system that has been called, and not without justification, 'an incoherent and uncertain theory, a theory so full of contradictions and lacunae that one is surprised to see to what extent it could convince the majority of the physicists of the eighteenth century.'[42]

Feyerabend labeled such divergences between scientists' philosophies of method with actual practice as 'ideology', with Newton the great and consequential example. Lakatos, somewhat more charitably, wrote that 'philosophies of rationality lag behind scientific rationality even today', but 'some of "the most interesting historical questions are the interactions of the two — the actual standards and the believed standards.'[43] Scientists can have poor theories of their own practices, while philosophers lack a privileged methodology or neutral standard. So, for Lakatos, what 'constitutes "false belief" (or "false consciousness"), depends on the rationality theory of the critic...But no rationality theory can ever succeed in leading to "true consciousness".' (HS 119n3). The duality of practice and ideology is itself unremarkable because sciences are historical and social practices. But as Feyerabend puts it, Newton's methodological legacy for centuries was his 'schizophrenic combination

[41] Feyerabend's mock Rule IV defense of the wave theory shows the rule's arbitrariness (*PE* 49).

[42] *PE* 46 with Feyerabend quoting Vasco Ronchi, *Histoire de Lumière*.

[43] Lakatos (1978a), p. 154.

of a conservative ideology and progressive practice.' That's again what's (sarcastically) meant by Feyerabend's 'classical empiricism'.

## 13.5 Galileo and the Trojan Horse

The examples above (except Newton) illustrate Feyerabend's criticism of research programmes taken as normative criteria for progress, beyond their role as interpretative schema. There's no time limit for a degenerating programme to stage a comeback (heliocentrism) or for a non-progressing programme to deliver some breakthrough (atomic-kinetic theory). Programmes generally can require tenacity and tolerance of anomalies and stalled progress. On Lakatos' own account, an observational 'basis' can be in flux, and so requiring 'hindsight' and more time to stabilize (Michelson-Morley). For programme participants, there's no guidance for what to improve or when to jump ship. As an historiographic perspective, theory choice can be guided by preferences for theoretical methods rather than research programme predictive power (*STR*). As Feyerabend observed for Einstein's 'principled' approach, predictive success as a dominant norm can imply *someone* acted 'irrationally' by taking up the 'wrong' programme.[44] Lakatos, for Feyerabend, enables his own methodological *anarchism*, as research programmes never need to be conclusively rejected, and a variety of theoretical adjustments or reinterpretations are always possible. That anarchism of method renders Lakatos a 'Trojan Horse'.[45] On Lakatos' behalf, research programmes can help *keep score* through their conceptual apparatus for outsiders, insiders, historians and philosophers to articulate programme progress, stagnation or decline. But as a normative theory, research programmes rule out little of substance as non-progressive or truly 'degenerating'. The positivist-Popperian 'demarcation' project runs aground when taken too far.

That normative role for research programmes creates a second Trojan Horse. Lakatos took it upon himself to use research programme analysis to lobby for publication rejection, with the elementary particle physics of the 1960s being his example of a degenerating programme. The particle-prolific 'standard model' was initially slow to develop but then took off in just the kind of situation envisioned by Feyerabend. The appalled historian of science Gerald Holton warned of

> ...the late philosopher Imre Lakatos who aimed to set up universal criteria for distinguishing between what he called 'progressive' as opposed to 'degenerating' research programmes. One of his candidates for the category of degenerating research programmes was contemporary elementary particle physics. He offered his help to funding agencies and editors of scientific journals, so that they might reject 'degenerating' work and 'lay down statute laws of rational appraisal which can direct a lay jury [for example, Congress] in

[44] See *PE* Ch. 10; *AM,* Ch. 16.

[45] *AM*, p. 200; and 'observational terms are Trojan Horses which must be watched most carefully', p. 75.

> passing judgment.' We can tell here that the fire in the stake is still smoldering, the stake on which all those were burned who ran afoul of an absolutist system.[46]

What is supposed to be historical reason can become dogma, and not as a philosophical quibble about descriptive history and normative guidance. Holton unfortunately misses the usefulness of research programmes for clarifying complex practices for history, education or public policy. Given that, Lakatos, the supposed defender of rationality in science, became an example of the elitist authoritarianism he wanted to guard against. Lakatos had advised Feyerabend to make the ideas resulting in *Against Method* as extreme as possible to highlight their underlying logic. Lakatos followed that advice himself. The perennial contradictions of normative versus descriptive history are a consequence of Lakatos' hyper-rationality, itself something of a caricatured gestalt like those making up *P&R*.

There's yet now a *third* option for Lakatos the Trojan Horse, and one aligned with Feyerabend's advice that new ideas in science may have to be introduced covertly, or through suggestive reframing of received ideas in new conceptual terms.

Feyerabend's illustrious example was Galileo, who explained that relative motion, understood from riding in a moving boat or carriage, can be applied to a moving earth circling the sun. Perceptions apparently refuting the earth's motion, like a stone dropped from a tower, get reinterpreted by Galileo as evidence for the earth's motion, not against it. Unlike the examples of relative motion, using newly-invented telescopes could not appeal to easily understood examples, much less any optical theory, As 'propagandist', Galileo worked to convince others of what's expected to be seen using telescopes pointed to the stars rather than a distant mountain. For present purposes, we only need this *idea* of Galileo, leaving aside its significance for his physics or its reception. That includes Galileo's political context as (dangerously) opposed to his Copernican views, and therefore *requiring* oblique arguments and persuasive rhetoric.

The third Trojan Horse is Lakatos as such a 'Galileo', by way of his Hegelian innovations and thorough-going historicism. The principal outcome of the philosophy of science debates spanning Popper, Kuhn, Feyerabend and Lakatos is a central role for history in understanding the sciences.[47] Lakatos' unorthodox 'rational reconstructions', by bracketing and cross-cutting narratives with their 'actual history', ostentatiously represent history as an uncertain interpretative ground for the sciences and other human institutions.

Most directly, Lakatos is explicit that all history of science is *theory-laden* by methodological categories,[48] with 'actual history', like theory-neutral observations, a chimera. The categories at work, whoever the philosopher of science, are 'theories',

---

[46] Holton (1988) p. 459.

[47] See Hacking (1983) Ch. 1 and p. 66. Popper's 'falsification' is Lakatos' pivot to history as noted above.

[48] '*Scientific "factual" propositions* are theory-laden: the theories involved are "observational theories." *Historiographic "factual" propositions* are also theory-laden: the theories involved are methodological theories.' HS 119n1 orig. emph.

'refutations,' 'phenomena', 'appearances', 'contradictions', 'inference', 'confirmations', 'programmes' and so on. That is modern, with Hegel explicitly identifying the inescapable role of categorical judgments in history writing of all kinds, among other observations on the craft.[49] Hegel, like Lakatos, created his histories of art, philosophy and world events to emphasize just this point. Lakatos, like Hegel, distinguished histories of discourse from events, with their histories of reason deliberately biased to the latter. The dependence of historical reconstruction on interpretive apparatus is also Lakatos' unwavering normative role for research programme comparisons. His view was that *all* histories of science are normatively biased through *some* methodological norms for what counts as science or progress, and just how.

Popper in *The Open Society and its Enemies* famously criticized Hegel as a historical determinist. But like *P&R*, Hegel's *Phenomenology* is only rational reconstruction: 'This way of looking at the matter is something contributed by *us,* by means of which the succession of experience through which consciousness passes is raised into a scientific progression.'[50] Lakatos also makes this 'way of looking' explicit through his caricatured 'gestalts' combined with the actual history corresponding to them as stereotypes of reason. Rational reconstruction in the history of science is similar to that of *P&R,* but minus the individual-historical pedagogy. Bohr's early quantum theory, the Michelson-Morley experiments, atomic and phenomenological theories of heat and others. *Historiographic* research programmes are themselves another expression of scientific rationality, including contradictions, anomalies, reconstructed history as 'predictions', refutations, progress and so on.[51]

Feyerabend didn't recognize the guiding historiographic motivation, but was on the right track. In the 1970 *essay* 'Against Method', he notes with elan that

> The only philosopher who imbibes the hidden brew of Leninism is Lakatos – and the results are evident in his magnificent work. All that is required now is that he confess his vices openly so that others may learn to delight and enlighten us in a similar way.... [O]ne must praise Lakatos for having made such excellent use of his Hegelian upbringing.. .One must perhaps also criticize him for not revealing his source of inspiration in a more straightforward manner...[52]

Feyerabend's error is to cite Vladimir Lenin, whose 1909 *Materialism and Empiriocriticism* was required reading in the USSR and its satellites, including Hungary. Instead, the source of Lakatos' Hegelianism is Lakatos' model in Hungary, Georg Lukács, himself a famous *critic* of Lenin and scientistic Marxism generally.[53] The overarching point of Lukács' classic 1920s' collection, *History and Class*

---

[49] See Hegel (1975/1830) p. 211 on categorical judgment; on language versus historical events, p. 135.

[50] Hegel (1977/1807), Introduction, p. 55.

[51] '*All methodologies function as historiographical (or meta-historical) theories (or research programmes) and can be criticized by criticizing the rational historical reconstructions to which they lead.*' HS 122, orig. emph.

[52] Feyerabend (1970), pp. 100–101, 115.

[53] See Arato and Breines (1979) on an intellectual tradition mostly unrecognized in Anglo-American philosophy.

*Consciousness,*[54] is the role of subjectivity, social realities and non-mechanistic historicism in Marx. Lukács conjectured there was an underlying and substantial Hegelian architectonic, with that suspicion confirmed by discoveries of Marx's 'economic-philosophical' manuscripts in the 1930s, along with an extensive rough draft (*'Grundrisse'*) for *Capital*. Knowing German philosophy well, Lukács recognized a latent and pervasive Hegel in Marx, principally the concept of alienation, concealed and repurposed to better fit the political economy of Adam Smith, Ricardo and others, and to which Marx oriented his philosophical-historical critique. Marx was the first Hegelian Trojan Horse. Lukács followed, through his sub rosa critiques of Stalinism in Russia and again in post-War Hungary.[55] Following WWII, Lukács continued to explain Hegel's confusing philosophical histories,[56] Hegel's *anti*-elitist pedagogical goals, and the *Bildungsroman* style of the *Phenomenology*. Anti-elitism was Lukács' veiled criticism of Stalinism and its personality cult.

Lukács brings us back to Feyerabend and the philosophy of modern science. Lakatos virtually 'deduces' his historiographic views via a self-application of research programme concepts to their own development out of positivism, Popper and, most of all, the use of historical 'data' to 'falsify' the latter. All history of science writing is a science of research programmes organizing methodological categories: Popper's conjectures and refutations, Kuhn's paradigms and revolutions, research programmes, inductive proofs. What seems a clever trick is a compact means for integrating history and its narratives as fallible, theory-laden, and possibly normative roles for how past science 'should have gone'.

Lakatos makes his own philosophy of science another historical product, like the evolving and transient 'series of theories' making up research programmes themselves. Lakatos developed his philosophy of science by historicizing Popper, and then historicizes *that*. This path to historicization isn't new. It was 1920s Lukács, in 'The Changing Function of Historical Materialism,'[57] where he explicitly 'applies' Marxism to itself, rendering it too as a passing historical ideology. For Lukács, 'orthodox Marxism' was solely about historical *method*, and not 'the 'belief' in this or that thesis, not the exegesis of a 'sacred' book.' Lukács and Lakatos both note their self-application is an ancient skeptical gambit, the 'turning of the tables', or 'peritrope'. For Lakatos, his 'Pyrrhonian machine de guerre', plus history, 'falsifies' Popper and all methodologies. All first-order scientific theories plus their second-order philosophies and historiographies are 'born refuted', 'progress in an ocean of anomalies' and so on.

So now back to Feyerabend, who made skeptical method explicit in *Against Method*. The title resonates with Sextus Empiricus' *Outlines of Pyrrhonism,* with

[54] Ropolyi (2002) argues Lakatos' (lost or stolen) Debrecen thesis applies Lukács' *History and Class Consciousness* to philosophy/sociology of science. See Long (1998), Bandy (2010) on Lakatos' long hidden Hungarian past.

[55] Principally in *The Young Hegel* and *The Destruction of Reason*: Lukács (1976, 1981).

[56] See my (2001) p. 37 and notes. Hacking (1983), p. 125 and many others misunderstood Lakatos' historiography.

[57] Lukács (1971/1923), Chapter 5.

chapters 'Against the physicists,' 'Against the logicians', 'Against the musicians' et al. as recognizable dogmatic targets for skeptical tropes. For Feyerabend, arguments and conclusions are framed, as for a skeptic, *within* a debate claiming some absolute criterion for truth or progress, including research programmes, critical rationalism and varieties of positivism. The skeptic practices an 'immanent critique', with undecidability ('*isostheneia*') or contradictions *assuming* the proposed standard. Many confuse Feyerabend's Pyrrhonism with *negative dogmatism*, or *Academic skepticism*, typically by taking his 'anything goes' as a positive philosophy when it's only a conditional conclusion following demands for a *universal* standard.[58] Feyerabend also favored cross-cultural epistemic comparisons to neutralize any one of them as fundamental, with such contrasts also being a standard trope (tenth in a list Sextus attributes to Aenesidemus). The archaic Homeric world is Feyerabend's poster child for an anthropological-contrastive approach he attributes originally to Xenophanes.

Both Lakatos' and Feyerabend's philosophies of science thereby enter the history of philosophy as innovative versions of mitigated or constructive skepticism. Even Feyerabend, as a champion of the sciences, and someone sensitive to human suffering, is not a *completely* passive or neutral Pyrrhonist, though he comes close. He once quipped he was 'Pyrrho himself', but that's not *quite* right. Lakatos, like Hegel in philosophy, would hope his historicism amounts to a *mitigated* or *constructive* skepticism as found in modern science: as put by Pierre Duhem, 'we have an impotence to prove, not conquered by any dogmatism', and 'an ideal of truth not conquered by any Pyrrhonian skepticism.'[59] That's a third way between the dogmatism of certain knowledge and the confused Academic skepticism of 'no knowledge'.[60] The skeptic needs a 'negative language' to state their conclusions, with that cautiously positive project taken on by Hegel and Lakatos.[61] In skeptical practice, the peritrope may easily lead to some kind of contradiction. Here, Popper's falsification criterion is falsified by the history of science, when 'applied' (i.e. interpreted) to itself, and not an unreasonable demand. Because research programme historiography is an historical tradition, the peritrope creates no problem. The historicism of Hegel's *Phenomenology* is similarly self-consistent.[62] Lakatos had planned a book, *The Changing Logic of Scientific Discovery*; there's no book, but a 'changing logic' is implied by research programmes as historiographic theories. Feyerabend could similarly safely 'apply' historicist-anthropological standards to himself, while Lakatos is more 'mitigated', systematic and provisionally positive in the spirit of Duhem and other scientists.

---

[58] 'Conceptual schemes' eliminating *any* inter-communication is another expression of negative dogmatism. See Feyerabend (1999b), pp. 123ff.

[59] A seventeenth century version is Pierre Bayle's; see Popkin (1979), p. 213, and dedicated in memory of Lakatos. Duhem is quoting Pascal.

[60] The trope 'from relation' makes things relative to perceivers or other entities; see (*P&R* 99) and Kappa *above*. On Feyerabend and skepticism see (Neto 1991; Sankey 2011).

[61] On 'negative language', see Popkin (1979), p. 48.

[62] On the *Phenomenology* as Pyrrhonism become historiography see my (2001), Ch. 3.

Therefore skepticism, variously utilized and expressed, is a deep positive bond for Lakatos and Feyerabend, with both acutely aware of Hume's problem of 'how the skeptic lives their skepticism'. Both agreed on the influential role of Pyrrhonism and mitigated skepticism in early modern science. Both framed their accounts of Newton's Rule IV as internalizing skeptical-dogmatic debates, especially as these appeared through the Reformation, Martin Luther and challenges to Church dogma. That's even as a constructive and mitigated skepticism would be implicit in the scientific *practice* of Newton and others.[63] Matteo Motterlini notes archival materials suggesting this epochal transformation of ideas would have figured highly in the *Changing Logic* with Newton as hero and anti-hero.[64] For Feyerabend, mitigated skepticism is implicit in non-foundational epistemologies in the nineteenth century as found in John Stuart Mill and Ludwig Boltzmann, and then as a response to positivism in Popper's 'negative progress' and finally Lakatos. Feyerabend and Lakatos fundamentally agreed that *no* knowledge has certain foundations, with each providing an answer for how to live their skepticism as philosophers of science and mathematics. Their planned book, *For and Against Method,* could have ended there, perhaps a draw, or suspension of judgment, between the two.

Except for one last comparison of these two lives in dialogue. Our story began with Feyerabend's praise for Lakatos as a fallen philosophical hero, like the 'best of the Achaeans' memorialized in the Homeric epics. That acclaim can include Lakatos as excelling at the rhetorical and covert methods Feyerabend attributed to Galileo. But more is involved than the clever dialectics of a brilliant and ambitious émigré intellectual. Lakatos found himself in an Anglo-American philosophical world, including Karl Popper in particular, as wholly inhospitable to Hegel,[65] if not abjectly repulsed. Hence Lakatos' need to dissemble his Hegelian-Marxist legacy. Yet that was nothing compared with the survival skills needed in the hyper-closed and history-falsifying society of Hungarian Stalinism from which Lakatos fled following the failed 1956 Hungarian uprising. The revolt was a revolution of truth, led by many journalists, writers and intellectuals. For many, 'my crime was to have believed in yours' in a pervasive life of lies and deceit—the Hungarian *Lebenslüge*.[66] Ever resourceful, Lakatos applied skills honed under Hungarian Stalinism in the Cambridge PhD thesis which is now a philosophical classic. Caught between open and closed societies, Lakatos' cross-cultural legacy is a masterclass in how one lies to survive. Lakatos rose to the top, spreading his gospel of historicist rational reconstructions, 'actual history', contradictions, hindsight and more. Imre Lakatos was the Hegelian avenger taking the Popperian citadel from within, a Trojan Horse *par excellence*. Paul Feyerabend has the last word on his great and cunning friend, this 'Galileo' among us.

[63] See *PE* Ch. 2 and Popkin, Ch. 7.

[64] Lakatos and Feyerabend (1999), pp. 393–395 (Lakatos-Feyerabend correspondence).

[65] See my (2001), p. 319n8 including Popper's 'quilt quotations' of Hegel.

[66] See Aczél and Méray (1958), including pp. 57–93 on Lukács and my (2001) Ch. 12, 'Hungary 1956 and the Inverted World'.

## References


Aczél, Tamás, and Tibor Méray. 1958. *The Revolt of the Mind*. New York: Praeger.

Arato, Andrew, and Paul Breines. 1979. *The Young Lukács and the Origins of Western Marxism*. New York: Seabury Press.

Bandy, Alex. 2010. *Chocolate and Chess (Unlocking Lakatos)*. Budapest: Akadémiai Kiadó.

Clark, Peter. 1976. Atomism Versus Thermodynamics. In Howson (1976), 41–106.

Feyerabend, Paul. 1970. Against Method: Outline of an Anarchistic Theory of Knowledge. In *Analyses of Theories and Methods of Physics and Psychology: Minnesota Studies in the Philosophy of Science IV*, ed. M. Radner and S. Winokur, 17–130. Minneapolis: University of Minnesota Press.

———. 1975a. Imre Lakatos. *British Journal for the Philosophy of Science* 26: 1–18.

———. 1975b. *Against Method*. 1st ed. London: New Left Books. (*AM*).

———. 1981a. *Realism, Rationalism and Scientific Method: Philosophical Papers 1*. Cambridge: Cambridge University Press. (*RR*).

———. 1981b. *Problems of Empiricism: Philosophical Papers 2*. Cambridge: Cambridge University Press. (*PE*).

———. 1999a. *Knowledge, Science and Relativism: Philosophical Papers 3*, ed. J. Preston. Cambridge: Cambridge University Press.

———. 1999b. *Conquest of Abundance: A Tale of Abstraction versus the Richness of Being*, ed. B. Terpstra. Chicago: University of Chicago Press.

———. 2016/2009. *Philosophy of Nature*, ed. H. Heit and E. Oberheim, trans. D. Lotter and A. Cross. Cambridge: Polity Press.

Gentner, Dedre, and Susan Goldin-Meadow, eds. 2003. *Language and Mind: Advances in the Study of Language and Thought*. Cambridge: MIT Press.

Hacking, Ian. 1983. *Representing and Intervening*. Cambridge: Cambridge University Press.

Havelock, Eric. 1963. *Preface to Plato*. Cambridge: Harvard University Press.

Hegel, G.W.F. 1977/1807. *Hegel's Phenomenology of Spirit*. Trans. A. Miller. New York: Oxford University Press.

———. 1975/1822. *Lectures on the Philosophy of World History, Introduction: Reason in History*. Trans. H. B. Nisbet. Cambridge: Cambridge University Press.

———. 1975/1830. *Hegel's Logic*. Pt. 1 of Hegel's *Encyclopedia*. Trans. W. Wallace. New York: Oxford University Press.

Hirosige, Tetu. 1976. The Ether Problem, the Mechanistic Worldview, and the Origins of the Theory of Relativity. *Historical Studies in the Physical Sciences* 7: 3–81.

Holton, Gerald. 1988. *Thematic Origins of Scientific Thought: Kepler to Einstein*. rev. ed. Cambridge: Harvard University Press.

Howson, Colin, ed. 1976. *Method and Appraisal in the Physical Sciences: The Critical Background to Science 1850–1900*. New York: Cambridge University Press.

Kadvany, John. 1991. Dialectic and Diagonalization. *Inquiry* 34: 3–25.

———. 2001. *Imre Lakatos and the Guises of Reason*. Durham: Duke University Press.

Lakatos, Imre. 1970. Falsification and the Methodology of Scientific Research Programmes. In Lakatos (1978b) Ch. 1. (FM).

———. 1971. History of Science and Its Rational Reconstructions. In Lakatos (1978b) Ch. 2. (HS).

———. 1976. *Proofs and Refutations: The Logic of Mathematical Discovery*, ed. J. Worrall and E. Zahar. Cambridge: Cambridge University Press. (*P&R*).

———. 1978a. Newton's Effect on Scientific Standards. In Lakatos (1978b) Ch. 5.

———. 1978b. *The Methodology of Scientific Research Programmes: Philosophical Papers 1*, ed. J. Worrall and G. Currie. Cambridge: Cambridge University Press.

———. 1978c. *Mathematics, Science and Epistemology: Philosophical Papers 2*, ed. J. Worrall and G. Currie. Cambridge: Cambridge University Press. (*MSE*).

Lakatos, Imre, and Paul Feyerabend. 1999. *For and Against Method*, ed. M. Motterlini. Chicago: University of Chicago Press.

Levinson, Steven. 2003. *Space in Language and Cognition: Explorations in Cognitive Diversity*. Cambridge: Cambridge University Press.
Long, Jancis. 1998. Lakatos in Hungary. *Philosophy of the Social Sciences* 28: 245–311.
Lukács, Georg. 1971/1923. *History and Class Consciousness: Studies in Marxist Dialectics*. rev. ed. Trans. R. Livingstone. Cambridge: MIT Press.
———. 1976/1948. *The Young Hegel: Studies in the Relations Between Dialectics and Economics*. Trans. R. Livingstone. Cambridge: MIT Press.
———. 1981/1954. *The Destruction of Reason*. Trans. P. Palmer. Atlantic Highlands: Humanities Press.
Morrison, Margaret. 1995. Scientific Conclusions and Philosophical Arguments. In *Scientific Practice: Theories and Stories of Doing Physics*, ed. J. Buchwald. Chicago: University of Chicago Press, Ch. 9.
Neto, J.R.M. 1991. Feyerabend's Scepticism. *Studies in History and Philosophy of Science Part A* 22 (4): 543–555.
Nye, Mary Jo. 1972. *Molecular Reality*. New York: American Elsevier.
Popkin, Richard. 1979. *The History of Scepticism: From Erasmus to Spinoza*. 3rd ed. Berkeley: University of California Press.
Robb, Kevin, ed. 1983. *Language and Thought in Early Greek Philosophy*. LaSalle: Hegeler Institute.
Ropolyi, László. 2002. Lakatos and Lukács. In *Appraising Lakatos: Mathematics, Methodology, and the Man*, ed. G. Kampis et al. Dordrecht: Springer.
Sankey, H. 2011. Epistemic Relativism and the Problem of the Criterion. *Studies in History and Philosophy of Science Part A* 42 (4): 562–570.
Snell, Bruno. 1953/1948. *The Discovery of the Mind: In Greek Philosophy and Literature*. Trans. T.G. Rosenmeyer. New York: Dover reprint 1982.
Szabó, Árpád. 1978. *The Beginnings of Greek Mathematics*. Trans. A. M. Ungar. Dordrecht: D. Reidel.
Whorf, Benjamin Lee. 1956. *Language, Thought and Reality: Selected Writings of Benjamin Lee Whorf*, ed. J. Carroll. Cambridge: MIT Press.
Williams, Bernard. 1993. *Shame and Necessity*. Berkeley: University of California Press.
Zahar, Elie. 1976. Why Did Einstein's Programme Supersede Lorentz's? In Howson (1976), 211–276.

# Chapter 14
# Paul Feyerabend in Retrospect

**Joseph Agassi**

**Abstract** In this paper, Agassi reflects on his own interaction with Feyerabend, as it evolved over the years, and especially on Feyerabend's intellectual legacy as a philosopher of science and a teacher. Whereas he appreciates Feyerabend's early work and endorses some of his criticism of Popper, he regards Feyerabend's later change of view, and his turn from rationalism to relativism, as unserious. Also, he ascribes the enduring success Feyerabend's works still enjoy among his peers, more to his provocative style than to their actual content.

**Keywords** Paul Feyerabend · Karl Popper · Imre Lakatos · Rationality · Relativism

Paul Feyerabend said in his autobiography that we were friends of sorts. Let me tell you of what sorts. We enjoyed each other's company tremendously in private while in public he said about me things that were less friendly, telling me in private to ignore them since he did not mean them. It was too much for me, I am afraid, and I disengaged; he expressed regret, repeating that he did not mean what he said of me in public. Let me get off my chest one item, the very worst, and do it at once, and then be relatively comfortable.

Feyerabend wrote about the Führer, he declared that the monster was "eben ein intelligenter Mensch, intelligenter als miesten kritische Rationalisten, mit einem klaren Blick für die Komplexität historischer Prozesen"—an intelligent man, more intelligent than most of the critical rationalists, with a clear view of the complex

Joseph Agassi died before publication of this work was completed.

J. Agassi (✉)
Tel-Aviv University, Tel Aviv, Israel

York University, Toronto, ON, Canada

© The Author(s), under exclusive license to Springer Nature Switzerland AG 2024

S. Gattei, R. Corvi (eds.), *Feyerabend in Dialogue*, Boston Studies in the Philosophy and History of Science 346,
https://doi.org/10.1007/978-3-031-71938-7_14

historical processes.[1] He did not think the monster was more intelligent than most critical rationalists. He could not possibly find anything that indicates that the monster had a clear view of anything, let alone the complexity of historical processes. The Catastrophe that the monster has caused hardly indicates intelligence or a clear view of historical processes; nor does the monster's *Mein Kampf* testify for any. In a tragic error, it is well known, Jews took that book for a joke. Feyerabend could not consider the monster more intelligent than critical rationalists like himself. Nor could he consider Popper "just a tiny puff of hot air in the positivist teacup". If Popper responded to all this, he did so only obliquely, and only in a private letter (to Hans Albert: "Heute ist er Anarchist; morgen ist er Fascist"—today he is an anarchist, tomorrow he is a Fascist).[2]

Feyerabend was an academic star. As a teacher and a speaker, he always improvised, and in addition to much learning and interesting original thinking that he displayed in all of his talks, he had charisma. He joins the many remembered stars whose contributions were major or minor, light or heavy, whose reputation and mass appeal were ephemeral or lasting, vain or justifiable. All of these categories are occupied so that the criterion known as the test of time is a default option, not always reliable, although the light, vain reputation is usually less likely to be durable. I say "usually", since some durable human weaknesses perpetuate some vain reputations. One of these human failings is power worship and so some cruel tyrants have managed to retain renown for generations on end. The same holds for the seriousness of a contribution. The exceptions here are some empty works durable due to their pomposity. Some folly may supplement a serious contribution to our heritage and may be better remembered. My standard example here is Imre Lakatos, friend and colleague of the both of us, known less for his lasting, significant contributions to the philosophy of mathematics and more for his prattle about the philosophy of science, expressed for the purpose of gaining immediate fame, in a manner similar to that of Feyerabend. They became close friends. As to Feyerabend's fame, it is too early to know how durable it is. Hopefully it will depend on the value of his contribution. How valuable is it?

The organizers of a famous conference in memory of Feyerabend in Berlin in 2012 have openly asked this question. His provocative proposals, they say in response, continue to excite an extensive assortment of heated debates across a range of disciplines; these have proven to be of interest to a broad audience around the world. Thomas Nagel says, he was "consistently outrageous".[3] He was more than that. What? The Berlin conference, its organizers say, should encourage research on his philosophical proposals in efforts to do justice to the notorious complexity of his ideas and their relevance to ongoing discussions. They have offered a question, not an answer—which is fair. Let me follow them.

[1] Paul Feyerabend, *Erkenntnis für freie Menschen* (1979, 92).

[2] Hans Albert-Karl Popper *Briefwechsel*, 2005, 207.

[3] Thomas Nagel, The New Republic, October 12, 1998, 32–38, II.

In 2006 Friedrich Stadler and Kurt R. Fischer edited a volume about Feyerabend, called *Paul Feyerabend: ein Philosoph aus Wien.*[4] The book is full of expressions of praise of him as a serious thinker; yet all of its authors but one admit explicitly that he had contributed no interesting idea. The exception is Julie Floyd. She has ascribed to him two or three ideas that are allegedly important and allegedly novel. One is that observation is theory-laden. This idea is very important. Galileo already stated it very clearly. Otherwise, he said, as one walks in a moonlit street one would see the moon jumping from rooftop to rooftop like a cat. Feyerabend learned this idea not from Galileo but from Popper.

Another new idea that Julie Floyd has ascribed to him is pro-science pluralism. Although pluralism is not new, pro-science pluralism is. Contrary to Floyd's claim, however, Feyerabend had never expressed it; Popper had. Feyerabend did support pluralism but not any version of it that can possibly count as pro-science. He was anti-science, and even with some justice. For, the classical idea of science was the set of proven assertions; Popper's great breakthrough was his having replaced proof with openness to empirical criticism. It is this that opened the road to pro-science pluralism. Popper developed pluralism before Feyerabend and I together sat at his feet. Admittedly, his *Logik der Forschung* of 1935 is not pluralist. On the contrary: he introduced there his theory of degrees of testability (§30) just in order to explain the alleged unanimity in physics. The unanimity in question is limited, as researchers make efforts to break it.[5] Popper presented his pluralism in his magnificent "Towards a Rational Theory of Tradition" (1949),[6] where he offered a theory that puts the diversity of tradition, legislation and scientific hypotheses on a par. Marcello Pera has offered a penetrating analysis of that paper.[7] In allusion to Feyerabend's exaggerated, bombast slogan, "anything goes", Pera contrasts Popper's pluralism with the assumption of "a very thin moral framework, within which almost everything goes" (p. 279). Pera is a conservative, and he presents Popper as one, in a manner that commands the attention of those who do not see him that way. Nevertheless, Popper's reformism that rejects both extreme social philosophies, radicalism and strict conservatism, allows for all options from near-radicalism to near-conservatism.[8]

The conclusion of my review of Stadler and Fischer is, the book must be mistaken: Feyerabend surely did make a serious contribution. The fault is in the tradition that credits novelty only to a discovery of a new fact or of a new theory. Editors of periodicals and of encyclopedias, for a conspicuous example, contribute novelty in ways still unacknowledged and still unstudied. What then was his contribution?

---

[4] See my review of it *in Philosophy of the Social Sciences* (38, 2008, 303–5).

[5] See my "Scientific Schools and Their Success", in my Science and Society, 1981, 164–91.

[6] Reprinted in his *Conjectures and Refutations* (1962, 120–35).

[7] Marcello Pera "Karl Popper's 'Third Way': Public Policies for Europe and the West", in Ian Jarvie, Karl Milford, and David Miller (eds.), *Karl Popper, A Centenary Assessment*, 2006, Vol. 1, 273–81.

[8] John Wettersten, "On Two Non-Justificationist Theories", in Joseph Agassi and Ian C. Jarvie (eds.), *Rationality: The Critical View*, 1987, 339–41.

It was of another kind of contribution that we all recognize and still do not acknowledge: that of a public teacher, of an author who shows the reading public how to apply new techniques or new ideas. There are good and bad teachers. Feyerabend was both. He taught good and bad things and things that were both. Like a sage, like a guru, like a zadik, he offered some obviously outrageous proposals intending his audience to think and find what interesting proposals he had in mind.

Example. The philosophy of science is taught regularly as ideas about rational belief. So Feyerabend taught it by calling Galileo a mountebank and by teaching scientific lies. Another example. Dr. Helmut Mayer's review of the first volume of Feyerabend's posthumous works, *Naturphilosophie*, edited by Eric Oberheim and Helmut Heit, cites Feyerabend to say, "The Odyssey begins with Parmenides". This aphorism is both profound and jocular. It answers the question, what is peculiar to our way of thinking? (Remember: the original *Odyssey* is the story of a long, arduous homeward journey.) It begins with Parmenides' doctrine of rationality as proof and ends with the recognition of the impossibility of scientific proof. Third example. In response to my having called Feyerabend a Popper disciple he said this: he was not one, although he did listen to Popper's lectures, sat in his seminars, visited his home, and talked to his cat. Were the cat able to respond verbally, it would have queried your report: how come Agassi both called you a Popper disciple and begged you on his knees to be one?

The name of one of Feyerabend's books is *Farewell to Reason*; it is obviously a declaration of irrationalism, yet Oberheim has objected to this as a "popular misconception" of "Feyerabend's later philosophy according to which Feyerabend dropped *rationality* as the explanation of scientific advance, arguing that science's development is primarily guided by *power, propaganda, and prejudice.*" We can easily clarify the misconception by taking seriously the book's tirade against "the alleged friends of Reason": it is not against the true friends of reason. Feyerabend says there, "if this is reason, then farewell to reason". He does not say, "farewell to reason". It only seems so.[9]

Ambiguity runs systematically through Feyerabend's later writings. He said, "I can hold opinions without either having or giving reasons." This looks like license for unlimited arbitrariness; it is the (true) observation that we cannot avoid all arbitrariness. Most rationalists will mistakenly disagree and disclaim all arbitrariness. (Claiming that one is free of prejudice is humbug, said Bertrand Russell.) Deviation from classical rationalism may be critical rationalism (or some other, still unknown version of rationalism) and it may be irrationalism (of all sorts). The terms, non-justificationism (that Willard Quine or William Bartley invented) and non-foundationalism or anti-foundationalism (that Richard Rorty invented) are near-synonyms; as it happens, they are used differently. "Non-foundationalism"

[9] A view of Feyerabend that Cardinal Ratzinger, later Pope Benedict XVI, has expressed is cited in a famous paper by Angelo Matera, "The Death of Irony: Benedict and the Enemies of Reason", *The National Catholic Register*, Jan 29, 2008. The Cardinal has attributed to Feyerabend exaggerations rightly and irrationalism wrongly.

denotes irrationalism; "non-justificationism" denotes critical rationalism. Feyerabend's assertion, "I can hold opinions without either having or giving reasons" when understood as license, is careless license to turn philosophy into sophisticated cocktail-party chat and license to be seriously critical of any view.

Thus, Dr. Oberheim is right: Feyerabend complained that many scientists make science propaganda and prejudice; he did not praise them. The cause of the popular misconception that he complains about, then, is due to his style. It is one thing to write ambiguously — which (*pace* Wittgenstein) is unavoidable to some extent — and quite another thing to introduce the systematic ambiguity that George Orwell called double-speak. Traditional religious double-speak is prevalent, on the popular (false) assumption that critical thinking is only for the educated. One of the reasons for my having given up my religious background is my unconditional dislike of double-speak; I admit I do not particularly like reading the older Feyerabend.

Consider his claim that he used the services of quacks. He did consult them, but never used their advice. Consider then his claim that he never was a Popperian. In his book on Feyerabend's contribution, Oberheim has commented on this. To say that Feyerabend distanced himself from the Popper school is one thing, and to say that he took Popper's ideas seriously in a pluralist spirit and examined them is quite another. Oberheim ascribes to Feyerabend the second option, which is Popperian, since his criticism is pluralist and Popper proposed to take seriously and examine critically any possible answer to any interesting question. Many earlier thinkers wished to present this very idea consistently. Feyerabend cited Helmholtz, Boltzmann, Mach and Duhem as examples. They all failed, as they had endorsed the traditional equation of rationality with proof. Feyerabend glossed over their failure, chiding Popper for not having paid them homage. He may still be right.

Feyerabend contributed some gems that appear in his three volumes of essays (Cambridge University Press). Let me single out his "Problems of Empiricism" and "How to be a Good Empiricist". His thesis, as he has put it (Vol. 3, ix-x, 104–5), is,

$$\text{criticism} \rightarrow \text{proliferation} \rightarrow \text{realism}$$

Popper said,

$$\text{criticism} \rightarrow \text{realism} \rightarrow \text{proliferation} \rightarrow \text{criticism}$$

(realism makes explanations competing and invites their critical examination).

True, Feyerabend ascribes all this to John Stuart Mill. But even were his ascription correct, it would scarcely belittle Popper.

Feyerabend's *Knowledge without Foundations* is terrific. John Preston rightly views it (in his introduction to Feyerabend's third volume of essays that he has edited) one of his two most Popperian works. There are then two versions of Feyerabend, the young mature one and the old immature one. I appreciate young Feyerabend's work and have endorsed some of criticism of Popper. In his introduction to his first two volumes he refers to his own change of view, and rather than explain it, he observes that all ideas can be defended. This is true, yet only good ideas can be defended well. Popper began his thinking by taking seriously the fact that good ideas may be false: Newton's mechanics was his paradigm case. Classical rationalists took for granted the truth is absolute and that only true ideas are defensible. Disappointment with classical rationalism led to relativism about truth rather than about good defense—against what younger Feyerabend and I deemed commonsense.

To return to Parmenides—to the Parmenides of Karl Popper and of Arpad Szabo,[10] to be precise. The unique thing about western philosophy is its Greek origin; the unique thing about Greek philosophy is its equation of the distinction between reality and appearance with that between durable substance and fleeting phenomena. To be precise, since all distinctions are valid, we should say, the dichotomy between the absolutely stable and the utterly ephemeral is the peculiarly Greek: everything is either unchanging substance or elusive mirage. The ephemeral, Parmenides concluded, is but a dream.

A subtle difference between Greek and non-Greek worldviews invites highlighting. That things are not always what they look is obvious. Parmenides was the first to express explicitly the great dichotomy as a principle on which his worldview rested: the world splits into dreamlike appearances and immobile reality. Real things may be made of atoms and they may be made of the Four Elements of Aristotle, or of the Five Elements of the *Wu Xing*. This desk in front of you is unreal; what looks like a desk is really something else. It may be a swarm of atoms or a mix of the four elements or of the five elements. This puts the Greek and the Chinese cosmologies on a par: both fit the great dichotomy; but the dichotomy is Greek, not Chinese. And there lies the difference: pushing a familiar idea to the limit leads to the theory of substance, and this has no equal in cosmologies that are not heirs to the traditional Greek. Objections to this theory have led to its overthrow in 1905; Feyerabend instructs us to live without it.

Example. Feyerabend felt strongly about the cleavage between art and science.[11] Since art was traditionally not subject to the classical standards of rationality, pluralism in art was permitted; writings in the critical spirit proliferated in art criticism, especially in literary criticism. Feyerabend tried to link the histories of art and of science. Were the two ready to learn from each other, he proposed, were historians of science pluralists, they might contribute to rendering science partake as much as art in the heritage of the common educated modern citizen. This will also establish the study of the interaction between art and science. I will not elaborate here, having done so elsewhere.[12] Let me only observe that the Enlightenment view of art as inferior, as a handmaid of science, is an immediate corollary to the great dichotomy that dismissed diversity. Feyerabend went so far as to defend diverse magical

[10] Arpad Szabo, *The Beginnings of Greek Mathematics*, 1978.

[11] "On the Improvement of the Sciences and the Arts and the Possible Identity of the Two"; *Proceedings of the Boston Colloquium for the Philosophy of Science, 1964-1966. In Memory of Norwood Russell Hanson*. 1967, 387–415.

[12] Joseph Agassi and Ian C. Jarvie, *A Critical Rationalist Aesthetics*, 2008, Ch. 5 (a).

worldviews—including medieval magic and reading tea-leaves.[13] He did not mean it as he admitted in his response[14] to my severe criticism of his *Against Method*. His study of Brunelleschi's discovery/invention of perspective[15] is remarkable. Brunelleschi solved a problem that is on the borderline: it belongs firmly and conspicuously to both the story of art and the story of science, blurring the demarcation between them although the grand dichotomy requires it to be sharp.

The review of *The Conquest of Abundance* by Stephen R. L. Clark[16] is impressive. Clark suggests that Feyerabend had a vision of putting an end to "the long-lasting battle between philosophy and poetry" (Plato, *Republic* 607b), namely, between science and art. That is impossible: they can never unite; but they may cooperate, and this is vision enough. This battle is at the center of the greatest philosophical cleavage today, with young Ludwig Wittgenstein and his cohorts sustaining one side and Martin Heidegger and his cohorts sustaining the other.[17] The slightest agreement with Feyerabend is a sharp dissent from both camps. Why then did he (and others) pretend to follow Wittgenstein? Perhaps because he viewed as tragic Wittgenstein's failed effort to overcome the narrowness of his first book without withdrawing its message. (This, Yaakkko Hintikka and Gregory Landini agree, characterizes all the later, posthumous writings of Wittgenstein.)

The Greek view of proof as the criterion for rationality is anti-pluralist. It was a glorious idea, as it broke the hold of mythical-magical thinking and the xenophobia that it enhances. Popper broke new ground when he opposed the new irrationalism as the return to magic and tribalism and identified rationality with criticism (*The Open Society and Its Enemies*, 1945, Chapter 24). Did Feyerabend agree? It is hard to say, since he wished the multitude to read him one way and the *cognoscenti* quite a different way. Steven Toulmin made a point similar to that of Feyerabend. In his *Return to Reason,* 2001, he suggested refrain from applying the traditional criterion of rationality stringently as it is injurious. "The chief task of this book", he declared (p. 13), "is to show what is needed if we are to treat that injury, and reestablish the proper balance between Theory and Practice, Logic and Rhetoric, Rationality and Reasonableness." Recognizing "the collapse of Foundationalism" (p. 137), of the theory of rationality as proof, he demanded not to replace it but to soften its application. Did Feyerabend agree? He equivocated; I think he did not.

The same goes for Rorty. What renders philosophies that advocate the diminution of reason attractive? Julien Benda has raised this question between the two

[13] *Against Method*, 1975, 14, 44, 65, 93, 205, 298, 304.

[14] Joseph Agassi, *The Gentle Art of Philosophical Polemics: Selected Reviews and Comments*, 1988, 405–16.

[15] Paul Feyerabend, *Conquest of Abundance: A Tale of Abstraction Versus the Richness of Being*, 1999, Ch. 4.

[16] Stephen R. L. Clark, Review of Paul Feyerabend, *The Conquest of Abundance*, *Inquiry*, 45, 2002, 249–68.

[17] In Eli Friedlander's *Signs of Sense: Reading Wittgenstein's Tractatus*, 2001, Wittgenstein changes allegiance.

world wars.[18] It invites discussion of taste and is thus psychological. And tastes diverge. They may however converge by institutionalized means that evoke similar emotions in large publics: the excessive demand to avoid error and take responsibility puts excessive pressure that makes it reasonable to purchase books that may offer a reduction of the pressure. This is better achieved by replacing the demand to avoid error with the demand to avoid irresponsible error. The laws of civilized countries recognize this; many moralists still do not.

Feyerabend never tired of condemning the violence and tyranny that some people practice in the name of reason and their ascriptions of intellectual hegemony to some people—priests, scientists or teachers. The traditional opinions about rational methods, Feyerabend observed with some justice, boost their ascriptions. So he condemned all methods. As Oberheim has noted in his review of Feyerabend's *Conquest of Abundance*,[19] Feyerabend "emphasizes that common sense and science tend to disguise the fact that worldviews are not forced on us but are formed by individual decisions, and accordingly he … challenges the status ascribed to contemporary science as sole provider of truth, rejecting science as an objective process of discovering nature's secrets in favor of a worldview according to which 'art is a product of nature as a work of art'." Since Popper was not engaged in such campaigns, the claim of Feyerabend for deviation from Popper is somewhat vindicated this way. The question arises, was Popper right to advocate science without warning against its dangers? I do not know; the choice of a cause to fight for is a matter of personal decision, since every volunteering for a fight of this kind is supererogatory. To the extent, however, that these matters are also open to public debate, Popper was right: irrationalism is politically more dangerous than rationalism. Feyerabend disagreed. He compared the advocates of the Vietnam War with the Nazis and blamed me for siding with these advocates. That was too much for me. But now we are here to praise Feyerabend, not to bury him. His last works were written in the spirit of conciliation, suggesting that no cultural differences are too big to overcome and no disciplinary differences are so big as to separate one discipline form another. Descriptively this is notoriously false: Popper cited the Nazi adage, "shoot first and talk later". But this may be a sermon and a proposal to aim at conciliation. If this proposal refers to the Nazi culture too then it is obviously excessive: there was no way to reconcile to its atrocities.

Feyerabend's work opened up debates to many new publics. The appraisal of the value of this invites preliminary research. Effort to render Feyerabend's proposals useful has to lead to rewording them more carefully and more honestly. Hopefully, they will then be more conciliatory—but not all the way: there is a limit to toleration. Feyerabend denied this, and he supported this denial by odd arguments that point at the positive value of violence ("Violence is necessary to overcome the impediments erected by a well-organized society […], and it is beneficial for the individual, for it releases one's energies and makes one realize the powers at

[18] Julien Benda, *La trahison des clerks*, 1927.

[19] Eric Oberheim, review of Feyerabend's *Conquest of Abundance*, *Isis*, Vol. 92, pp. 597–598.

one's disposal").[20] This repelled both Gellner[21] and Popper. But, again, we come to praise Feyerabend, not to bury him.

All conciliations raise new conflicts as Georg Simmel has taught[22] and as Popper has stressed. The abolition of all conflict is an impossible, dangerous utopian dream. This is not to preach against utopian dreams: Ian Jarvie rightly criticized Popper on this point:[23] he agrees that dreams may be dangerous, but, like conflicts, risks are unavoidable. In sum, those seriously intending to engage in Feyerabend's project—of showing all cultures to be one and all disciplines to be one—should note that creating unity carries the seeds of a new diversity: pluralism is here to stay.

Bas C. van Fraassen views Feyerabend's study of Brunelleschi as the peak of his output,[24] a part of his efforts to put art and science on a par, in the hope that science will thereby cease to be esoteric. Now, science progresses, not art. As van Fraassen notes, Feyerabend was fascinated by the dramatic aspect of both the arts and the sciences. We can and perhaps should generalize this point to cover all information relevant as background to scientific progress. For example, after Hans Christian Ørsted discovered electromagnetism, he reported, he was dazed for three months. The reason why this kind of information is ignored, not to say suppressed, is that it is judged irrelevant, and by obsolete criteria. Feyerabend opposed this passionately. He said, all idolization is harmful. In his story of Brunelleschi's output he tried to combine the history of art (including the theatre) and of science (mainly geometrical optics) with the histories of their social settings. To repeat, he meant to avoid harm, at least the harm of excess stereotyping. He certainly had a point, even if not a new one: already *Ecclesiastes* (1:18) complained about the pain that knowledge brings about. If Feyerabend made people see that even what they are sure is harmless is possibly harmful, if he made some of us more aware of our causing harm, then we must acknowledge this.

Let me conclude with Feyerabend's gift to me, his posthumous contribution to one of the *Festschrifts* dedicated to me, "Universals as Tyrants and as Mediators" that deals with scientific revolutions. He questioned there the view of research as the search for regularity rather than as muddling through: it does not bring happiness. What does? He did not know. Science is the search for the universal, and the universal is too uniform. How did it come about? Already Al-Farabi found this puzzling. Greek culture is special because it taught that no culture is special, he noted; he found this paradoxical. The search for the universal, said Feyerabend, is accidental: the motive for the search for generalities in mathematics, he said, is to satisfy the requirement of some ritual. Now it does not matter what the motive was. Popper's breathtaking classic "Back to the Pre-Socratics" describes the pre-Socratic revolution as the readiness that Thales showed to let his disciples dissent. Why,

---

[20] *Against Method*, p. 187.

[21] Gellner, *British Journal for the Philosophy of Science*, 26: 331–42, 340.

[22] Georg Simmel, *Conflict and the Web of Group Affiliations*, [1922] 1955.

[23] Ian Jarvie, "Utopia and the Architect", in J. Agassi and I. C. Jarvie (eds.), *Rationality: The Critical View*, 1987, 227–43.

[24] Bas van Fraassen, Feyerabend's *Conquest of Abundance*, *Times Literary Supplement*, 507, June 23, 2000, 10–11.

however, did they want to dissent? Perhaps because they did not like the Master's thesis. Their motive was clearly either pre-rational and thus pre-revolutionary, or rational and thus post-revolutionary. Either way, this misses the revolution. Did rationality appear whole, as Popper and Gellner say, or did it evolve? Was its motive rational or a-rational?

Did Feyerabend try to offer an alternative theory of the rise of Greek culture, science, rationalism? I do not know. He clearly did wish to return to the pre-rational its dignity, just as we have returned the dignity to medieval paintings (to his delight). Does all this deprive rationality of its dignity? Feyerabend's irrationalist followers say yes, the rationalist ones say, no. And, beautiful as the pre-rational is, surely the rational is no less beautiful and no less legitimate. Feyerabend loved it, perhaps too secretly but too well, and he complained only about its imperialism. We concede this point. The rest is history.

The question remains: how do changes of patterns of conduct occur? The rationality principle allows changes as due to changes of circumstances—material, individual and social. The changes can be imposed and humans may cause them according to their aims and circumstances. This leaves inexplicable some changes—say, those that Feyerabend discussed, from pre-rationalism to rationalism or even from mediaeval art to Renaissance art. How are these explicable? Jarvie and I address this question (end of our "The Rationality of Magic", 1967). We say, "The problem is not, then, 'how on earth can they believe in magic?'; it is rather 'can people with inefficient magical beliefs come to be critical of them, under what conditions and to what extent?' To us this seems the really urgent sociological problem posed by magic." Some profound changes may follow the change of circumstances. Brunelleschi's revival of ancient building methods (for building the copula of the Florence cathedral of St. Mary of the Flowers) made people realize that they can revive antiquity, contrary to their older view that only the Second Coming will.

How did Brunelleschi arrive at this idea? Antonio Manetti, his biographer, reports that this idea guided him from the start: Brunelleschi and Donatello went to Rome and acted as the first archeologists. Whatever the details of the story are, clearly, early circumstances led here to new conclusions. More generally, say Jarvie and I, ideological progress is explicable as the discovery that given ends and worldviews yield to other ends and worldviews as superior—rightly or not but rationally. Feyerabend found that the error was of taking the modern dismissal of mediaeval art as the paradigm was one of facile dismissal. He took today's ideological and intellectual strife as in dire need of a healing process. He hoped that his opposition to excessive stereotyping will help such a process even though he knew we cannot avoid stereotyping and can only hope to reduce the damage. Yet he was right in demanding that we should not dismiss as *passé* even past superstitions. As Jarvie has put it,[25] "It is only in relation to its past that a tradition of enquiry can regain a sense of progress."

**Acknowledgemnts** Yoav Ariel, Daniel Cohen, Ian Jarvie and Nathaniel Laor have helped me in the difficult task of writing this paper. They have my profound gratitude.

[25] Ian C. Jarvie, "Epistle to the Anthropologists", 1975.

# Index



© The Editor(s) (if applicable) and The Author(s), under exclusive license to Springer Nature Switzerland AG 2024

S. Gattei, R. Corvi (eds.), *Feyerabend in Dialogue*, Boston Studies in the Philosophy and History of Science 346,
https://doi.org/10.1007/978-3-031-71938-7

**SPRINGER NATURE**

# GPSR Compliance

*The European Union's (EU) General Product Safety Regulation (GPSR) is a set of rules that requires consumer products to be safe and our obligations to ensure this.*

*If you have any concerns about our products, you can contact us on ProductSafety@springernature.com*

In case Publisher is established outside the EU, the EU authorized representative is:

Springer Nature Customer Service Center GmbH
Europaplatz 3
69115 Heidelberg, Germany

The manufacturer's authorised representative in the EU is Springer
Nature Customer Service Centre GmbH, Europaplatz 3, 69115 Heidelberg,
Germany. If you have any concerns regarding our products, please
contact ProductSafety@springernature.com

Printed and bound by CPI Group (UK) Ltd, Croydon, CR0 4YY
30/09/2026
02225599-0007